Aseeva/Zaikov
Combustion of Polymer Materials

R. M. Aseeva/G. E. Zaikov

Combustion of Polymer Materials

Translated by T. Fabrikant

With 87 Figures and 23 Tables

HANSER PUBLISHERS

Hanser Publishers, Munich Vienna New York

Distributed in the United States of America by
Macmillan Publishing Company, New York
and in Canada by
Collier Macmillan Canada, Ltd., Toronto

Original Title: Gorenie polimernych materialov
© Izdatel'stvo Nauka, Moskva 1981

Translated by T. Fabrikant

Distributed in USA by
Scientific and Technical Books
Macmillan Publishing Company
866 Third Avenue, New York, N.Y. 10022

Distributed in Canada by
Collier Macmillan Canada Distribution Center,
539 Collier Macmillan Drive, Cambridge, Ontario

Distributed in all other countries by
Carl Hanser Verlag
Kolbergerstraße 22
D-8000 München 80

CIP-Kurztitelaufnahme der Deutschen Bibliothek
Aseeva, Roza M.:
Combustion of polymer materials / R. M. Aseeva ; G. E. Zaikov. Transl. by T. Fabrikant. - Munich ; Vienna ; New York : Hanser, 1985.
ISBN 3-446-13989-3
Einheitssacht.: Gorenie polimernych materialov ⟨engl.⟩
NE: Zaikov, Gennadij E.:

ISBN 0-02-948930-X Macmillan Publishing Company, New York

Library of Congress Catalog Card Number 86-060300

Printed in Germany (FRG) by Appl, Wemding

Contents

Foreword IX
Preface to the English Edition XI
Editor's Preface to the Russian Edition XIII
Authors' Preface to the Russian Edition XV
Introduction 1
References 9

Chapter 1 Origin of the Combustion Process:
Self-Ignition and Ignition of Polymers 11
General Concepts of Combustion 11
Theory of Self-Ignition and Ignition of Polymers 14
Ignition in the Condensed Phase 14
Heterogeneous Ignition 27
Ignition in the Gaseous Phase 33
Self-Ignition Temperature and Ignition of Polymers 44
Reaction between Polymers and Oxygen 47
References 56

Chapter 2 Polymer Combustion and its Characteristics 60
Multistage Combustion of Polymers 61
Mass Transfer in the Combustion of Polymer Materials 66
Heat Transfer during Polymer Combustion 72
Flame Temperature of Polymers 72
Changes in Heat Transfer Parameters and in Hydro-
dynamics during Polymer Combustion 79
The Rate of Combustion of Polymer Materials 90
The Spreading of a Flame over the Surface of a
Polymer Material 91
Combustion of Spherical Polymer Particles 103
Combustion of Polymer Materials for a Counterflow
of Gaseous Oxidiser 106
Stability Limitations on the Diffusion Flames of
Polymer Materials 108
References 113

Chapter 3 Chemical Processes in the Combustion of Polymeric Materials 119
Chemical Processes in the Condensed Phase 120
Pyrolysis of First-Group Polymers (Chain Scission/ Volatile Products) 130
Pyrolysis of Second-Group Polymers (Intramolecular Detachment of Atoms or Groups/Non-Volatile Products) 141
Heterogeneous Oxidation of Carbon 170
Chemical Processes in the Gas Phase 177
Diffusion Flames of Polymers 178
Smoke Formation during the Combustion of Polymers 194
Chemical Aspects of the Reduction in Combustibility of Polymer Materials and Smoke Liberation during their Combustion 212
Mechanism of Flame Retardant Action 214
Retardation of Smoke Formation 241
References 246
Addendum 259
The Effect of Oxygen on Polymer Decomposition 259
Chemical Processes in the Gaseous Phase 265
Chemical Aspects of the Reduction in Combustibility of Polymer Materials and Smoke Formation during their Combustion 267
Halogen-Containing Flame Retardants 268
Phosphorus-Containing Flame Retardants 270
Metallic Compounds 271
References 274

Chapter 4 Methods for the Study of the Combustion Process and Estimation of the Flammability of Polymer Materials 277
Experimental Methods for the Study of the Combustion of Polymer Materials. Determination of Combustion Characteristics 277
Rate of Combustion 278
Shape and Structure of Flame and Burning Surface 280
Temperature Profiles during Combustion 282
Concentration Profiles of Products inside a Flame 284

Gas-Dynamic Characteristics 287
Estimation Methods for the Flammability of Polymer
Materials 288
Determination of the Ignitability of Polymeric
Material 292
Determination of the Capacity of Materials to
Propagate Combustion...294
Determination of the Smoke-Formation Capacity of
Polymer Materials 300
Determination of the Toxicity of Polymer Combustion
Products 301
Modelling Methods for the Full-Scale Combustion
of Polymer Materials 304
Modelling at Atmospheric Pressure 304
Modelling at Elevated Pressure 305
References 306

Chapter 5 Reduction of the Flammability of Polymeric Materials 309
General Tendencies in the Design of Polymer
Materials with Reduced Flammability 310
Synthesis of Incombustible Polymers 313
Chemical Modification of Polymers 315
Use of Flame Retardants..317
Use of Fillers 322
Application of Fireproof Coatings 326
Criteria for Evaluating Flame Retardant Effectiveness 329
Polymer Materials with Reduced Flammability 338
Polyolefin-Based Materials 339
Polyvinyl Chloride-Based Materials 344
Materials Based on Styrene Polymers and Copolymers 347
Materials Based on High Molecular Weight Linear Polyesters 352
Materials Based on Polyamides 356
Materials Based on Carbamide Resins (Urea-formaldehyde,
Melamine-formaldehyde) and Phenol-formaldehyde Resins 359
Materials based on Polyurethanes 361
References 364
Subject Matter Index 373

Foreword

This book is a significant contribution to our understandings in regard to this important scientific subject - polymer combustion and the chemical and physical processes involved. Among other things, it is also a current reminder that science and its quest for knowledge can still transcend political boundaries.

The concern and interest in regard to combustion processes are universal and the authors have managed to bring together the concepts from combustion engineering, physics, and chemistry in an understandable cohesive manner. This very difficult task, in itself, makes it an unusual and valuable text. It also makes available a large amount of U.S.S.R. literature on this and related subjects not readily accessible to many scientists.

Much remains to be solved in improving the flammability behaviour of many polymeric materials. Such difficult problems will be solved only by multidisciplinary studies capable of taking a broader and less parochial view. The present authors' approach is a step in the right direction. This is certainly a universal problem area where future cooperative international efforts should be considered.

Eli M. Pearce
Dean of Arts and Sciences and
Director of the Polymer
Research Institute
Polytechnic Institute of NY
Brooklyn, NY 11201 USA

Professor Pearce was a member of the Committee for Fire Safety Aspects of Polymeric Materials appointed by the National Materials Advisory Board of the National Research Council/National Academy of Sciences.

Preface to the English Edition

The use of polymer materials in daily life and in industry is closely connected with the problem of fire safety. This is the reason why manufacturers and consumers pay serious attention to the flammability properties of polymer materials. The flammability of a material, in turn, is linked to its exposure to different stages of a complicated combustion process and also to the relationship between the combustion process and the conditions under which this takes place.

In this book we explain the principles of combustion and the transformation processes that occur during the combustion of polymer materials.

The book is intended primarily for chemists, researchers and engineers who develop, reprocess and utilise polymer materials. We have therefore tried to write about the scientific fundamentals of polymer combustion in a practice-oriented way and to show the practical conditions under which combustion initiates and terminates.

A large number of polymer materials are used every day. These materials have specific characteristics and properties that affect their behaviour in fire hazard situations. Since combustion itself, in the first instance, is a chemical reaction between the material and the oxidative atmosphere, the chemical aspects of the combustion of polymer materials are emphasised in this book. A review is also given of the interdependence of the combustibility properties and the chemical structure of the polymers.

From our point of view, this monograph differs from many other books on the same subject in that we emphasise a number of general characteristics and principles intrinsic to all polymers but do not disregard the individuality of polymer substances. A general approach to the reduction of polymer combustibility and to the efficiency analysis of combustibility reduction methods is also given special coverage.

This field of polymer science is developing exceptionally rapidly. Since the publication of the Russian edition a fairly large number of publications have appeared on theoretical and experimental research, both on the combustion process itself and on the action mechanism of retardants. We thus felt it necessary to add some new sections to the English edition. These additions illustrate recent progress in the field of fundamental research into the combustibility of polymer materials.

In conclusion to this preface we would like to quote the words of Donald J. Cram (UCLA) in his preface to M. Szwarc's monograph "Carbanions, Living Polymers, and Electron Transfer Processes," New York, Wiley-Interscience (1968).

> "As in all endeavours, chemistry has its scholars, investigators, craftsmen, prospectors, speculators, adventurers, entrepreneurs, statesmen and chroniclers whose combined functions give moment and substance to the science. In the early stage of the development of a branch of chemistry occasionally one person integrates some of these activities. But as the field expands, the hard economics of the time drive investigators into specialisation, both of subject matter and of function".

Many hundreds (if not thousands) of scientists in the world are working on combustion problems today. They are researching numerous subjects, employing a large variety of methods and approaches, ranging from synthetic organic chemistry to physics, from fundamental scientific problems to practical applications. On the basis of Professor Cram's classification, we feel that we are playing the role of "investigators" in this monograph and that the book may be useful to persons concerned with the practicalities of the development of specific polymer products and the determination of their combustibility. It could also be of assistance to those engaged in research on reducing the combustibility of polymer materials and products made from these.

In the middle of the first century A.D. the Roman scientist, Pliny, asserted that mankind owed all its achievements to the cockerel, since it was the cockerel's crowing every morning that woke everyone up. Whilst the role of the cockerel may indeed have been important in Pliny's time, it is safe to say that, at the end of the twentieth century, mankind owes many of its achievements to polymers.

One problem in this field, however, remains largely unsolved - how to make polymers noncombustible and thereby completely safe in use. It is precisely this problem which is engaging the minds of scientists investigating the reduction of combustibility of polymer materials.

R. M. Aseeva
G. E. Zaikov

Editor's Preface to the Russian Edition

This book is devoted to one of the current directions in the study of polymers, namely the combustion of polymers and means of reducing their combustibility.

In recent times there has literally been an "explosion" in the number of publications on this subject in many countries. The majority of these publications are of an applied nature, however, since the greater the number of polymers manufactured, the greater the need to reduce their combustibility becomes. Practical interest is stimulating fundamental studies in this area, however, together with attempts to draw scientifically-based inferences from the experimental results of polymer combustion. This is the approach that authors have adopted and this, therefore, distinguishes the monograph from others.

The combustion of polymers and the reduction of polymer combustibility constitute a complex problem. Its solution is linked not only to problems of polymer chemistry but also to questions of chemical kinetics (mostly problems of chain reactions, such as thermal and thermo-oxidising degradation, combustion and explosion) and to organic chemistry (the synthesis and properties of additives that lower polymer combustibility). The solution also takes in various branches of physics, as well as technological processes.

A major task is the creation of uniform tests to determine the combustibility of polymers and the effectiveness of added fire-retardants. The authors of the monograph thus focus their attention on this question, since the lack of uniform experimental methods has resulted in a situation where experiments carried out in different countries virtually "speak to each other in different tongues".

The authors of this monograph have succeeded in presenting a critical examination, in condensed form, of the experimental combustion results of different classes of polymers. The monograph reflects the changes in the theory and in the developmental paths followed by research in this field. It also correctly emphasises that there has been a lack of continuity in the development of this subject. This lack of continuity has been marked by a transition from a chiefly empirical approach to methods based upon current theory and experimental kinetic data.

The information presented and the ideas developed in the monograph of R. M. Aseeva and G. E. Zaikov will be of interest to all those concerned with polymer combustion and the reduction of polymer combustibility.

N.M. Emanuel

N. M. Emanuel until his untimely death in 1985 was a Member of the Academy of Sciences of the USSR and worked at the Institute of Chemical Physics in Moscow.

Authors' Preface to the Russian Edition

The study of the processes of ignition and burning in natural and synthetic polymers, and also in various composite materials, has taken on great scientific and practical significance. Interest in this rapidly evolving area of science stems from the growing need to establish scientific foundations for the synthesis of non-flammable polymers. There is also a need for a rational technology to manufacture fire-safe materials, to determine the conditions under which these may be used and to eliminate the possibility of fires emerging and spreading.

Once the mechanism and rate laws for the ignition and burning of polymers have been established, together with the reaction mechanism for the various ingredients that generally form part of a multi-component polymer system, it will then be possible to find effective means of suppressing or, as the case may be, accelerating these processes.

Among the most extensive theoretical and experimental investigations are those related to the combustion of condensed polymer systems employed as solid rocket fuels and as explosive materials. A.F. Belyaev, P.F. Pokhil, K.K. Andreev, O.I. Leipunskii and others have contributed greatly towards the formulation of a combustion theory for condensed systems of this type. A general combustion theory for condensed systems was developed on the basis of work by N.N. Semenov, Y.B. Zel'dovich, D.A. Frank-Kamenetskii and others. Its basic scientific aspects have also been applied to ordinary polymer materials.

In this monograph the authors have attempted to discuss modern concepts of the combustion of both natural and synthetic polymers. They have also tried to reflect the wide interest that exists in the creation of fire-safe materials and have endeavoured to systematise and generalise the theoretical and most significant experimental studies in polymer combustion by Soviet and other scientists. This information is not always easily accessible to chemists working on the synthesis of non-flammable or low-flammable polymers. The information is scattered over a large number of journals, chemical and non-chemical, and also appears in the proceedings of national and international conferences and symposia on combustion.

In the manufacture of polymer materials with reduced flammability, the empirical approach still predominates. The authors hope that their work will offer a solution to many of the practical problems encountered in synthetic chemistry and technology by focussing on the mechanism and principles of polymer combustion.

The compilation of a theory for the diffusion combustion of polymers is far from complete. Processes such as the high-temperature conversion of polymers which occurs during rapid heating (by hundreds of degrees and more in a second) under combustion conditions have been poorly studied.

A detailed examination is required of the mechanism and kinetics involved in the formation of the new phase at the burning surfaces (in particular, the carbonised layer or char) and of its effect on heat and mass transfer during combustion. The authors would like to focus attention on such gaps in our understanding of the diffusion burning of polymers so as to direct the efforts of experts and scientists towards their solution.

This book consists of an introduction and five chapters. The basic conceptions on the combustion process of polymers are presented in the introduction. Chapter 1 contains general theoretical and experimental studies on the ignition of polymers. The most important characteristics of polymer combustion are considered in detail in Chapter 2. Chapter 3 is devoted to the chemical aspects of polymer combustion, to problems of smoke formation during burning and to means of retarding these processes. Methods of determining combustibility and studies of polymer combustion are included in Chapter 4. Chapter 5 presents general material on methods of reducing the flammability of polymer materials.

The authors wish to thank N.M. Emanuel for his constant attention to their work on polymer combustion, S.A. Tsuiganov for discussing various problems covered in the monograph and for his valuable comments during preparation of the manuscript. The authors also acknowledge the help of M.I Artsis in the compilation of this monograph.

R.M. Aseeva
G.E. Zaikov

Introduction

Chemical science and industrial chemistry are among the leading branches of technology when it comes to furnishing scientific and technical progress. The production of polymer materials is one of the fastest-growing branches of the chemical industry. World production of plastics, chemical fibres, synthetic rubbers and other polymer materials currently totals about 100 million tons per annum, approaching the production of steel and cast iron.

Growth in the production and use of large numbers of polymer materials in various fields of engineering is somewhat hampered by a number of serious deficiencies, in particular by their flammability.

Fires caused by the ignition and combustion of polymer materials lead to extensive losses in industry, human lives and priceless historical monuments. The history of fire prevention shows that up to 80% of fires are caused by low-heat ignition sources (sparks, burning matches, cigarettes etc.). A reduction in the flammability and combustibility of polymers and the development of fireproof materials is thus an urgent problem calling for an immediate solution.

Many countries have special laws limiting or prohibiting the use of flammable polymer materials in civil and industrial construction, in means of transport (aircraft, automobiles, railway coaches, ships), in electrical and electronic engineering and in the manufacture of everyday products. These measures serve to intensify the search for a solution to this particular problem and the development and implementation of fireproof polymer materials in various fields.

A number of products made of polymer materials will be manufactured solely in fire-resistant grades in the near future [1]. Western European countries and Japan have seen an increase in the number of fire-resistant brands in total polymer material consumption [2].

The flammability of natural and synthetic polymer materials is determined by the ability of such materials to ignite and spread the combustion process. The word "combustion" has a very broad meaning. It includes a combination of complex physical and chemical processes. Combustion is defined as a fast, self-accelerating exothermal redox process that is able to spread in the environment and is accompanied by luminosity and the formation of a flame. Combustion may be of either a chain or a thermal nature, depending on the cause of initiation and development: accumulation of active particles in the system or the liberation of heat.

The combustion of common polymer materials is viewed as a heat-generating process, as in the case of the overwhelming majority of other substances. Initiation of the polymer combustion process is based essentially on the fact that the heat liberated as a result of redox reactions is not able to enter the surrounding atmosphere but heats the reacting system and increases the rate of reaction. The resulting heat then causes self-propagating combustion even after the ignition source has been removed. The phenomenon of a progressive auto-acceleration of a chemical reaction under the effect of liberated heat is called a thermal explosion [3-5].

The specific characteristic of a thermal explosion is the existence of critical conditions for its development. The chemical nature of the fuel and the oxidising substances differs, as does the mechanism of the combustion reactions. A pronounced relationship between the rate of heat liberation and the temperature is a major factor characterising the reaction occurring in heat liberation. When the rate of heat input is equal to the rate of heat consumed to maintain the process and the losses to the surrounding atmosphere then a steady-state combustion process has been established.

In common practice, the oxygen from the air is used as an oxidising agent in polymer combustion. The reaction between the oxygen and the polymer and the fuel products of polymer degradation and vaporisation is dependent on the reagent mixing and diffusion conditions. This is why the combustion of polymer materials in air is viewed as a process taking place in a diffusion regime, and the polymer flames are classified as diffusion flames in this case. The flame is defined as a gaseous zone of the combustion process in which fast exothermal reactions take place between the fuel products of polymer degradation and vaporisation and the gaseous oxidiser. The flame is usually characterised by visible radiation. The zone of the flame that has the highest temperature is called the flame front; beyond the front of the flame the concentration of combustible components is assumed to be equal to zero.

During the combustion of polymer materials inside and on the surface of the condensed phase (polymer matrix) complex physical and chemical processes also occur (phase transition, thermal and thermal-oxidative degradation, in some cases formation of a new phase, etc.). The mechanism of polymer combustion thus includes a combination of various physical and chemical processes through which the original substance is transformed into high-temperature combustion products. The spontaneously spreading volumetric zone in which these processes occur is called a combustion wave. The spreading of a combustion wave occurs as a result of the physical processes of heat transfer and diffusion.

The simplest structure of a combustion wave, which corresponds to an adiabatic transformation of the reagent into combustion products (i.e. when there are no losses of heat to the surrounding environment), is represented in Fig. 0.1. In this case the combustion wave consists of a narrow reaction zone and a heated system layer adjacent to it. Inside the combustion wave there is a sharp change in temperature from the initial value T_{in} to a maximum value T_{max}, which corresponds to the complete combustion of the reagent. The structure of a combustion wave for polymer substances is more complex in character and is a function of the nature of the polymer system and the combustion conditions.

The combustion and the flames of the condensed substances are classified according to different criteria (phase and aggregate state, character of oxidiser and fuel component mixing, aerodynamics of the reagents' flow motion, etc). The combustion of the condensed substances is subdivided into homogeneous and heterogeneous combustion as a function of the state of the components participating in the combustion process. In the case of homogeneous combustion there is a phase contact area between the reacting components of the system. In the case of heterogeneous combustion the process occurs at the phase contact area. The combustion of a large number of polymer systems, especially fire-resistant ones, takes in characteristics of both homogeneous and heterogeneous processes.

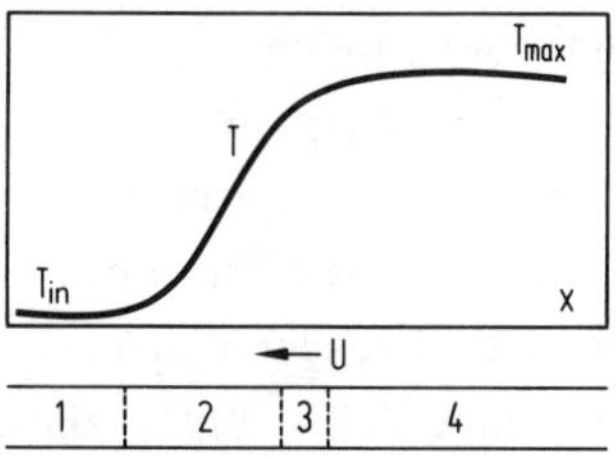

Fig. 0.1 Diagram of the simplest adiabatic structure of a combustion wave. (1) Original substance, (2) heated layer, (3) reaction zone, (4) combustion products. The arrow indicates the direction of combustion wave spread.

The process is characterised by the fact that the high-temperature degradation of polymers during combustion is often accompanied by the formation of a new phase called the carbonised layer. The carbonised layer is a result of the reaction between the gaseous oxidiser and the surface carbon. The rate of the heterogeneous chemical formation process of the carbonised polymer layer is determined by the rate of oxygen diffusion out of the gaseous phase towards the carbon surface. There are also other possible conditions under which a heterogeneous reaction between oxygen and the surface of the polymer will occur.

The combustion of metals is an example of true heterogeneous combustion. When A.G. Merzhanov studied the basic physical and chemical processes inside the combustion wave for condensed substance combustion and distinguished the reacting zone (which is regarded as "leading" or controlling the rate of combustion), he divided all condensed substances into two classes: nongaseous and vaporised during combustion.

The condensed substances of the first class do not form any smoke at all during combustion. We can include in this class various thermite mixtures which are produced as combustion products during the combustion of nonvolatile condensed substances (metal oxides). The overwhelming majority of condensed substances belongs to the second class. They are vaporised to begin with and this is followed by homogeneous combustion of the vaporisation products in the gaseous phase. Such substances are also subdivided [6] into

1. Evaporised volatile substances
2. Nonvolatile decomposing substances

Polymers degrade (decompose) at a temperature lower than their vaporisation temperature. They form gaseous products during combustion as a result of degradation. Dispersion processes play an important role during the combustion of the decomposing condensed substances.

The dispersion phenomenon was first discovered by A. Belyaev in the combustion of fulminating mercury. In this phenomenon the substance does not have sufficient time to react completely in the combustion zone and is carried away from the surface (by the mass of the gaseous products) into the plume of the flame in the form of fine particles. Since the dispersion depends on the aggregate state of the reaction layer, there are several sub-groups of substances (among the degrading condensed substances) that react in either a liquid or a solid phase.

As a result of their rapid degradation during combustion, polymers also display an inclination towards the dispersion of particles. Particles with a reticular structure are especially prone to dispersion [7]. The dispersion eventually increases the effective rate of polymer combustion. Multicomponent polymer materials may contain different additives (such as plasticisers), which can evaporate and thus affect combustion.

Mention has already been made of the fact that the combustion and flame of common polymer materials are viewed as diffusion processes which result from the way in which fuel and oxidiser are mixed in the reaction zone. It is, however, also possible for so-called mixed or premixed flames of polymer-containing systems to occur. In these flames, reagent diffusion is not the determining factor. Systems used for rocket fuels and as explosive substances constitute examples of polymer-containing systems that form premixed flames. In these cases the oxidiser and the fuel reagent are either a part of a single macromolecule (ballistic explosive substances, nitrocellulose type) or are premixed (mixed rocket fuels) [8-11].

Whether a diffusion regime or a kinetic regime of combustion prevails in the flame of similar polymer-containing systems will depend on the nature of the mixed fuel, the ratio of fuel to oxidiser, the component dispersion and the conditions of combustion [9, 10].

The flames are subdivided into laminar and turbulent flames according to the nature of movement of the oxidising agent and the fuel flow. Inside the laminar diffusion flames of condensed substances the flow motion is fluid and laminar; reagent mixing and heat transfer occur through molecular diffusion and natural convection. Inside turbulent flames, these occur through molecular and turbulent diffusion. Flow turbulence increases the rate of complete combustion and the rate of flame diffusion.

A change in the flow movement regime inside the flame will set off turbulent peturbations in the event of an increase in the centre of the combustion diameter. This is a common phenomenon and is observed in the combustion of substances that are in different aggregate states. The turbulence results from the development of natural convection. The interreaction between warmed up gaseous combustion products (which flow against gravity) and the surrounding atmosphere intensifies the turbulence effect. The larger the source of combustion is, the more pronounced the disturbance will be.

The critical diameter of the combustion source, above which there is a breach in the laminar flame structure, is a function of the nature of the fuel. It was discovered, for instance, that laminar flames of solid paraffin ($C_{26}H_{54}$), urotropin and ferrocene burning in air change to turbulent flames when the combustion source diameter is in the vicinity of 23-32 mm with very low critical Reynold's number values: $Re_{crit.} = 150$ [12]. The development of fluctuations inside the flame of carbon oxide and the change from a laminar regime into a turbulent regime has been detected at Re = 21 and with a jet (burner) diameter equal to 4 mm [13].

The Reynold's number, which determines the type of motion, is dependent upon the rate of flow motion U, its diameter d and the properties of the substance:

$$Re = Ud\rho/\mu$$

where ρ and μ are the density and viscosity of the substance respectively. In the general regime of gases and liquids, the flow will change at $Re_{crit.} \sim 2300$. A distinction can be drawn between two types of turbulence depending on whether the magnitude of the heterogeneities is larger or smaller than the width of the flame front: microturbulence and macroturbulence. With premixed flames, the acceleration of flame diffusion during microturbulence is caused essentially by an increase in the rate of heat transfer inside the front of the flame, whilst during macroturbulence the acceleration is caused by the change in shape of the flow front.

In the case of diffusion flames, the acceleration of flame spread is caused initially by an increase in the rate of reagent mixing. This leads to an acceleration in the redox process inside the flame front. The conditions under which a diffusion combustion regime changes into a kinetic combustion regime may then be ultimately attained. It is assumed that the turbulence of a flame does not essentially change the mechanism of combustion reactions or the approach adopted to its study - the methods used to study polymer diffusion flames are based on the instrumental technique formerly developed to study the combustion of gases and condensed types of fuel. A survey of the experimental methods applied in determining different flame properties may be found in the references [14, 15].

The study of flames is intended to reveal the basic characteristics of the physical and chemical processes that take place in gaseous phases during the combustion of substances. These characteristics are closely linked to the characteristics of processes that occur directly at the interface area and inside the condensed phase. It is extremely difficult and, at the present time, practically impossible to obtain complete information on the chemistry of polymer combustion. There are great difficulties involved in studying the heterogeneous reactions that take place inside the condensed phase under the influence of the high polymer heating rate prevailing during combustion. The development of a special instrumentation technique is required to quantify the chemical and physical transformation of a polymer substance during combustion in the condensed phase.

Natural and synthetic polymer compounds are highly complex systems. The detailed, kinetic mechanism of degradation is still not clear for the majority of polymers, even at a relatively moderate temperature, nor is the rate of thermal effects. It is known that the macromolecular nature of a substance influences its thermal oxidative composition. Molecular heterogeneity, superstructural organisation and the presence of admixtures and additives are all factors that play an important role in the vaporisation and decomposition of polymers, as are the conditions under which these processes occur.

In order to determine the relationship between the structural characteristics of polymer substances and the mechanism of their combustion it is necessary to know and understand the physical and chemical processes of transformation from the initial substance into the final combustion products at all stages of the transformation. This ultimate goal cannot be achieved without exploring chemical kinetics and the influence that physical factors have upon it.

During the diffusion combustion of polymer materials the chemical reactions of substance formation are diffusion controlled, which means that the transfer of mass and heat plays an important role. The theory of the diffusion combustion of substances is based on the macroscopic kinetics of chemical transformation [4]. Extensive research into the mass and heat transformation phenomena occurring during the combustion process is providing an opportunity to approach a solution to the problem of determining the actual kinetics and mechanism of the chemical transformation of substances during combustion. At present there is a tendency to accumulate information and broaden knowledge on the chemistry of the combustion process through an analysis of macroscopic characteristics.

The macroscopic combustion characteristics include the shape and size of the flame, the structure of the combustion wave, the temperature fields and the concentration profiles of the original, intermediate and final products of the reaction inside the combustion wave, the rate of the process, the critical ignition condition and critical combustion conditions, the heat liberation and the completeness of combustion.

The presence of the various critical phenomena that occur during the initiation and development of combustion is a specific characteristic of the combustion process. The determination and study of critical combustion conditions is considered to be one of the major tasks. Knowledge of the mechanism and the critical conditions of combustion for a polymer material will provide a scientific basis for determining actual flammability and for laying down fire-prevention standards when polymer articles are used in technical fields.

Different critical conditions of ignition and combustion have now been determined for polymer materials: a critical concentration of oxygen in the surrounding atmosphere (oxygen index); a critical temperature (temperature index) and pressure; a critical polymer sample size, above and below which extinction of the process is possible; concentrational limits of flame ignition and flame diffusion, and other critical characteristics.

A mathematical description of the combustion process includes differential equations for the chemical kinetics, heat conductivity and diffusion, these equations being derived from the laws of the conservation of matter, energy and motion for the reacting system in question. The theory of a diffusion flame has, in general, been developed for a three-dimensional state, and a system of corresponding differential equations in three-dimensional steric coordinates is applied.

This type of mathematical analysis is very difficult. In order to solve the flame equations and check them experimentally in practice, use is made of unidimensional combustion models, which simplify the problems. With an analysis of the macroscopic characteristics of quasi-unidimensional flames it is possible to calculate the rate of the process and obtain data on the combustion mechanism.

The diffusion combustion of polymers is studied mostly on the basis of quasi-unidimensional and two-dimensional systems, such as the combustion of a round flat flame with a counterflow of oxidiser, candle-like combustion (analysed along the vertical stationary axis of the flame) and also a flame that spreads over the plane surface of a polymer sample.

In order to analyse the complex combustion process and quantify it, a study is made of the structure of a combustion wave, and the so-called leading zones are determined. These zones determine the general rate of combustion.

The idea of elementary combustion models is a very fruitful one [6]. Modelling the combustion process in order to quantify it is the most promising approach in the development of combustion theory. The validity of the models used can be checked by a comprehensive experimental study of the macroscopic combustion characteristics.

Modelling full-scale and large-scale fire tests could be of great interest for predicting the potential fire hazards of polymer materials and for creating effective fire-prevention measures [16, 17]. The modelling helps in establishing a qualitative relationship between the potential flammability of materials and the results of rapid, small-scale laboratory studies of the processes of polymer combustion. Various approaches have been developed in the modelling of the combustion process. The one most frequently applied is physical modelling, which is based on the similarity theory. The use of computers also plays a large role in the solution of many applied problems of combustion theory.

The study of the mechanism and properties of polymer combustion is at an elementary stage at present. In order to achieve a scientifically sound approach to the problem of flammability reduction and create incombustible polymer materials, it will be necessary to combine the efforts of polymer technologists, physicists and chemists working in the field of combustion, on such problems as the high-temperature degradation of polymers under conditions close to combustion conditions. A study should also be made of the relationship between chemical structure, polymer substructure and the mechanisms of ignition and combustion, and also of the effect of ageing in polymer materials on combustibility. These efforts must be directed towards establishing the mechanisms of the flame-quenching effects of various ingredients and towards the development of methods for a qualitative evaluation of their effectiveness.

References

1. Chem. Age Intern., 1975, vol. 110, no. 2904, pp. 12-17.
2. Chem. Age Intern., 1975, vol. 108, no. 2888, pp. 9-14.

3. Semenov N. N., Usp. Phys. Nauka, 1940, vol. 23, no. 3, pp. 251-265.
4. Frank-Kamenetskii D. A., Diffuziya i teploperedacha v Khimicheskoy Kinetike, 2nd ed., elaborated and edited, Moscow: Nauka, 1967.
5. Merzhanov A. G., Dubovitsky F. I., Usp. Khim., 1966, vol. 35, no. 4, pp. 654-683.
6. Merzhanov A. G., Combust. and Flame, 1969, vol. 13, no. 2, pp. 143-156.
7. Shteynberg A S., Ulihbin V. B., Dolgov E. I., Manelis G. B., in Gorenie i vzriv, Moscow: Nauka, 1972, pp. 124-127.
8. Sarner S., Khimiya reactivnih topliv. V. A. Ilyinsky, tr. from Eng., M.: Mir, 1969, p. 488.
9. Summerfield U. M., Sutherland G., in Issledovanie raketnih dvigateley na tvejordom toplive. Summerfield, tr. from Eng., ed. M.: Izd-vo Inostr. lit., 1963, pp. 104-129.
10. Bakham N. N., Belyaev A. F., Gorenie condensirovannih geterogennih system. M.: Nauka, 1967.
11. Novikov S. S., Pokhil P. F., Ryazantsev U. S., Phys. goreniya i vzriva, 1968, vol. 4, no. 4, pp. 469-473.
12. Maltsev V. M., Maltsev M. I., Koshporov L. Y., Osnovnie Kharacteristiky goreniya. M.: Khimiya, 1977.
13. Maklakov A. I., Zh F. Kh., 1956, vol. 30, ed. 3, p. 708.
14. Fristzom R. M., Vestenberg A. A., Struktura plameny, Metallurgiya, M.: 1969.
15. Pokhil P. F., Maltsev V. N., Zaytsev V. M., Metody issledovania protsessov goreniya i detonatsii. M.: Nauka 1969.
16. Deris J., J. Appl. Sci., Appl. Polymer Symp., 1973, no. 22, pp. 185-193.
17. Kanury A. M., 15th Symposium (Intern.) on Combustion, Pittsburg: Combust. Inst., 1974, pp. 193-202.

Chapter 1

Origin of the Combustion Process: Self-Ignition and Ignition of Polymers

General Concepts of Combustion

The burning of a substance is always the result of a self-accelerating chemical reaction. This progressive self-acceleration, in turn, results either from heat or from chemical reactions in the substance. In practice, there are many different conditions that initiate the combustion process. These conditions are dependent on variations in the chemical nature and the aggregation of flammable substances and on external conditions (the environment).

A distinction can be drawn between two major types of combustion initiation on the basis of the aggregate state of the substance and the self-acceleration of the chemical reaction:

1. A self-accelerating chemical reaction taking place in the whole of the system
2. A self-accelerating chemical reaction not taking place in the entire system but only in a small part (on the surface layer, for instance). Once ignition commences in this particular area the rest of the system ignites without any external influence.

These two regimes (1 and 2) correspond to the self-ignition and forced-ignition process of the system.

Self-ignition is frequently termed either forced ignition or simply ignition. A special case of forced ignition is so-called focus ignition. Focus ignition occurs under the mechanical influence (either friction or impact) of explosives.

There is no clear-cut borderline between self-ignition and forced ignition. Forced ignition is a more complex phenomenon and is characterised by the nonstationary development of the process in time and space.

As self-ignition progresses, the heat coming from outside is distributed rapidly throughout the whole system, and the T^0 of the system becomes equal to the T^0 of the surroundings. The difference in temperatures simply determines the period of thermal inductance (the time the system needs to reach the temperature of the environment).

In spite of the fact that the external heat source may not be present, the reaction will take place simultaneously in the whole system. The rate of reaction is exponentially dependent on the temperature. If the heat of the chemical reaction is not dissipated rapidly in the environment, this will lead to a random increase in temperature and to self-acceleration of the reaction.

Ignition commences at the centre of the system, where heat transfer is inefficient. The size of the system determines the rate of heat transfer and plays an essential role in the self-ignition process. Since the reaction occurs over the entire volume, self-ignition is sometimes called volume ignition.

With forced ignition, the size of the system is unimportant. The external heat is not distributed throughout the full volume of the system. A self-accelerating chemical reaction commences in the local section to which the heat is being applied, and ignition takes place as a result. Forced ignition requires an external impetus.

The system will only ignite when it attains the conditions of the stationary combustion process. If these conditions are not present then either ignition will not take place or the flame will not spread and will die out. The latter condition is typical of flashing. In practical terms, the moment of self-ignition and forced ignition can be identified by the appearance of a flame, by a glowing or a smouldering of the condensed phase or by an increase in pressure or in the system's exothermicity to a level where a stable burning process is ensured.

An examination of the experimental criteria produced concepts for specific temperatures:

T^0 flash
T^0 self-ignition
T^0 ignition

As has already been mentioned, however, these temperatures should not be taken as specific physical constants. They are simply to be regarded as useful parameters. The temperatures only describe a very specific system and the conditions of that system.

When the conditions are kept constant, these parameters are useful for a comparative evaluation of different substances. The time between the moment when the source of ignition starts to take effect and the moment when the flame occurs (when the first self-glowing of the system appears) is called the induction period or delaying period of self-ignition and ignition. There are a number difficulties involved in experimentally defining ignition and self-ignition times. The difficulties are caused by specific physico-chemical processes relating to the origin of combustion.

The combustion of a polymer system may be the result of:

1. Exothermal self-accelerating oxidation reactions that occur in the condensed phase
2. Heterogeneous reactions on the surface of the system
3. Exothermal reactions between gaseous products of polymer degradation and gaseous oxidisers

Specific physico-chemical processes are responsible for the ignition of combustion in a large number of polymer substances, and there are many theories dealing with these specific problems, mostly in the field of solid fuel flammability. It is important to determine the precise quantitative criteria for ignition. Another important problem to be solved is the definition of the critical conditions at which stable combustion develops.

Very often forced ignition of polymers causes the initiation of burning. In this case, the critical conditions of ignition are largely dependent on the ignition source conditions.

So-called chain-reaction self-ignition and ignition (where active particles or atoms cause the self-acceleration of the reaction) has been found to occur in a number of gaseous systems only under pressure. In this case there is practically no critical heating due to the chemical reaction. An explosion-type reaction under isothermal conditions is a very rare phenomenon.

The origin and development of the combustion process for common polymer materials is mostly a result of the ignition of gaseous polymer degradation products resulting from thermal degradation. The following diffusion phenomenon and its subsequent effect on the self-acceleration of the exothermal reactions is very important.

Let us take a brief look at the existing theories which give a mathematical description of the ignition initiation process of polymers and examine some examples of experimental research carried out on specific systems.

Theory of Self-Ignition and Ignition of Polymers

Ignition in the Condensed Phase

The process of self-ignition and ignition in the condensed phase can best be described using classical thermal theory. We assume that:

1. There is no mass transfer due to diffusion
2. The reactions in the condensed phase are not sensitive to the conditions of the gaseous environment (for instance, pressure). Nitrocellulose could be taken as an example.
3. Oxidising groups are present in the molecular structure of the substance.

Thermo-oxidative, exothermal decomposition reactions occur in the solid phase. This process may take place either in the stationary phase or in the self-ignition and ignition phases.

According to the stationary theory of thermal explosions, the system in the initial state is presumed to be inert. The reaction starts after an increase in temperature has occurred. The velocity of the reaction ω changes in accordance with the Arrhenius law:

$$\omega = k_0 \exp(-E/RT)\, \varphi(\eta), \tag{1.1}$$

where

k_0 preexponential factor
E activation energy
R universal constant
T absolute temperature
$\varphi(\eta)$ rate law for the reaction under isothermal conditions

The stationary distribution of temperature during the time of heat liberation can be described by the following equation

$$\lambda \left(\frac{\partial^2 T}{\partial x^2} + \frac{\partial^2 T}{\partial y^2} + \frac{\partial^2 T}{\partial z^2} \right) + Q\omega = 0$$

or (1.2)

$$\lambda \nabla T = -Qk_0 \exp(-E/RT)\, \varphi(\eta),$$

where

λ heat conductivity factor of the system
Q heat of reaction

The stationary distribution of temperature within the system constitutes a criterion of self-ignition. The essential assumptions made in the calculation of the critical conditions are:

1. A zero order of reactions
2. Neglection of substance burning during the induction period

If δ is the dimensionless Frank-Kamenetskii parameter [1], then:

$$\delta = (Q/\lambda)\,(E/RT_0^2)\,r^2 k_0 \exp(-E/RT_0), \qquad (1.3)$$

where

r active area of system

T_0 temperature of environment

This parameter expresses the relationship between the heat absorbed and the heat removed (cooling). The parameter also allows the conditions of explosion to be determined.

A thermal explosion takes place when $\delta = \delta_{crit.}$. The value of $\delta_{crit.}$ was calculated by Frank-Kamenetskii for the symmetrical regions of the geometric forms

1. Plane-parallel: $\delta_{crit.} = 0.88$
2. Cylindrical: $\delta_{crit.} = 2.00$
3. Spherical: $\delta_{crit.} = 3.32$

These values of $\delta_{crit.}$ for gases were derived under stable vessel wall temperature conditions.

The conditions of stability correspond to the point at which the Biot number, Bi, tends to infinity. The Biot number determines the ratio between the external and internal transmission of heat:

$$\mathrm{Bi} = \alpha r/\lambda$$

where

α heat transfer coefficient

There are a number of different phases for condensed systems, from $\mathrm{Bi} \to 0$ (self-ignition phase) to $\mathrm{Bi} \to \infty$ (combustion phase). There is an approximate formula for the calculation of $\delta_{crit.}$ in condensed systems [2]:

$$\frac{1}{\delta_{crit.}} = \frac{1}{\delta_n} + \frac{e}{(1+n)\,\mathrm{Bi}}, \qquad (1.4)$$

where

n form (configuration) characteristic ($n = 0, 1, 2$)

δ_n critical value, computed by Frank-Kamenetskii for each form under consideration.

There is no dependence on the temperature distribution over the volume of the system when the more versatile nonstationary theory of thermal combustion is used. Only the mean value of the temperature is employed. Initial heating, its variations in time and the degree of matter conversion are the parameters analysed in the theory. It is assumed that the initial temperature is equal to the environment temperature T_0. Using Θ as the dimensionless variable of temperature and τ as the dimensionless variable of time:

$$\theta = (E/RT_0^2)\,(T - T_0) \qquad (1.5)$$

and

$$\tau = (Q/c)\,(E/RT_0^2)\,k_0 \exp\,(-E/RT_0)\,t, \qquad (1.6)$$

where

c specific heat capacity

t time

In the simplest case, when $\Psi(\eta) = 1$, the rate of heat absorption as a result of exothermal reaction can be described by the following equation:

$$\frac{d\theta}{d\tau} = \exp\theta - (1/\mathrm{Se})\,\theta. \qquad (1.7)$$

where

$d\Theta/d\tau$ heat liberation

$\exp\Theta$ heat absorption

$(1/\mathrm{Se})\Theta$ heat removal

The initial conditions of self-ignition are:

$\tau = 0, \ \Theta = 0$

Thus the Semenov number Se [3] is another dimensionless parameter and serves as an equivalent to parameter δ:

$$\mathrm{Se} = \frac{t_-}{t_+} = \frac{QV}{\alpha S}\,\frac{E}{RT_0^2}\,k_0 \exp\left(-\frac{E}{RT_0}\right), \qquad (1.8)$$

where

t_- heat (thermal) relaxation time

t_+ adiabatic reaction time

V system volume

S system surface (at which the heat removal from the system is taking place)

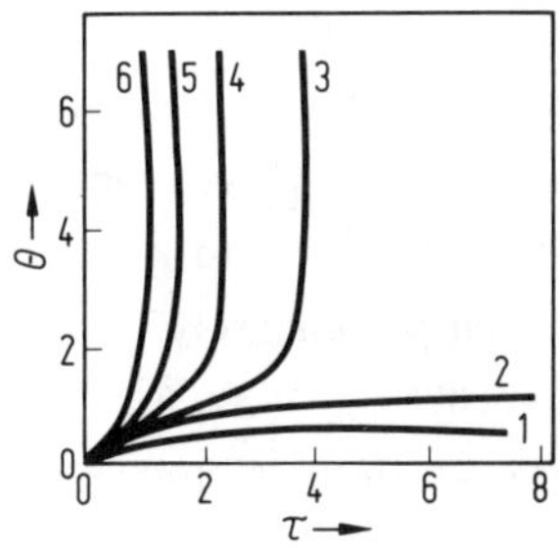

Fig. 1.1 Relationship between dimensionless parameter of self-ignition heating θ and dimensionless parameter of time τ. Se: 1 = 0.31, 2 = 0.367, 3 = 0.407, 4 = 0.49, 5 = 0.82, 6 = 6.1.

Thermal self-acceleration of the process starts when the values of the Semenov numbers exceed $Se_{crit.} = e^{-1} \sim 0.368$ (Fig. 1.1).

The thermal theory of ignition makes it possible to calculate the nonstationary temperature distribution in the substance and also such parameters as:

1. Ignition time delay
2. The amount of heat supplied by the ignition source (by the moment of ignition)
3. The width of the chemical reaction zone (in the compound)

When the values of critical temperatures and induction periods are taken into account, it is possible to calculate the kinetic parameters of the exothermal reaction in the condensed phase. The nonstationary temperature distribution inside the reacting system has been analysed by the authors of this work. Taking into account the rate of heating in the system, they established the existence of the upper limit of the self-ignition regime $\delta'_{crit.}$ [4].

The initial heating of the entire system occurs virtually simultaneously at $\delta \leq \delta'_{crit.}$ when system heating is due to chemical reaction. The maximum temperature prevails in the centre of the system throughout the entire heating stage. When δ reaches $\delta'_{crit.}$ the initial heating occurs near the surface but the heat wave has sufficient time to reach the centre of the system. The ignition of the system takes place in the same way as in the absence of a heat wave.

When $\delta > \delta'_{crit.}$ the heat wave does not have time to reach the centre of the system. A transitional region can be observed. It is interesting that at $\delta_{crit.} < \delta < \delta'_{crit.}$ the self-ignition regime can exist within a very narrow domain of parameters. In the case of the cylindrical samples, for instance, variation of the Frank-Kamenetskii number δ to $\delta'_{crit.}$ corresponds to an enlargement of sample diameter of only 2.4 times or to an increase of just a few degrees in the temperature (by approximately $1.8RT_0^2/E$). Thus a self-ignition regime exists when $2.07 < \delta < 12.0$.

Figure 1.2 shows the region of δ values in which variation of the origin of the combustion region takes place (in other words, the space-dimensionless ignition parameter $\xi = x/r$ shifts from the centre of the system). Combustion is taking place at δ equals 100 and more to reveal the basic principles of the ignition process in the condensed phase. The simple model of forced ignition has been taken as an example [5]. There are several simplifications assumed in this model:

1. Stable physical properties of matter throughout the process
2. Unidimensional heat transfer
3. A zero order of chemical reaction

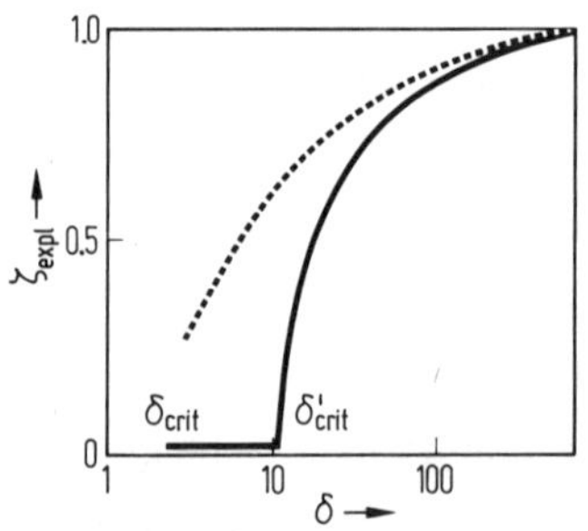

Fig. 1.2 Relationship between dimensionless spatial coordinate of ignition $\xi_{expl.}$ and Frank-Kamenetskii parameter. Dotted line indicates combustion regime.

The work of a heating source is expressed in relation to the temperature conditions on the surface of the condensed phase. Two limiting conditions are usually examined:

1. Surface temperature $T_s = T_0$ = const.; heat flow near the surface fluctuates with time
2. The heat flow supplied to the surface, $q_0 = -\lambda(\partial T/\partial x)_s$ = const., and the surface temperature increases with time

The task is to solve the following equation:

$$\frac{\partial T}{\partial t} = a\frac{\partial^2 T}{\partial x^2} + \frac{Q}{c\rho} k_0 \exp\left(-\frac{E}{RT}\right), \tag{1.9}$$

where

$a = \lambda / c\rho$ thermal conductivity factor of a substance

ρ density of the substance

The most important point of the analysis is the selection of the ignition criterion. Thus the concept of scale temperature, T_*, is introduced. T_* characterises the self-acceleration of the chemical reaction [6].

The ratio of the rate of heat supply from the thermal source to the rate of heat liberation in the chemical reaction is taken as the basis for the selection of T_*:

$$Qk_0 [\exp(-E/RT_*) - \exp(-E/RT_{in})]\, \sigma x_0 = \begin{cases} q_0, \\ \alpha (T_0 - T_*), \end{cases} \tag{1.10}$$

where

$$x_0 = [(RT_*^2/E)(\lambda/Qk_0)\exp(E/RT_*)]^{1/2} \tag{1.11}$$

x_0 the extent of the chemical reaction zone (when combustion occurs)

σ dimensionless heat flow parameter, in the case of $T_* \sim T_1$, $\sigma = 4.2$

T_{in} initial temperature

The essential similar property in the fundamental ignition theory is the following magnitude:

$$\theta_i = \frac{E}{RT_*^2}(T_* - T_i), \tag{1.12}$$

where

T_i temperature of the substance

The T_* values are not adequate for different heating conditions and different heat transfer mechanisms.

In the case of ignition at a constant surface temperature, the surface temperature is taken as a scale temperature, which determines the course of a chemical reaction.

When combustion is caused by a stationary heat flow, temperature T_* correlates with temperature T_1 at the inflection of the curve of surface temperature increase over time (Fig. 1.3). The ultimate case of adiabatic self-ignition takes place at $T_* = T_{in}$, $q_0 = 0$, $\alpha = 0$. The ignition performance temperature is assumed to be either close to or equal to the scale temperature by the method of approximation.

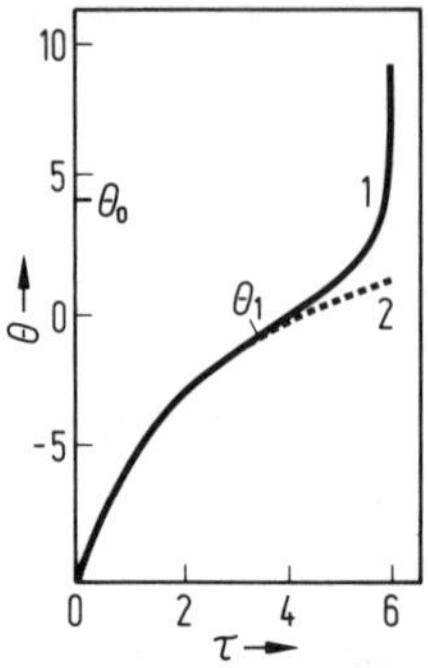

Fig. 1.3 $\theta(\tau)$ relationship. Substance is being heated by a steady heat flow. $\sigma_0 = 4.2$. 1 = chemically active substance, 2 = chemically inert substance, θ_0 = temperature of heat flow, θ_1 = temperature at inflection point.

The heating of a substance in order to ignite it may be carried out by

1. A hot gas (convective heat transfer)
2. An incandescent material (conductive heat transfer)
3. A flow of radiant energy

If the ignition of a substance is caused by a flow of hot gas at a constant temperature $T_0 > T_{in}$, then the heat exchange on the surface is described by Newton's law:

$$q_S = -\lambda \frac{\partial T}{\partial x}(0, t) = \alpha (T_0 - T_S), \tag{1.13}$$

where

q_s and T_s are the heat flow and surface temperature respectively.

Using dimensionless temperature parameters it is possible to present this equation as

$$\sigma_S = \frac{\theta_0 - \theta_S}{\theta_0}, \tag{1.14}$$

Parameter θ_0 characterises the conditions of heat transfer to the surroundings:

$$\theta_0 = \frac{E}{RT^2_{ign.}}(T_0 - T_{ign}) = \frac{\lambda}{\alpha x_{ign.}}. \tag{1.15}$$

where

$T_{ign.}$ ignition temperature

$x_{ign.}$ specific cross-section of the chemical reaction

The calculation of the ignition temperature $T_{ign.}$ is reduced to the solution of equation 1.9 at the boundary conditions (1.14).

The ignition time delay $\tau_{del.}$ is determined from the ratio of the dimensionless parameters:

$$\frac{\theta_0}{\theta_0+\theta_i}=\varphi\left(\frac{\sqrt{\tau_{del.}}}{\theta_0}\right). \qquad (1.16)$$

An analysis of the ignition mechanisms in the solid phase shows that the ignition time delay is a function of the heat exchange coefficient α:

$$\tau_{del.} \sim \alpha^{-m}$$

and of environment temperature T_0. In a certain narrow range of T_0 temperatures

$$\tau_{del.} \sim \exp(A/RT_0)$$

where A = const.

The A values are only close to the actual value of activation energy E in extreme cases, when the surface temperature is constant and equal to the environment temperature.

Small values of temperature coefficient A for the convectional heat exchange could be consolidated by the fact that initiation of the reaction actually starts at the lower temperature [5]:

$$(T_{ign.} < T_0)$$

When incandescent material causes ignition, such as a hotplate with constant temperature inducing the heat exchange, it is assumed that the temperature of the superficial layer momentarily takes on a certain T_S value and remains constant during the combustion process.

Taking the specific temperature of an incandescent state of semi-infinite size and a ratio of the thermal activities of the contact bodies of $k_\xi = (\lambda_1 c_1 \rho_1/\lambda_2 c_2 \rho_2)^{1/2}$ it is possible to arrive at condition at which the surface temperature equals the temperature of ignition [7]:

$$T_{ign.}=T_S=T_{in}+(T_0-T_{in})\frac{k_\xi}{k_\xi+1}, \qquad (1.17)$$

where

T_{in} initial temperature of the ignitable substance.

When the ignition of the substance is caused by a radiant energy flow it is possible, as a function of the optical properties of the matter, i.e. absorption coefficients, for a certain portion of the energy to be absorbed by inner layers and cause additional warming of the substance. This amount of absorbed energy is proportional to the incident flow q_0 and to the absorption coefficient μ. The amount of absorbed energy decreases exponentially over the depth of layer x. In this case, equation 1.9 needs to be modified by the inclusion of the term $q_0\mu \exp(-\mu x)$, allowing inclusion in the analysis of a heat source of a nonchemical nature within the heat source.

Reviews of ignition in condensed phase models are given in the literature [5, 8].

Two variants are observed:

1. A constant incident radiant energy flow q_0 (this correlates with the ignition experiments in arc reflection furnaces)
2. An incident radiant energy flow that varies with time, in accordance with a given law

In both cases there is an analysis of:

1. The conditions at which heat loss to the environment is absent
2. The conditions at which the heat supplied to the surface of the heated body may be partially lost to the surroundings as a result of either thermal conductivity or forced convection.

Both the self-ignition and forced ignition of polymers in condensed phase theories have been examined in detail taking nitrocellulose as an example. The following method has been employed for the self-ignition study: spherically shaped particles are injected into the stream of heated gas which has

- A low value Reynold's number, $\mathrm{Re} \leq 1$
- A low value Grashof's number:

$$\mathrm{Gr} = (gd^3/\nu^2)\,\gamma\Delta T < 1$$

where

g acceleration due to gravity

d diameter of a particle

γ coefficient of cubic expansion due to temperature increase

ν kinetic viscosity of gas

The Grashof number determines the type of motion in the case of natural convection. With small Re and Gr numbers the heat exchange between the particles and the gas takes place by thermal conductivity.

Heat transfer (heat emission) coefficient $\alpha = \lambda_g / r$

where

λ_g thermal conductivity of gas

r radius of a particle

A schematic drawing of the assembly used is shown in Fig. 1.4.

The time between the moment of injection and the moment of ignition (time delay $t_{del.}$) was determined both visually and with the help of a special photograph and film recording system. In this experiment $t_{del.}$ was determined in relation to the size of the particles, the temperature and the gas composition [9, 10].

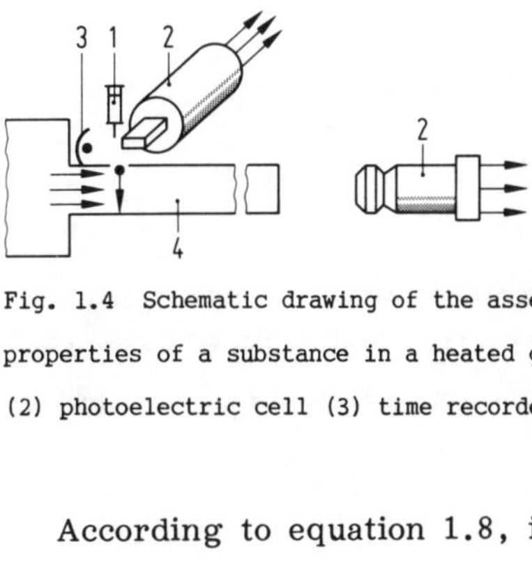

Fig. 1.4 Schematic drawing of the assembly used for the determination of self-ignition properties of a substance in a heated gas: (1) syringe for injecting the particle, (2) photoelectric cell (3) time recorder, (4) reaction vessel.

According to equation 1.8, it is possible to express the critical self-ignition condition of a particle (thermal explosion) in the form

$$\ln (\alpha S T_{0,crit.}^2 / V) = \ln [Q k_0 \exp (E/R)] - (E/R)(1/T_{0,crit.}). \qquad (1.18)$$

The average value of the temperature of explosive and nonexplosive decay of nitrocellulose has been taken as the critical temperature $T_{0,crit.}$. The experimental results correlate well with the straight line, which is expressed in semilogarithmic coordinates of equation 1.18 (Fig. 1.5).

The effective kinetic parameters of nitrocellulose degradation were calculated from the gradient of the straight line and the segment cut off from the coordinate axis. In the interval 223-225°C:

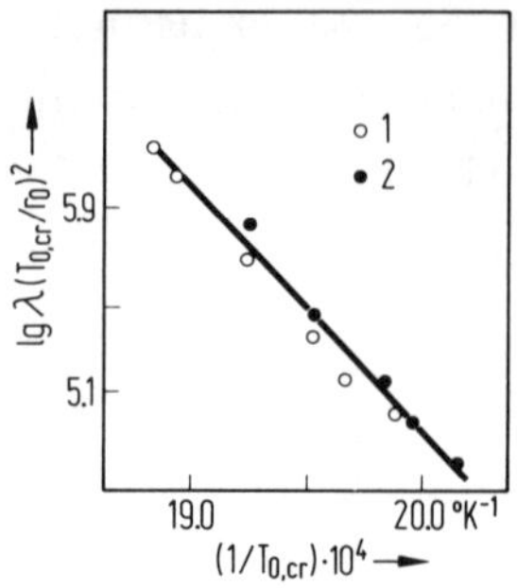

Fig. 1.5 Relationship between the critical condition for thermal explosion of nitrocellulose spherical particles and the reciprocal temperature of heated gas (see equation 1.18).
1 = air, 2 = argon.

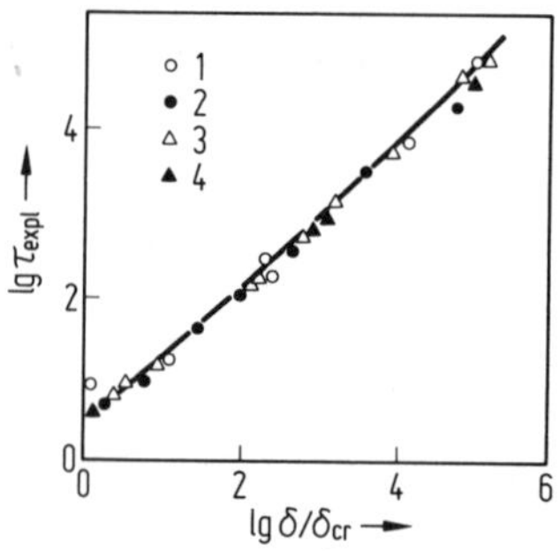

Fig. 1.6 Logarithmic relationship between the dimensionless parameter of thermal explosion of time $\tau_{expl.}$ and the ratio $\delta/\delta_{crit.}$ for spherical nitrocellulose particles in a heated gas. Radius of particles in microns: 1-50, 2-100, 3-125, 4-150. The dotted line is the calculated relation.

E = 48,500 cal/mole, $Qk_0 = 5.5 \times 10^{21}$ cal/cm$^2\cdot$sec

These kinetic parameters are used to calculate the self-ignition time delay of the nitrocellulose particles which may be compared with experimentally obtained values of $\tau_{expl.}$ (Fig. 1.6).

Dimensionless ignition time delays $\tau_{expl.}$ were calculated according to a formula which takes in the following parameters:

- The heating time needed for the particle to reach the temperature of the gas τ_h
- The induction period $\tau_{ind.}$ (pre-explosive heating time for a particle)

$$\tau_{expl.} = \tau_h + \tau_{ind.}, \tag{1.19}$$

where

$$\tau_h = \{0{,}819\,\theta_{in}^{0{,}27} + \frac{1}{1{,}32 - 0{,}04\,\theta_{in}}[(1 + 4\chi\beta)\Delta -$$
$$- 1{,}06]^{0{,}914}\Delta^{-0{,}01}\}(1 - 1{,}6\beta);$$
$$\tau_{ind.} = \frac{\Delta^{0{,}7}}{(\Delta - 1)^{0{,}7}}, \quad \Delta = \delta/\delta_{crit.},$$
$$\beta = RT_0/E, \quad \chi = \varepsilon\sigma T_0^2 E r_0/\lambda_g R, \quad \theta_{in} = (E/RT_0^2)(T_{in} - T_0).$$

The magnitude of $\tau_{expl.}$ may be expressed as a function of a single parameter $\delta/\delta_{crit.}$, since Θ_{in}, β and X remain practically unchanged during the entire experiment.

The concurrence of estimated and experimental $\tau_{expl.}$ data confirms that the theory of thermal explosion is applicable to the self-ignition of the polymer used in the condensed phase.

The ignition of spherically shaped particles is possible in combustion models either when the temperature of the heated gas is high or when the particles are of a large diameter. Studies of forced ignition of nitrocellulose (the forced ignition was caused by a current of heated gas) have been carried out on cylinder-shaped samples [11-13].

It was found that the ignition time delay for nitrocellulose $t_{del.}$ is a function of the following parameters:

1. The heat transfer coefficient α at T_{in} = const., T_0 = const.:

   $t_{del.} = \alpha^{-m}$, $m = 1.64$

2. The temperature of flow of gas T_0 (in the interval 290-350°C) at α = const., T_{in} = const.:

   $t_{del.} \sim \exp(A/RT_0)$, A = 8300 cal/mole, $A \leqq E$

3. The initial temperature of a sample T_{in} at α = const. and T_0 = const.:

   $t_{del.} \sim (T_0 - T_{in})^n$, $n = 2$

From the forced ignition experiments and a knowledge of $t_{del.}$ (α, T_0,

T_{in}) relationships, it was possible to determine the kinetic parameters of nitrocellulose degradation.

The method of successive approximations was one method used. Equation 1.10 at T_{in} = const. appears as:

$$\alpha (T_0 - T_*) = 4{,}2\,[\lambda Q k_0\,(RT_*^2/E)\exp(-E/RT_*)]^{1/2}. \tag{1.20}$$

The ignition time delay for the samples is calculated according to the following formula [6]:

$$t_3 = 0{,}46\,\frac{c\rho\lambda}{\alpha^2}\left(\frac{T_* - T_{in}}{T_0 - T_*}\right)^2\left(1 + \frac{12}{(T_* - T_{in})\,E/RT}\right). \tag{1.21}$$

The subsequent steps are:

1. Assumption of a reasonable value for E
2. Determination of T_* (α, T_0) from equation 1.21
3. Determination of a new value for E from equation 1.20

The final results of E and Qk_0 are thus determined after a number of approximations.

The second approximation method is based on the following relationship [12]:

$$\lg\frac{t_{del.}T_0^{2+m}}{(T_0 - T_{in})^n} = \text{const} + 0{,}4343\,\frac{E}{R}\left(1 - \frac{m}{2}\right)\frac{1}{T_0}, \tag{1.22}$$

where m and n equal 1.64 and 2 respectively.

The value of E for nitrocellulose, determined from the gradient of the line (coordinates taken from equation 1.22) is equal to 48,000 cal/mole in the temperature interval T_0 = 290-350°C (Fig. 1.7).

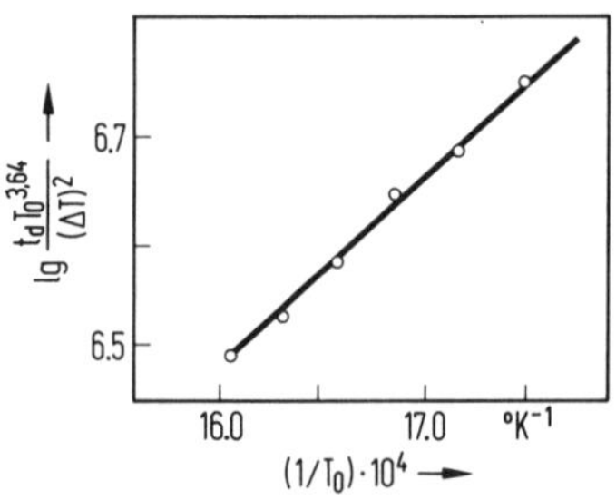

Fig. 1.7 Interdependency of ignition characteristics of nitrocellulose in semilogarithmic coordinates of equation 1.22.

Heterogeneous Ignition

In the case of heterogeneous ignition of flammable solids, the self-accelerating exothermal reaction takes place at the interface between the solid fuel and the gaseous oxidising agent. The two types of heterogeneous ignition under consideration [14] involve either:

1. No internal heating; ignition is initiated by a spontaneous exothermal surface reaction at room temperature immediately after the oxidising agent comes into contact with the surface of the solid phase, or
2. The heterogeneous reaction between the gaseous oxidising agent and the solid fuel is initiated and sustained by internal heating. In the initial state the system is assumed to be inert.

The first type of heterogeneous ignition is called hypergolic. The ignition of polymer substances in an atmosphere of highly active gaseous oxidants (F_2, ClF_3) at a common initial temperature is representative of the first type of reaction.

The ignition of high-molecular-weight compounds in an atmosphere of less active gaseous oxidising agents such as O_2[1)] is representative of the second type of reaction.

A process that involves an exothermal self-accelerating reaction between the gaseous degradation products of the fuel and the gaseous oxidant is not possible. There are a few examples of similar heterogeneous ignition of high-molecular-weight substances. These reactions include the ignition of polymethylmethacrylate in an atmosphere of heated oxygen at a high pressure and the ignition of carbon fibres and carbon particles in an atmosphere of either hot air or hot oxygen.

The initiation of combustion as the result of a heterogeneous reaction on the surface of a substance is determined totally by the kinetics of the reaction - by the conditions of the heat exchange and mass exchange of the fuel and oxidant on the surface.

The mathematical description of this nonstationary process includes a system of nonlinear differential equations. These equations take into account the diffusion of gaseous oxidant and degradation products of the

[1)] The rate of a heterogeneous reaction at an initial low temperature (room temperature) is negligible. It is necessary to raise the temperature of oxidising or the surface temperature to make the reaction rate noticeable.

substance, which can also act as oxidising agents. This latter factor complicates the analysis and solution of the mathematical problem.

While analysing a thermal regime on the surface during a heterogeneous reaction, Frank-Kamenetskii discovered that the heterogeneous ignition of solid combustible materials is inevitably related to the transition of the reaction from a kinetic regime to a diffusion regime.

The magnitude of the surface reaction and the conditions of the heat exchange with the surroundings are the two factors that determine the heterogeneous regime of ignition: self-ignition or combustion. Both regimes can be observed in hypergolic ignition and in heterogeneous ignition with external heating.

This conclusion has been established from the analysis of a substantially simplified thermal model of heterogeneous ignition without allowance for the diffusion process [15].

When the process of self-ignition is progressing, the heat that the substance receives from a heat source and from the reaction surface is able to spread evenly throughout the entire area of the presurface layer. Furthermore, the presurface layer correlates with the specific size of the sample.

With a combustion regime, the thickness of the heated layer in which the heterogeneous reaction occurs is less than the specific size of the sample.

The heterogeneous ignition theory is based on the application of unidimensional models. These unidimensional models consider gaseous and condensed phases in the form of semi-infinite regions with invariable, isotropic properties.

Figure 1.8 shows a model of heterogeneous ignition of a polymeric substance with external heating [16]. The thermal flow on the surface is constant (q_0 = const.). The transmission of heat from the heated oxidant

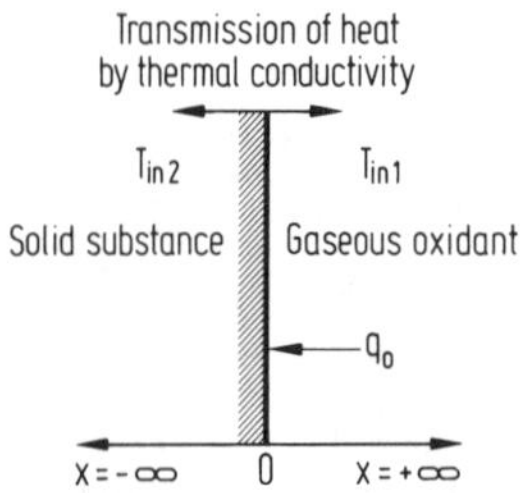

Fig. 1.8 Schematic drawing of a heterogeneous ignition model for a polymeric substance.

(initial temperature $T_{in\ 1}$) to the surface of the solid substance (initial temperature $T_{in\ 2}$) occurs as a result of thermal conductivity. According to this model, the heterogeneous ignition of the substance is described by the thermal conductivity equations for the fuel and the oxidant and by the diffusion equation for the oxidant. The temperature distribution in the gaseous state ($y>0$) is described by the equation

$$\frac{\partial\theta_1}{\partial\tau}=\frac{\partial^2\theta_1}{\partial y^2}, \tag{1.23}$$

and in the solid state ($y<0$) by the equation

$$\frac{\partial\theta_2}{\partial\tau}=\frac{a_2}{a_1}\frac{\partial^2\theta_2}{\partial y^2}, \tag{1.23a}$$

where

$\theta = (T - T_*)(E/RT_*^2)$ dimensional variable of temperature

$\tau = t/t_0$ dimensionless variable of time

$t_0 = r_0^2 c_1 \rho_1/\lambda_1$ time scale

$y = x/r_0$ dimensionless spatial coordinate variable

$r_0 = (RT_*^2/c_0^{\nu} EQk_0)\exp(E/RT_*)$ length scale

$a_i = \lambda_i/\rho_i c_i$, i = 1, 2 (indices 1 and 2 belong to an oxidant and a solid substance respectively)

The diffusion of a gaseous oxidant towards the surface of a solid substance ($y>0$) is described by the equation

$$\frac{\partial c}{\partial\tau}=\mathrm{Le}\,\frac{\partial^2 c}{\partial y^2}, \tag{1.24}$$

where

$\mathrm{Le} = D/a$ Lewis parameter.

The object is to solve this system of equations for the initial and boundary conditions:

$$\theta_1|_{y=\infty}=\theta_1|_{\tau=0}=\theta_1^0, \qquad \theta_2|_{y=-\infty}=\theta_2|_{\tau=0}=\theta_2^0, \tag{1.25}$$

$$\theta_1|_{y=0}=\theta_2|_{y=0}, \qquad m\left(\frac{\partial\theta_2}{\partial y}-\frac{\partial\theta_1}{\partial y}\right)\bigg|_{y=0}=\sigma+c_0^{\nu}\exp\frac{\theta_0}{1+\beta\theta_0}, \tag{1.26}$$

$$c|_{y=\infty}=c|_{\tau=0}=1, \quad \frac{\partial c}{\partial y}\bigg|_{y=0}=\frac{\gamma}{\mathrm{Le}}c_0^{\nu}\exp\frac{\theta_0}{1+\beta\theta_0}. \tag{1.27}$$

where

c_0 initial concentration of oxidant

ν the order of a heterogeneous reaction according to the oxidant

$\sigma = (q_0 / Q k_0 c_0^{\nu}) \exp(E/RT_*)$ dimensionless thermal flow

$\gamma = \rho_1 c_1 R T_*^2 / Q E c_0$; $Le = D/a_1$; $\beta = RT_*/E$; $m = \lambda_2/\lambda_1$

The scale temperature T_* is represented in this case by the temperature of the interface at the initial moment for which

$$T_* = (T_{in1} + k_\xi T_{in2})/(1 + k_\xi), \tag{1.28}$$

where $k_\xi = \varepsilon_2/\varepsilon_1 = (\lambda_2 c_2 \rho_2 / \lambda_1 c_1 \rho_1)^{1/2}$ is the relative coefficient of thermal activity.

The conditions at which ignition occurs are determined by the time relationship between the temperature of an interface $\Theta_0(\tau)$ and the concentration of oxidising agent at the interface $c_0(\tau)$.

Figure 1.9 shows the variation of Θ, the dimensionless parameter that is a function of surface temperature when a first-order chemical reaction occurs, for different conditions. The order of the chemical reaction is determined by the oxidant ($\nu = 1$).

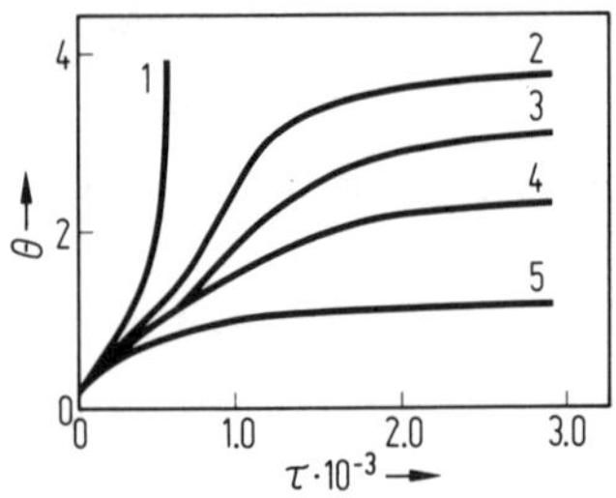

Fig. 1.9 $\Theta(\tau)$ relation at $\sigma = 0$, $\beta = 0.03$, $k_\xi = 40$, $\nu = 1$ and different $\bar{a}$. (1) $\bar{a} = 0.0014$, (2) 0.0057, (3) 0.0071, (4) 0.0085, (5) 0.0141.

The transition of a heterogeneous reaction on the surface from a kinetic to a diffusion regime is taken as an ignition criterion (ignition number).

$$\left. \frac{d^2\Theta_0}{d\tau^2} \right|_{\tau=\tau_*} = 0, \tag{1.29}$$

where

τ_* induction period for the surface self-heating.

The following critical effects have been discovered during heterogeneous ignition:

1. Concentration ignition range
2. Limitations that are due to heat loss to the surroundings
3. So-called thermokinetic limitations, which depend on the thermophysical properties of a system and on the kinetic parameters of the surface reaction [16-18].

Figure 1.9 illustrates the situation when heating of the surface depends on the change in a thermokinetic parameter $\bar{a}$ [16].

$$\bar{a}_* = c_1\rho_1 RT_{*\text{lim}}/QE. \tag{1.30}$$

where

$\bar{a}$ thermokinetic limit

for the ignition conditions, when $\bar{a} \leq \bar{a}_*$:

$$\tau_* = \frac{\pi(1+k_\xi)^2}{2[\pi+2(\sigma+1)(1-2\beta)]}. \tag{1.31}$$

The induction period of ignition increases as the relative coefficient of thermal activity increases and diminishes with the growth of a thermal flow, temperature T_*.

An experimental study has been conducted of the heterogeneous ignition of polymethylmethacrylate (PMMA) in an oxygen atmosphere [19]. The experiment was carried out on high-pressure apparatus [P = (50-100) x 10^5 N/m²]. This apparatus allows the temperature of a polymer sample and an oxidising gas to be adjusted. The chemical drawing of the apparatus is shown in Fig. 1.10. The application of pressure suppresses the ignition of decomposition products in the gaseous state. Under these conditions, the molar concentration of the monomer in the diffusion layer (above the surface of the sample) is much lower than its concentration at the ignition limit.

The temperature of an oxidant near the surface ($T_{\text{in }1}$) and the temperature of a solid sample ($T_{\text{in }2}$) have been measured using a thermopile. The temperature of the contact area T_* has been calculated from equation 1.28 at various experimental values of $T_{\text{in }1}$, $T_{\text{in }2}$ and k_ξ.

The induction period of heterogeneous ignition diminishes exponentially with the increase of T_* (Fig. 1.11).

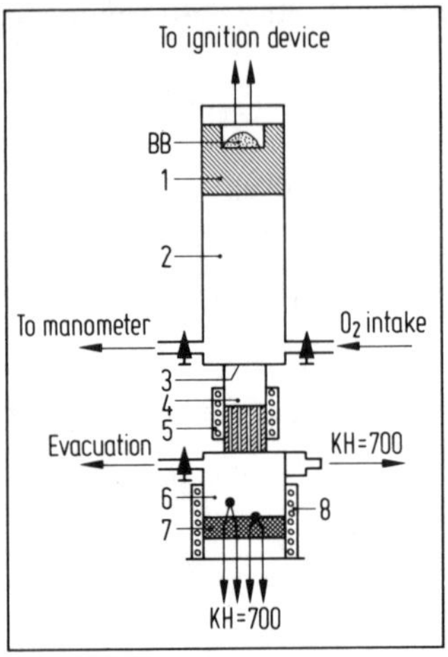

Fig. 1.10 Schematic drawing of a high-pressure unit used for the experimental study of heterogeneous ignition of polymer materials. (1) Piston (plunger), (2) compression chamber, (3) diaphragm, (4) tubing, (5) cellular honeycomb, (6) combustion chamber, (7) sample, (8) heater.

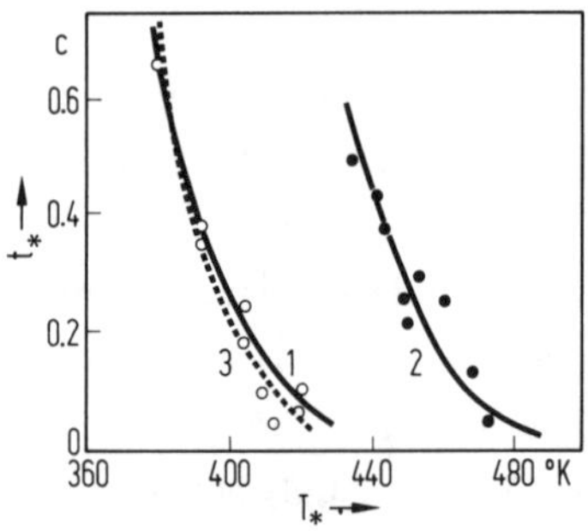

Fig. 1.11 Relationship between induction period t_* and the surface temperature of PMMA in an oxygen atmosphere. (1) k_ξ = 13, (2) 15, (3) computed data.

If there is no thermal exchange with the surroundings (α = 0), then using the dimension magnitudes, equation 1.31 will take the form:

$$t_* = \varepsilon_1^2 \left(\frac{RT_*^2}{c_0^{\nu} E Q k_0} \exp \frac{E}{RT_*} \right)^2 \frac{\pi (1 + k_\xi)^2}{2 [\pi + 2 (1 - 2RT_*/E)]}, \tag{1.32}$$

where

ε_1 coefficient of thermal activity of oxidiser

Q enthalpy of the reaction

ν the order of the reaction with respect to the oxidant (ν = 1)

c_0 initial volume concentration of the oxidant

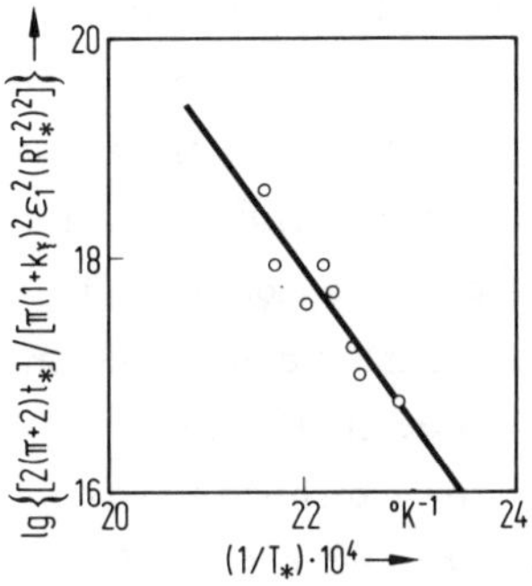

Fig. 1.12 t_* (T_*) relationship for PMMA in oxygen atmosphere at k_ξ = 15. Coordinates of equation 1.32 are used.

The diagram, which is constructed by using the semilogarithmic coordinates of equation 1.32, permits the kinetic parameters of the heterogeneous reaction of PMMA with oxygen to be determined (Fig. 1.12).

In the temperature ranges $T_{in\ 2}$ = 380-470°K and $T_{in\ 1}$ = 480-550°K, at P = (50-80) x 10^5 N/m², the following effective values have been obtained for the activation energy: E = 50,000 J/mole, Qk_0 = 8.4 x 10^{10} J/m²·sec.

The heterogeneous ignition of carbon in an atmosphere of oxygen, which is not complicated by any ignition of volatile products within the gaseous phase, has been studied by Frank-Kamenetskii and co-workers [1]. Gaseous products which could ignite within the gaseous phase were removed from this zone as the velocity of the oxidiser flow increased.

Ignition in the Gaseous Phase

In the theory of the ignition of a polymeric substance in the gaseous state it is assumed that the substance is degraded as a result of an external influence. The gaseous degradation products generated diffuse into the surrounding gaseous oxidant and react with it. Thus the self-accelerating exothermal reaction, which leads to the initiation of the flame, occurs in the gaseous phase.

The gaseous phase theory is based on the examination of a unidimensional ignition model of a semi-infinite combustible plate in the atmosphere of a gaseous oxidant. As in the case of the theories discussed earlier, the most important aspect of this theory is the selection of an ignition criterion.

Different models of ignition in the gaseous state have been discussed in the work of Prais and his associates [8]. The mathematical descriptions of all the models of the ignition of a condensed substance in a gaseous atmosphere presume a dependence on a number of assumptions, including the stability of the thermophysical properties of the gas and the condensed phase; all the gaseous components are defined by the same mass diffusion coefficient and thermal conductivity. It is assumed that, in the gaseous phase, heat is liberated as the result of a second-order exothermal reaction. The gasification of a polymer material occurs as the result of a single-stage, zero-order endothermal reaction.

The conditions on the surface of a gaseous combustible material depend on the source of ignition of the heated gas and radiant flow. The model of ignition in the gaseous phase of a polymeric substance is shown in a schematic diagram in Fig. 1.13. As is apparent from the Figure, combustible degradation products enter the gaseous oxidant; the exothermal reactions start in the vicinity of the sample's surface.

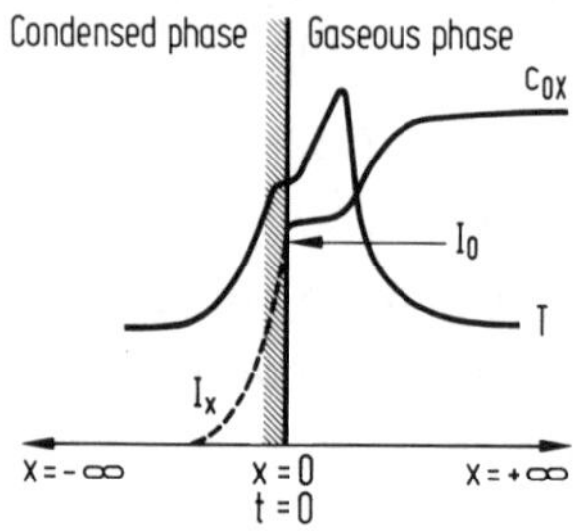

Fig. 1.13 Schematic drawing of ignition in a gaseous state model for polymer materials.

When heating is initiated by radiant flow, the ignition of a polymeric substance in the gaseous phase is described by the following system of equations [20].

The variation of a temperature in the solid phase is:

$$\frac{\partial T}{\partial t} = a_c \frac{\partial^2 T}{\partial x^2} - v_c \frac{\partial T}{\partial x} - \left(\frac{Q_{pyr} k_{0,}}{c}\right) e^{-E_c/RT} + \left(\frac{1}{\rho_c c_c}\right) \frac{\partial I_x}{\partial x} + \left(\frac{\mu Q_{pc.} I_x}{\rho_c c_c}\right), \quad (1.33)$$

and in the gaseous phase is:

$$\frac{\partial T}{\partial t} = a_g \frac{\partial^2 T}{\partial x^2} - v_g \frac{\partial T}{\partial x} + \left(\frac{Qk_{0,g}}{\rho_g c_g}\right) c_{ox} c_f e^{-E_g/RT}. \tag{1.34}$$

The variation of the concentration of fuel in the gaseous phase is:

$$\frac{\partial c_f}{\partial t} = D \frac{\partial^2 c_f}{\partial x^2} - v_g \frac{\partial c_{ox}}{\partial x} - k_{0,g} c_{ox} c_f e^{-E_g/RT}. \tag{1.35}$$

The boundary conditions appear as:

$$T(x, 0) = T_0, \quad T(\pm\infty, t) = T_0, \quad T_g(0, t) = T_c(0, t),$$
$$\lambda_g\left(\frac{\partial T}{\partial x}\right)_{0,t} = \lambda_c\left(\frac{\partial T}{\partial x}\right)_{0,t}, \quad c_f(x, 0) = 0, \quad c_f(+\infty, t) = 0,$$
$$\dot{m}_f = c_f v_g - D\left(\frac{\partial c_f}{\partial x}\right).$$

where

μ thermal emission absorption coefficient

I_x thermal emission absorption function

$\dot{m}_f$ mass fuel consumption

v_c and v_g are convective thermal exchange coefficients for the condensed and gaseous phases respectively. (The terms that express conventional energy transfer are usually neglected.)

$Q_{pyr.}$, $Q_{pc.}$ and Q are the heat of a pyrolysis reaction of a polymer and the heat of a photochemical degradation reaction in a condensed phase and a gaseous state respectively.

Equation 1.33 contains terms that take into account the pyrolysis of a polymer, the heating of its interfacial layer as a result of radiation absorption and the heat liberation as the result of photochemical degradation of the polymer.

Equation 1.35 takes into account the complete combustion of a gaseous fuel as a result of its reaction with the oxidant.

In order to determine the ignition parameters (the induction period of ignition and the energy needed for ignition) it is necessary to solve this system of nonlinear differential equations. The analytical task of expressing the ignition parameters in terms of the thermal flow rate, pressure, oxidant concentration and physical and chemical properties of a combustible material and oxidising agent is very difficult.

Approximate solutions in respect of the induction period of ignition produce the following expression [20]:

$$t_{ign.} \simeq t_{CD} + \frac{1}{\mu}\left(\frac{\rho c \Delta T^*}{I_0(1-r)}\right) + \frac{\pi}{4} a \left(\frac{\rho c \Delta T^*}{I_0(1-r)}\right)^2, \quad (1.36)$$

where

μ absorption coefficient

r reflection coefficient on a solid substance

I_0 incident thermal (radiant) flow

a thermal conductivity coefficient

c specific heat of a solid fuel

ρ density of a solid fuel

$\Delta T^* = T^* - T_0$

T^* surface initiation temperature at which intensive pyrolysis of a polymer material commences

t_{CD} initiation time required for the initiation of the chemical reaction and for the diffusion of the reagents: $t_{CD} \sim 1/P^n$

Either term of equation 1.36 will predominate, as a function of the rate of thermal flow. If the radiant thermal flow is very intense, there will be instantaneous heating of the surface up to the initiation temperature of T^* and $t_{ign.} \approx t_{CD}$.

With an average rate of radiant flow, thermal retardation occurs as a result of the inner layers of the substance absorbing radiation (the ignition induction period is basically determined by the second term of equation 1.36).

Finally, with a low rate of thermal radiant flow, the last term of equation 1.36, which characterises the thermal retardation due to the heating through of the substance by thermal transfer, plays a dominant role.

In the case of convectional thermal flow

$$t_{ign.} \simeq t_{CD} + (\pi/4)a(\rho c \Delta T^*/I_0)^2. \quad (1.37)$$

Parameter I_0 represents a convectional thermal flow from heated gas towards the surface of a solid combustible material for relatively low thermal flow values:

$$(\pi/4)a(\rho c \Delta T^*/I_0)^2 \gg t_{CD}.$$

It is possible to determine the minimum energy required for ignition from the thermal radiant flow reaction time:

$$E = I_0 t_{ign.} \quad (1.38)$$

According to the theory of ignition in a gaseous phase for solid combustible materials, the ignition induction period is a function of the pressure at a constant molar concentration of oxidant in the gaseous phase ($t_{ign.} \sim P^n$) and also of the concentration of the oxidant at a constant pressure ($t_{ign.} \sim c_{ox.}^{-m}$) [8].

In the models of ignition in the gaseous phase for polymer-containing systems, it is assumed that ignition occurs when the surface temperature reaches a critical magnitude at which intense gasification of the polymer is possible [16, 20], or when a sufficiently high rate of temperature increase occurs.

According to reference [21], the polymer material ignites when the speed of liberation of the pyrolised combustible gases reaches its critical value.

The course of a self-accelerating ignition reaction follows the change in direction of the temperature gradient in the gaseous state close to the surface. The ignition point corresponds to the following condition: heat liberation due to an exothermal reaction in a gaseous phase.

The ignition criterion has been derived from an analysis of the equations that describe the laws of energy and mass concentration in a gaseous phase and on the surface of the polymer under the influence of a radiant thermal flow. This ignition criterion, however, is general and restricted to a specific reaction in a gaseous phase:

$$I_0|_{x=0} = 2Q_{pyr.}\, \dot{m}_f|_{x=0} - \lambda_c \frac{\partial T}{\partial x}\bigg|_{x=0}. \qquad (1.39)$$

For ignition of a polymer material to occur, the radiant thermal flow reaching the surface must be equal to twice the thermal flow necessary for pyrolysis of the polymer, minus the heat losses (through thermal conductivity) in a condensed phase.

Using a diffusion combustion model and assuming that the ignition of a solid fuel takes places when the diffusion flame (gasification by-products plus gaseous oxidiser) occurs near the surface, Librovitch [22] demonstrated the need for simultaneous application of two ignition criterion conditions:

1. The surface temperature of a solid fuel must reach its vaporisation (gasification) temperature
2. The temperature gradient near the surface must be less than a

critical magnitude. This critical magnitude depends on the kinetics of the chemical reaction in a diffusion flame and on the rate of supply of an oxidising agent to the surface of a solid fuel during the combustion process.

A number of regimes for heating the solid fuel by radiant thermal flow or by a heated gas have been analysed.

1. Insufficient heat supply regime:

   The heat flow towards the surface of a solid fuel when the surface temperature reaches the gasification temperature is less than the thermal flow that corresponds to a critical gradient.

2. Very strong heat supply regime:

   The temperature gradient near the surface may exceed the critical gradient. In this case, a stationary solid fuel gasification process occurs. The diffusion flame does not appear. Thus in the model of ignition in a gaseous phase the selection of an ignition criterion is ambiguous.

All things considered, the theoretical study of the ignition of a polymer material should be oriented towards the development of a relationship between ignition characteristics, the individual physico-chemical properties of the substance and external conditions.

This problem is highly complex, especially in view of the fact that the physical and thermal properties of polymer materials have not been investigated over a broad range to date. The chemistry of the high-temperature decomposition process is not clear.

The unidimensional ignition models are based on a large number of assumptions, which include invariance of the properties of the gaseous and condensed phases. Even for the unidimensional models which describe the ignition process more completely, however, there is still no solution with regard to ignition characteristics.

The answer to the question of how objectively the simplified models reflect the real situation can only be given on the basis of an experimental study of the ignition process of polymers.

Despite the fact that this is a very important problem, there are only a few experimental studies that have established the fundamental interrelationships between the ignition characteristics, the individual properties of polymers and the conditions of external thermal influence. The efforts of experimentalists have in the main been directed towards the determination of characteristic ignition and self-ignition temperatures. We shall return to this problem later.

In this section it is expedient to take a look at the experimental results that have determined such ignition characteristics as the induction period (time delay) of polymer ignition, its temperature dependence and the limitations on ignition.

Self-ignition and forced ignition might also be distinguished within the ignition process in a gaseous phase. The formation of a flame as a result of combustion during contact between the sample and the hot gaseous oxidant, which is initiated by an additional source of ignition (flame, spark), is called self-ignition. The self-accelerating exothermal reaction occurs in the entire layer of degradation products surrounding the polymer sample.

Initially, this gaseous layer quickly warms up to the temperature of the environment. The process of polymer ignition in a gaseous state takes in the following stages:

1. Heating the sample up to its breakdown temperature (temperature of degradation)
2. Destructive degradation of the sample after the breakdown temperature (gasification temperature) is reached and generation of fuel gases [23].

The time until the appearance of the flame must include time t_1, which was used in heating the sample to a gasification temperature, and time t_2, used in attaining saturated concentrations of the combustible products in the gaseous layer (Fig. 1.14).

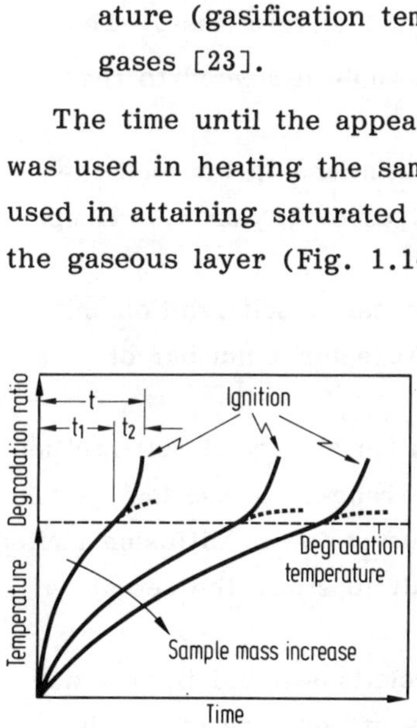

Fig. 1.14 Diagram of ignition process development of polymer materials in the gaseous phase.

It is assumed that the time spent on the self-heating of a gaseous layer due to the oxidation of combustible products is small in comparison with the time spent on heating and accumulation of fuels in the gaseous layer.

Experimentally, the pre-ignition time (the time in which ignition occurs) is registered by using a thermocouple near the surface of the sample when intense self-heating commences.

The two following factors play an important role in achieving ignition in a gaseous phase:

1. The rate of combustible product generation
2. The speed of combustible product diffusion through the surrounding space

If the rate of generation of gaseous products from the fuel is small, the gaseous products diffuse quickly and spread into the surrounding space. As a result, the saturation concentration will not be reached near the sample surface, and ignition will not occur.

An increase in the rate of an oxidiser flow also contributes to the ignition process. The time required to heat the polymer sample to the ambient temperature is a function of the mass of the polymer. To eliminate the influence of this factor it was suggested [23] that the so-called initiation ignition time could be derived by extrapolating experimental values of ignition time delay to a value that corresponded to zero mass. This time correlates with the situation in which the sample is heated to the temperature of the surrounding space.

Figure 1.15 shows, by way of example, the relationship between the induced self-ignition time and the mass of a polyamide (nylon 6,6) sample in heated air at 600-750°C.

Figure 1.16 shows the variation in the reciprocal of self-ignition initiation time ($1/t_{s\text{-}ign.}$) as a function of temperature for a number of polymers.

The computed values for the apparent activation energy of self-ignition of the polymers are found in the 8-10 kcal/mole range. These low values are linked to the fact that reactions in a gaseous state are diffusion limited for fuel gases. This makes it especially difficult to study the actual ignition mechanism in a gaseous phase.

Ignition induction periods and temperature relations found in this way give no idea as to how the individual properties of polymers affect the self-ignition process.

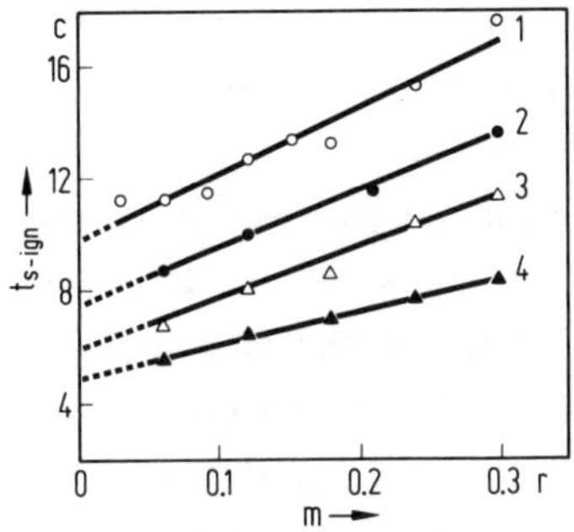

Fig. 1.15 Relationship between self-ignition temperature $t_{s\text{-}ign.}$ of polyamide (nylon 6,6) and mass of a sample *m*. Air temperature: (1) 600°C, (2) 650, (3) 700, (4) 750°C.

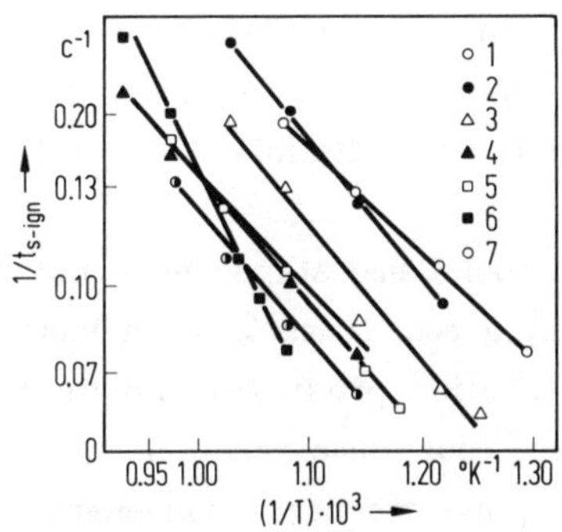

Fig. 1.16 Relation between self-ignition performance temperature of polymers and the temperature of air. (1) Cellulose acetate, (2) cotton, (3) polyacrylonitrile, (4) polyethylene terephthalate, (5) polypropylene (6) polyamide (nomex) (7) nylon 6,6.

During the study of the self-ignition of unsaturated oligoether-based polymers with a reticular structure in a pure oxygen atmosphere, the author, together with B.B. Serkov, discovered that the effective activation energy of a precombustion process in a temperature range close to the self-ignition temperature has a low value (8-10 kcal/mole). However, as the oxygen flow temperature increases (>450°C) the $E_{ef.}$ of self-ignition increases to 20 kcal/mole.

A sudden increase in the temperature coefficient of a precombustion process during the temperature rise is characteristic of the ignition process of many hydrocarbon gases [24]. This specific property is apparent when the concentration of oxygen in the gaseous mixture is increased. The observed effect is related to the multistage evolution of an ignition process of gases, a specific kinetic scheme of individual stages, and the presence of the so-called cool-flame stage, which initiates the development of a high-temperature process [24].

In a cool-flame zone, the oxidation of fuels does not proceed to completion. The intermediate products of oxidation (peroxides, aldehydes) play a large role in the kinetics of this process. A study of the effect of conditions on the self-ignition process for either of the precombustion process stages will give useful information about the self-ignition nature of polymers. The work of Delfosse [25-27] is of interest in this connection.

It was noticed that some polymers undergo all stages of the reaction with oxygen, from a low oxidation reaction through to single-stage, high-temperature self-ignition. With polyolefines, for instance, the oxygen pressure and temperature produce limitations. The slow oxidation reaction, the cool-flame reaction and the two-stage and, finally, single-stage, high-temperature self-ignition can occur within these limitations (Fig. 1.17) [26].

For polystyrene and polyvinyl chloride, the cool-flame stages have not been observed [27]. According to experiments, a cool-flame ignition has to appear in the polymers that develop as degradation products, and these products themselves are capable of undergoing a cool-flame reaction.

As for gaseous systems, the self-ignition of polymers has a concentration dependence. In this case the "concentration" of a polymer is represented by a ratio, namely the mass of a polymer sample inserted into the test vessel to the sum of the oxygen and polymer masses in the closed system [26].

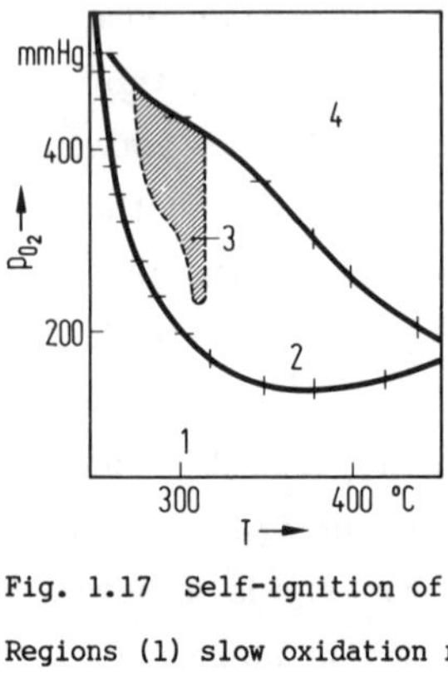

Fig. 1.17 Self-ignition of polypropylene (PP) in an oxygen atmosphere (C_{PP} = 25 mass%). Regions (1) slow oxidation reaction, (2) cool-flames, (3) "multiple" secondary cool-flames, (4) two-stage ignition. The single-stage self-ignition region is not shown.

The study of polymer ignition under the influence of a radiant thermal flow indicated that ignition induction periods are a function of the nature of the radiation source, the rate of the thermal flow and the optical and thermophysical properties of the polymers [28, 29]. The ignition induction period of polymers is adequately expressed by the semi-empirical formula:

$$t_{ign.} = \frac{160\,(\lambda\rho c)^{0,75}\Delta T_{ign.}}{(\bar{\mu} I_0)^2}, \qquad (1.40)$$

where

μ average emission absorption coefficient of a sample, this being a function of the optical properties of the sample and of the spectral properties of the radiation source

λ thermal conductivity coefficient

ρ polymer density

c thermal capacity of polymer (specific heat)

$\Delta T_{ign.} = T_{ign.} - T_0$

I_0 intensity of incident radiant flow

The $T_{ign.}$ function was derived from a mathematical analysis of temperature distribution within the sample on the basis of Fourier's law. Radiation from a benzene flame and from a high-temperature tungsten lamp, however, were used as the radiation source. The rate of radiation flow was varied over a range of 0.5-3.5 cal/cm^2·sec.

The polymer samples were shaped into 10 cm^2 sheets with a thickness of 1.27 cm. Optically transparent and nontransparent samples were used. In the studies cited, the ignition time delay, polymer density, incident radiation flow intensity and absorption coefficients were determined directly for each sample under consideration. The thermal conductivity coefficient, specific heat and ignition temperature data are taken from other literature sources.

Experimental results in coordinates of equation 1.40 were obtained for 12 polymers. The authors noted that it is especially difficult to determine the ignition temperature and that this is the main source of inaccuracy.

The comparison of the semi-empirical expression 1.40 with equation 1.36 shows that experimental results of the determination of $t_{ign.}$ reflect the complex combination of the second and third terms of equation 1.36.

Self-Ignition Temperature and Ignition of Polymers

The minimum temperature of a gaseous oxidant flow (air) at which either spontaneous initiation of fuel gases, flame combustion or decomposition (incomplete combustion) of a condensed phase is observed has been assumed to be the self-ignition temperature. The minimum temperature at which the sample releases a fuel gas in a quantity sufficient for the initiation of flame combustion when an ignition source (flame, spark, or glowing substance) is brought near the sample surface has been called the polymer ignition temperature.

Determination of these initiation temperatures is carried out by a method that has been developed in the US [30]. A similar method is being used in the USSR [31]. Standard sample sizes and optimum rates of gaseous oxidation at atmospheric pressure have been adopted.

In theoretical studies the concepts of ignition, initiation and self-ignition temperatures are used. Yet it is not possible to derive the $T_{ign.}$ and $T_{s\text{-}ign.}$ values adequately using standard methods. This is perfectly understandable. The method for determining the surface temperature of the ignitable polymer is far from easy.

Direct determination of the surface temperature at the ignition point using contact methods (thermocouples) and noncontact methods (IR detectors, pyrometers) is difficult. This difficulty is caused by many factors, the chief one being the nonstability of the process.

The reproducibility and accuracy of temperature determination in a stable and steady combustion regime depends on the sensitivity and durability of the detector recording the initial point of the run of a self-accelerating reaction, for reactions taking place either in a condensed phase or as a result of a heterogeneous reaction.

When using contact methods, it is important not only to maintain the contact for a long time but also to shield the ignition process from any catalytic or inhibiting effects that may result from the detector material.

When noncontact methods are employed [32], the result of the determination is markedly influenced by the colour variation and absorptive (or reflective) power of the polymer, resulting from its degradation, which commences even prior to the point of ignition. These difficulties involved in the recording of a surface temperature at a point of ignition have led to a practical solution to the problem.

It is assumed that the surface temperature of a polymer at the moment of its ignition is equal to the ambient temperature. The standard method of determining $T_{ign.}$ and $T_{s\text{-}ign.}$ is based on this assumption. We pointed out earlier that a real situation may not fulfil this requirement. Nevertheless, self-ignition and ignition initiating temperatures give useful information on how relatively readily-flammable the polymer materials are and on how different ingredients of composite materials affect the ignition itself.

Tables 1.1 and 1.2 give a number of $T_{s\text{-}ign.}$ and $T_{ign.}$ data for different polymer materials. As has been shown, flame retardants can affect $T_{s\text{-}ign.}$ in a very complex manner [36]. Figure 1.18 shows how flame retardants (chloroparaffin and Sb_2O_3) affect the self-ignition temperature of polypropylene. The data indicate that $T_{s\text{-}ign.}$ and $T_{ign.}$ give an indication of how the molecular structure affects the ignitability of polymers and also permit the study of the effect of additives and of the action of flame retardants and their efficiency.

Naturally, the initiation temperatures depend on the concentration of oxidant in the surrounding space (see Table 1.2) and on the rate of its flow and pressure.

As was discovered by Alvares and Martin [37], when α-cellulose is irradiated by a radiant thermal flow, its surface temperature drops by several hundred degrees with increasing concentration.

Additional detailed studies should reveal deeper interrelationships between these parameters and the individual properties of polymer materials.

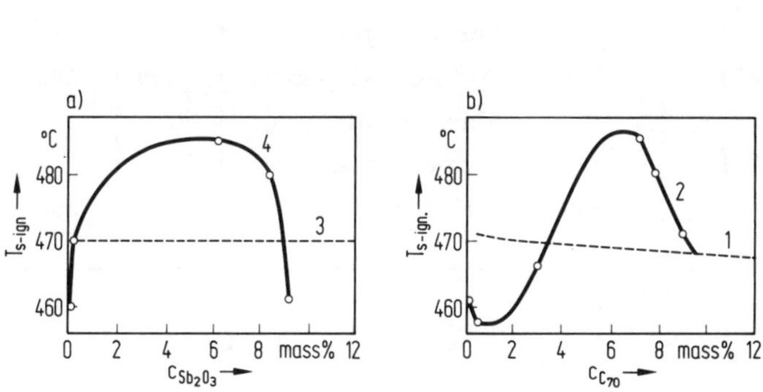

Fig. 1.18 Relationship between self-ignition temperature of polypropylene and the contents of flame retardants. (1) Chloroparaffin C_{70}, (2) Sb_2O_3 + C_{70}, (3) Sb_2O_3, (4) Sb_2O_3 + C_{70}.

Table 1.1 Ignition and self-ignition temperature of polymers [30, 31, 33, 34].

Polymer Material	$T_{ign.}$ °C	$T_{s\text{-}ign.}$ °C	Polymer Material	$T_{ign.}$ °C	$T_{s\text{-}ign.}$ °C
Polymethylmethacrylate	280-300	450-462	Polytetrafluoro ethylene	-	530
Polyethylene	341	349	Rigid element, polyether base, foam polyurethane	310	416
Polypropylene	325-343	325-388	Foam polyurethane (FPU)	-	528
Polystyrene	345-360	488-496	FPU with 50% Cl	-	528
Copolymer of styrene with acrylonitrile	366	464	FPU with 5% P and 50% Cl		554
Copolymer of styrene with methylmethacrylate	329	484	FPU with 10% P		638
Foam polystyrene	346	490			
Ether of cellulose and ethyl alcohol	290	297	<u>Reinforced Materials</u>		
Acetyl cellulose	305	375	Phenolformaldehyde plastics with paper filler base	300	430
Cotton	254	254			
Polyvinyl chloride	391	454			
Polyvinylidene chloride	532	538	Phenolic glass-reinforced plastic	520-580	571-580
Polyamide (nylon 6,6)	421	424	Silicon-containing, organic glass-reinforced plastic	490	550
Polycarbonate	522	550			

Table 1.2 Self-ignition temperature of unsaturated oligoether-based polymer materials with a reticular structure [35].

Oligoether	$T_{s\text{-}ign.}$, °C		Oligoether	$T_{s\text{-}ign.}$, °C
	Air	Oxygen		Oxygen
TGM-3	420	295	MP (dimethylacrylate propanediol-1,4)	282
OKM-8	439	330		
OKM-11	441	340	MB (dimethylacrylate butanediol-1,4)	288
OKM-12	411	350	OKM with phosphorus-containing flame retardant B-3 admixture	395

Reaction between Polymers and Oxygen

The reaction between polymer materials and a gaseous-oxidising agent (oxygen) is an exceptionally complex one. The dynamics and mechanism of the reaction depend not only on the nature of the substance of which the materials are composed but also on the physical characteristics of the surrounding environment, on the source of heating, on the size of samples and articles and on their position in relation to the ignition stimulant. All these factors together also determine the position of the self-supporting exothermal oxidation reaction that controls the ignition and combustion of polymer materials. Such a reaction is either of a homogeneous nature or localises in a gas phase near the surface (or at some distance from the surface).

Both flame and flameless combustion are fire hazards. The initiation of incomplete combustion, as in gas phase ignition, occurs as the result of a preliminary heating of the system. Smouldering always occurs under milder conditions (relatively low temperature, slow heating) than the flaming reaction. Smouldering is characterised by degradation of the polymer material, liberation of smoke and a visible glow.

In most cases incomplete combustion is found in polymer materials that have porous structures and are inclined to carbonisation (fabrics, foamed plastics, insulating materials based on cellulose), and the density of the materials plays a very important role in the process [38].

The mechanism of incomplete combustion of polymer materials is linked to the carbonisation of the polymer during pyrolysis and to the reactions between the char layer and oxygen, which diffuses towards the carbonic surface from the surrounding medium. Heat generated as a result of oxidation of the char layer maintains the polymer pyrolysis [39-41].

Incomplete combustion begins when the critical values of temperature $T_{ign.}$ and the rate of heat liberation (heat liberation is provided by a supply of oxygen for the reacting surface) are attained [39]. If the rates of char layer oxidation and pyrolysis are balanced then a stationary, incomplete combustion process occurs. If a violation of these equilibrium conditions occurs then the regime becomes a stable one [40, 41].

The temperature in the zone of incomplete combustion depends on the partial pressure of oxygen near the surface and on its mole fraction in the atmosphere [40]. In the case of incomplete combustion of cellulose materials the temperature reaches 900-1100°K; in the case of foamed polyurethane the temperature will reach 550-650°K [41]. The threshold concentration of oxygen required for a stable incomplete combustion regime of cellulose material is then substantially lower than in the case of polyurethane. Figure 1.19 illustrates the conditions of incomplete combustion of foamed polyurethane in the stationary regime and conditions of a transition to either quenching or flame combustion.

Gas phase ignition occurs under more drastic conditions, as is shown in Fig. 1.19. The temperature of incomplete combustion affects the diffusion rate of flameless combustion. This effect is indirectly linked to the kinetic parameters of the pyrolysis reaction of polymer materials and to their thermophysical properties. It was discovered with cellulose materials [42] that the geometry of an ignition source has a pronounced influence on the initiation temperature of incomplete combustion.

In order to obtain the minimum temperature required for initiation of incomplete combustion in cellulose materials there was assumed to be a

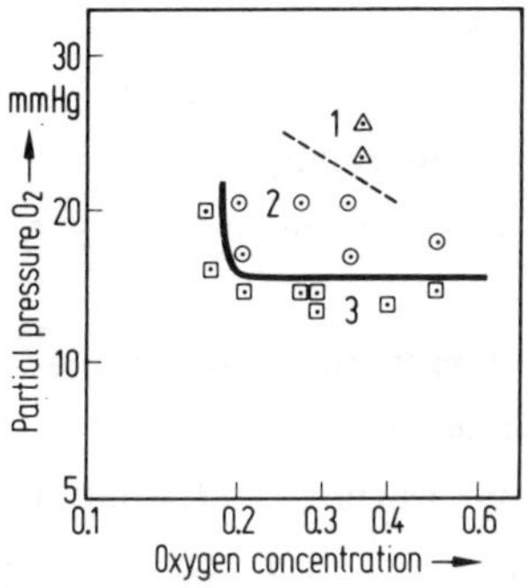

Fig. 1.19 Effect of the partial pressure of oxygen near the surface of foamed polyurethane and the concentration of oxygen in the atmosphere on a combustion regime. (1) Flaming combustion, (2) incomplete combustion, (3) absence of combustion.

phase of thermal oxidation degradation of polymers in the 280-370°C range. This phase was determined using a thermal analysis method.

The chemical mechanism was established [42]. With the assumption that heat transfer occurs through heat conduction and is described by experimentally established kinetic parameters and the heat of the decomposition reaction, the magnitudes of $T_{ign.}$ were derived, and these values are close to the values observed in practice. The theoretical model of incomplete combustion of foamed materials with a cellular structure has been described in [43]. This model of incomplete combustion of foamed polyurethane is based on two chemical processes: the formation of char during pyrolysis, followed by oxidation. It was discovered that heat transfer through radiation during the incomplete combustion of such materials must not be neglected. The incomplete combustion process is limited by the oxidation of the char layer.

It is easier to quench incomplete combustion by lowering the concentration of oxygen than by reducing the rate of char formation. The minimum energy $E_{min.}$ required for the initiation of stable incomplete combustion was proposed as a characteristic of the pyrolysis resistance of polymer materials.

This characteristic is expressed as the minimum time for which the heat flow of a given magnitude affects the material [44]. Before the stationary smouldering regime is attained, however, the polymer system passes through a transitional state. This state is characterised by the fact that the heat from the ignition source is sufficient to initiate incomplete combustion but the generated heat does not compensate for the general losses of heat from the reaction zone, and when the ignition source is removed the incomplete combustion dies. Unstable smouldering is indicated by an inflection of the temperature profile curve, which is recorded by a thermocouple.

It should be mentioned that, as the amount of energy supplied to the material increases, the difference between the minimum time for establishment of a stationary regime $t_{sp.}$ and the time for the initiation of unstable smouldering $t_{tr.}$ also grows [45]. It was discovered that when the energy supply is constant, different admixtures (phosphates, borates, mixtures of these with boric acid or with aluminium salts) enhance the initiation period for incomplete combustion; they affect $t_{sp.}$ values more than $t_{tr.}$ values.

As the concentration of additives increases, the difference ($t_{sp.}$ - $t_{tr.}$) also increases [44]. The magnitude $t_{sp.}$ - $t_{tr.}$ may be used as a measure of the rate of effectiveness, to show how the admixture affects the retardation of smouldering as a function of the chemical mechanism of the admixture effect [45]. Parameter $t_{tr.}$ is assumed to serve as a measure of the physical influence of the rate of addition on the development of the incomplete combustion process (due, for example, to its influence on the heat conductivity of the material and, accordingly, on the rate of heat loss).

The transition to flame combustion is dependent on temperature conditions in the polymer material and on the concentration of oxygen in the surrounding atmosphere. In this respect, the effect of the fluctuating flame or so-called secondary cool-flame should be mentioned. This effect is a result of the heterogeneous oxidation of polymer materials [46, 47]. Peroxide macroradicals aggregate on the surface and when their concentration reaches a critical value intensive decomposition of the surface polymer layer and volatilisation of fuel products occurs. Detachment of the oxidation products from the molten polypropylene takes place periodically as a series of superficial microruptures.

Oxidation of the liberated volatile products is followed by heating of the gas phase and by initiation of a cool-flame. The moving force acting on the oscillating flame is the fluctuation that occurs in the superficial layer of the oxidised polymer in the direction of the temperature gradient, resulting from competition between heterogeneous oxidation and gas-phase oxidation reactions. The frequency of fluctuations and the amplitude of the first fluctuation are directly proportional to the mole fraction of the oxygen in the oxidising flow. The effect of an oscillating cool-flame was detected during the oxidation of a number of polyolefines that contained hydrogen atoms in their structure; these hydrogen atoms were localised near the tertiary carbon atoms.

This flame was detected in the case of polyethylene and polyvinyl chloride. Thus the oscillating flame effect is a specific one and it combines characteristics of both heterogeneous and gas phase oxidation. When the temperature continues to increase, the oscillation process of polymer self-ignition degenerates and the formation of normal cool and hot flames is detected.

The study of ignition limits and the effect of different admixtures on these processes is especially useful for understanding the ignition mechanism. In the case of the gas phase ignition of polymers it is assumed that fuel gases are formed as a result of the thermal degradation of the polymer under the influence of heat supplied to the sample. Since, in the case of large-scale combustion, the transfer of heat from the flame towards the surface of a polymer material occurs mainly through radiation, much attention has been focussed on the study of the principles of ignition under the effect of this type of heat transfer. Ohlemiller and Summerfield discovered previously that the time delay for polymer ignition in an unheated atmosphere under the effect of radiant heat flow is greater than the ignition time for an identical heat flow coming from a heated gas [48]. It was assumed that the ignition is caused by two factors:

1. The transmission coefficient for the gaseous phase is such that there is slower heating of the solid fuel surface.
2. The unheated air slows ignition. Kashiwagi [49] studied the mechanism of reactive ignition of PMMA in an atmosphere of unheated air by CO_2 laser analysis of self-ignition under these conditions. An analysis of the pilot ignition of PMMA degradation products using a gas jet and heated coil also showed that the temperature of the surface is approximately 385-410°C in all cases.

The minimum radiation flow required for self-ignition (I_0 ~ 16 W/cm²), however, is almost twice that for pilot ignition (I_0 ~ 8 W/cm²), and so pilot ignition reduces the rate of thorough heating of the gaseous phase.

During self-ignition, a portion of the heat radiation is absorbed by the gaseous products of polymer degradation. This result was confirmed by Kashiwagi in small-scale experiments and in experiments with large samples of polymer materials that had been placed in various positions vis-à-vis the incident radiant heat flow [50]. The general tendency is for the amount of energy absorbed by the gas phase degradation products to be smaller when the samples are positioned vertically.

The ignition time delay and the minimum flow needed for ignition of the materials are thus larger than in the case of horizontally oriented samples. The dynamics of the heating of the polymer surface and the gaseous phase near it (0.15 cm above the surface) should also be mentioned.

Before PMMA starts to decompose, the temperature of the gas phase is lower than the surface temperature but after degradation has commenced the temperature of the gaseous phase starts to exceed the temperature on the surface of the PMMA. At the moment of ignition, the difference in temperatures reaches almost 100°C. The temperature of a vertical surface in the case of PMMA ignition was 460°C (t_s for horizontal ignition ~ 400°C).

The t_s for PMMA remained virtually constant during ignition, regardless of the energy of the incident radiant heat flow, I_0. For carbonised-type materials t_s increased as I_0 decreased. The type of heat transfer and the intensity of the heat flow towards the material determine the rate of retention and the level of energy supply in the condensed phase and, accordingly, the rate and mechanism of polymer material degradation and fuel product heating in the gaseous phase.

The ignition time delay (when all other conditions are the same) in fact increases linearly with the mass [51] or the width of the sample [52]. It decreases as the intensity of the heat flow increases:

$$\tau_{\text{ign.del.}} \sim \frac{1}{I_0^{\,n}}$$

where n fluctuates in a range of 0 to 2 as a function of I_0. Usually it is close to 2 for I_0 ~ 1-200 W/cm².

This data confirms that, under real conditions, a significant portion of the ignition period is taken up with warming the material. Japanese scientists [53] discovered that when the radiant flow is above 60 J/cm·sec, there is a sharp decrease in the ignition time delay for PMMA, a transition

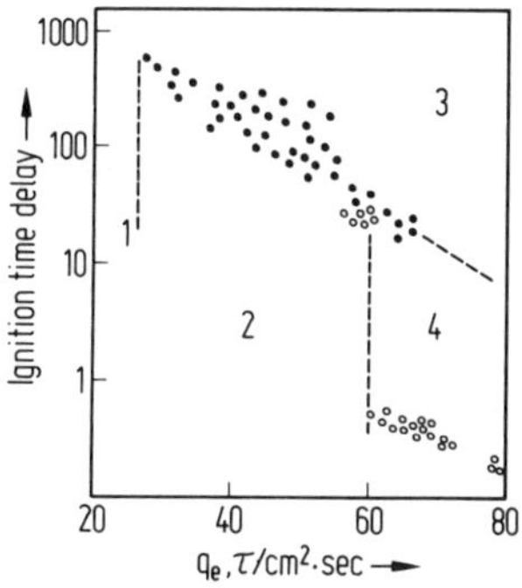

Fig. 1.20 Relationship between ignition time delay for PMMA and energy of the radiant flow. (1) Ignition limit, (2) no ignition, (3) stable flame, (4) transitional zone.

to an unstable regime, and a change in the localisation of the exothermal reaction zone (Fig. 1.20).

An analysis of the concentration and temperature profiles in the gaseous phase leads to the conclusion that ignition is localised in the zone where the concentration of fuel lies within the ignition limits and where the temperature exceeds the temperature of spontaneous ignition. Violation of this condition leads to the destabilisation of ignition. There is a clear-cut connection between the intensity of the heat flow, the time for which it acts on the material after ignition and the establishment of a stable flaming combustion regime [50].

When I_0 has high values after the ignition source has been removed (immediately after ignition of the polymer material), the flame is unstable. When I_0 is low ($<$13-11 W/cm^2 for PMMA), then ignition leads to a stable combustion regime, even if the heating source is removed immediately after ignition occurs. This is why both factors, namely the energy of the ignition source and its exposure time, play such an important role in the objective evaluation of the flammability and ignitability of materials. As usual, determination of the ignition criteria is a major problem in the theoretical analysis of gas phase ignition because the transition to a combustion regime is an asymptotic process and the momentum of ignition cannot be determined mathematically.

According to Kanury [54], the most objective criterion is the minimum mass flow of pyrolysis products into the gaseous phase. This flow is related to the lower concentration limit of fuel ignition in a specific atmosphere. A numerical analysis has been conducted for a gas phase model of polymer ignition by radiant flow [55]. Six different criteria were used. Numerical values were chosen for the gas phase reaction as follows:

General rate of reaction (1)

Local rate (2)

Maximum local temperature (3)

Acceleration of the rate of reaction (4)

Rate of heat liberation (5)

Temperature gradient (6)

All the values indicate the conditions under which the rate of heat liberation is either equal to or in excess of the heat losses.

Figure 1.21 shows that the selection of a criterion essentially affects the qualitative representation of the relationship between ignition time delay and the percentage of oxygen in the surrounding atmosphere (in other words, between the ignition limit and the oxygen concentration). For the first time it was possible to theoretically show the existence of a lower and upper ignition limit, related to the values of the kinetic parameters (activation energy) for the pyrolysis of a condensed fuel and gas phase reaction. By suitably combining these parameters it is possible to narrow the ignition limits and thereby decrease the flammability of materials. The nature of an inert diluent of the gas phase affects the ignition limits to a marked extent.

The application of the asymptotic method to analyse the ignition of polymer materials represents an important step [56]. With the aid of this method, conditions for conjugation were discovered and equations were solved that describe the conditions of the solid and the gaseous phase over

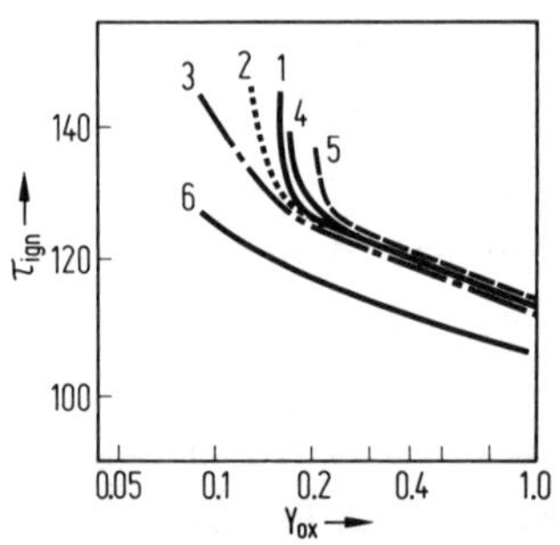

Fig. 1.21 Relationship between computed time of ignition delay (in dimensionless units) and concentration of oxygen in atmosphere at different ignition criteria [38 - 43].

a period of time at the different stages of the ignition process. The process of polymer vaporisation during the action of a stationary radiant flow received much attention. The process of PMMA vaporisation was conditionally subdivided into stages:

1. Inert heating
2. Transitional state (in the transitional state a rise in temperature was detected)
3. Diffusion controlled stationary vaporisation with a constant surface temperature

It was discovered that ignition occurs either during the transitional or the stationary stage of pyrolysis as a function of the relative rate of pyrolysis and gas phase oxidation reactions [56]. If it occurs at a later stage then the ignition time delay does not correlate with the energy absorbed by the material. During ignition in the transitional period of polymer vaporisation, the ignition time delay is dependent on the absorptivity of the C phase. As the absorption coefficient $\bar{\mu}$ diminishes, the ignition time delay increases.

The determination of the criteria affects the calculated ignition limits worked out on the basis of the energy of the gas phase reaction. The results of calculations of ignition time delay for the condensed system obtained with the asymptotic method tally better with the numerical solutions [55] for large heat flows.

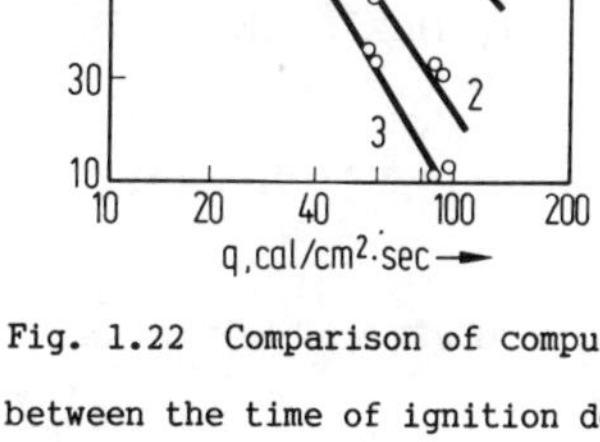

Fig. 1.22 Comparison of computed (unbroken lines) and experimental (dots) relationships between the time of ignition delay for polystyrene in an oxygen atmosphere and the power of a radiant heat flow and absorptivity of samples, varied by adding soot (1) 0, (2) 4, (3) 20% of soot.

Experimental data given by Ohlemiller [57] was used in order to check the basic aspects of the theory. The kinetic parameters of the surface pyrolysis reaction of polystyrene, which had been determined independently, were used in the calculation. The parameters of the gas phase reaction were assumed on the basis of the observed relationship between $t_{ign.del.}$ and the pressure of oxygen (the preexponential multiplier must be inversely proportional to $\sqrt{Po}$ and the activation energy of the gas phase reaction should be greater than the pyrolysis activation energy by a factor of two). If one considers the absorptive and reflection coefficients of the surface then good agreement is obtained between experiment and theory in terms of how I_0 and $\bar{\mu}$ affect the ignition time delay (Fig. 1.22) [57].

Hence it can be seen that theory opens up new ways for determining more detailed characteristics of the ignition of polymer materials on the basis of data from independent experiments.

References

1. Frank-Kamenetskii D. A., Diffuziya i teploperedacha v Khimicheskoy Kinetike, 2nd ed., elaborated and edited, M.: Nauka, 1967.
2. Barzikin, V. V., Merzhanov A. G., Dokl. AN USSR, 1958, vol. 120, no. 5, pp. 1271-1274.
3. Merzhanov A. G., Dubovitsky F. I., Usp. khim, 1966, vol. 35, no. 4, pp. 654-683.
4. Merzhanov A. G., Abramov V. G., Gontokovskaya V. T. Dokl. AN USSR, 1963, vol. 148, no. 1, pp. 156-160.
5. Merzhanov A. G., Averson A. E., Combust. and Flame, 1971, vol. 16, no. 1, pp. 89-94.
6. Averson A. E., Barzikin V. A., Merzhanov A. G., Dokl. AN USSR, 1966, vol. 169, no. 1, pp. 158-162.
7. Carslow G., Eger D., Teploprovodnost tvyordih tel., M.: Nauka, 1964.
8. Prais E., Bradley G., Deoritti G., Ibiritsu M., Raketnaya tekhnika i cosmonautika, 1966, no. 7, pp. 3-41.
9. Grigor'ev U. M., Maksimov E. I., Merzhanov A. G., Nauchnotechnicheskiye problemi goreniya i vzriva, 1965, no. 1, pp. 93-100.

10. Grigor'ev U. M., Maksimov E. I., Merzhanov A. G., Dokl. AN USSR, 1964, vol. 157, no. 6 pp. 1427-1431.

11. Rozenland V. I., Barzikin V. V., Merzhanov A. G., Phys. goreniya i vzriva, 1968, vol. 4, no. 2, pp. 171-177.

12. Lisitsky V. I., Merzhanov A. G., Nauchnotechnicheskiye problemi goreniya i vzriva, 1965, no. 2, pp. 62-67.

13. Grigor'ev U. M., Lisitskyi B. I., Merzhanov A. G., Phys. goreniya i vzriva, 1967, vol. 3, no. 4, pp. 512-516.

14. Anderson R., Brown R. S., Shannon L. T., AIIA J., 1964, vol. 2, no. 1, pp. 52-60.

15. Averson A. E., Barzikin V. V., Martem'yanova T. M., Phys. goreniya i vzriva, 1974, vol. 10, no. 4, pp.498-501.

16. Grishin A. M., Ignatenko I. A., Phys. goreniya i vzriva, 1971, vol. 7, no. 4, pp. 510-514.

17. Grishin A. M., Kuzin A. Y. in Gorenie i vzriv, M.: Nauka, 1972.

18. Grishin A. M., Ignatenko E. M., in Vsesouznyi simposium po goreniu i vzrivu. Autoref. dokl., Tchernogolovka: OIKhF AN USSR, 1969, pp. 15-20.

19. Isakov G. N., Grishin A. M., Phys. goreniya i vzriva, 1974, vol. 10, no. 2, pp. 191-196.

20. Ohlemiller T. J., Summerfield M., AIAA J., 1968, vol. 6, no. 5, pp. 134-140.

21. Deverall L. J., Lai W., Combust. and Flame, 1969, vol. 13, no. 1, pp. 8-11.

22. Librovitch V. B., Zh.PMTF, 1968, no. 2, pp. 36-42.

23. Miller B., Martin J. R., Meiser C. H., J. Appl. Polymer Sci., 1973, vol. 17, no. 2, pp. 629-638.

24. Sokolik A. S., Samovosplameneniye plamia i detonatsiya v gazah. M.: Izd-vo AN USSR, 1960.

25. Baillet C., Delfosse L., Antonik S., Lucauin M., C. r. Acad. sci. (Paris), 1972, vol. 274, pp. 146-149.

26. Baillet C., Delfosse L., Lucquin M., in Combustion Institute European Symposium. L.: Acad. Press, 1973, pp. 148-152.

27. Delfosse L., J. Macromol. Sci. Chem., 1977, vol. 11A, no. 8, pp. 1491-1501.

28. Hallman J. R., Welker J. R., Slipcevich C. M., SPE Journal, 1972, vol. 28, no. 9, pp. 43-48.

29. Hallman J. R., Welker J. R., Slipcevich C. M., in Society Plastic 30th Annual Technical Conference. Chicago, 1972, pt. 1, p. 283.
30. Setchkin N. P., J. Res. NBS, 1949, vol. 43, no. 6, p. 591-605.
31. Pozharnaya opasnost veshchestv i materialov pimenyaemih v khimicheskoy promishlennosti. Under ed. of I. V. Riabov, M.: Khimaya, 1970.
32. Smith W. K., King J. B., J. Fire Flammability, 1972, no. 1, pp. 272-276.
33. Ueno K., Kobunsy Kako, 1971, vol. 20, pp. 526-530.
34. Einhorn J. N., Fire Res. Abstrs and Revs, 1971, vol. 13 no. 3, pp. 236-242.
35. Aseeva R. M., Ushkov A. V., Andrianov R. A., Zaikov G. E., in Nehorlavost polymernych materiálov. Bratislava: Dom Techn. SVTS, 1976, pp. 3-5.
36. Delfosse L., Baillet C., Lucquin M., Spilda I., Ibid., pp. 22-25.
37. Alvares J. Martin S. B., in 13th Symposium (Intern.) on Combustion. Pittsburgh: Combust. Inst., 1971, pp. 905-919.
38. Anderson R. W., Fresichel W., J. Thermal Insulation, 1978, vol. 2, p. 7.
39. Kinbara T., Endo H., Sega S., in 11th Symposium (Intern.) on Combustion, Pittsburgh: Combust. Inst., 1967, pp. 525-531.
40. Moussa N. A., Toong T. Y., Garris C. A., in 16th Symposium (Intern.) on Combustion, Pittsburgh: Combust. Inst., 1977, pp. 1447-1457.
41. Ortiz-Molina M. G. et al. in 17th Symposium (Intern.) on Combustion, Pittsburgh: Combust. Inst., 1979, pp. 1191-1200.
42. Ohlemiller T. J., Combustion Science and Technology, 1981, vol. 26, pp. 89-105; 1980, vol. 24, pp. 129-137; 139-152.
43. Ohlemiller T. J., Bellan J., Rogers F., Combust. and Flame 1979, vol. 36, p. 197-215.
44. Day M., Suprunchuk T., Wiles D. M., J. Consum. Product Flammability, 1980, vol. 7, p. 3-14; 1979, vol. 6, pp. 233-243.
45. Day M., Suprunchuk T., Wiles D. M., in Proced. 5th Intern. Conference on Fire Safety, 14-18 Jan. 1980, p. 3.
46. Baillet C., Delfosse L., Lucquin M., Europ. Polymer J., 1981, vol. 17, pp. 779, 787, 791-195.
47. Rychly J., Matisova-Rychlá L., Spilda J., Europ. Polymer J., 1979, vol. 15, pp. 565-570.

48. Ohlemiller T. J., Summerfield M., in 13th Symposium (Intern.) on Combustion, Pittsburgh: Combust. Inst., 1971, pp. 1087-1094.

49. Kashiwagi T., Combust. and Flame, 1979, vol. 34, p. 231.

50. Kashiwagi T., Combust. and Flame, 1982, vol. 44, pp. 223-245.

51. Miller B., Martin J. R., in Flame Retardant Polymeric Materials, Lewin M., Atlas S. M., Pierce P. M., N.Y.-L: Plenum Press, 1978, vol. 2, pp. 63-101.

52. Matonis V. A., J. Fire and Flammability, 1978, vol. 9, pp. 14-29.

53. Muton N., Hirano T., Akita K., in 17th Symposium (Intern.) om Combustion, Pittsburgh: Combust. Inst., 1979, p. 1183.

54. Kanury A. M., Combustion Science and Technology, 1977, vol. 16, p. 89.

55. Kashiwagi T., Combustion Science and Technology, 1974, vol. 8, pp. 225-236.

56. Kindelan M., Williams F. A., Combustion Science and Technology, 1975, vol. 10, pp. 1-19; 1977, vol. 16, pp. 47-58; Acta Astronautica, 1975, vol. 2, pp. 955-979.

57. Niioka T., Williams F. A., in 17th Symposium (Intern.) on Combustion, Pittsburgh: Combust. Inst., 1979, p. 1163.

Chapter 2
Polymer Combustion and its Characteristics

Multistage Combustion of Polymers

The combustion of polymer materials, like the combustion process of any other fuel material, is a combination of complex physico-chemical processes, which include the transformation of initial products into combustion products. This whole conversion process may be divided into stages, with specific physical and chemical processes occurring in each of these stages.

In contrast to the combustion of gases, the combustion process of condensed substances and, in particular, of polymer substances, has a multi-phase character. Each stage of the initial transformation of a substance correlates with a corresponding value (combustion wave) with specific physical and chemical properties (state of aggregation, temperature range, concentration of the reacting substances, kinetic parameters of the reaction etc.). Experimental studies of the combustion of solid fuels conducted by Pokhil and his research team [1] made a significant contribution to the development of the multi-phase combustion concept.

It was established that exothermal chemical reactions resulting from the combustion of the condensed substances are not only concentrated in a narrow front flame zone. Vigorous chemical reactions also occur in the condensed phase of a fuel at a relatively low temperature.

The structure of the solid fuel combustion wave can be described by a unidimensional diagram (Fig. 2.1) [2, 3]. The heat that is absorbed in the condensed phase increases the temperature of the solid fuel to the point at which either a change of phase or chemical reactions commence.

As the temperature continues to rise, the condensed phase is vaporised as a result of sublimation, the evaporation from either the exothermal or the endothermal decomposition of a substance. The resulting gaseous products react in a gaseous state and become combustion products. The burning surface is an interfacial boundary between reaction zones in condensed and gaseous phases. The three zones to be distinguished are:

1. Vapour-smoke gaseous zone adjoining the surface (so-called hissing surface zone)
2. A high-temperature flame reaction zone
3. A combustion products zone

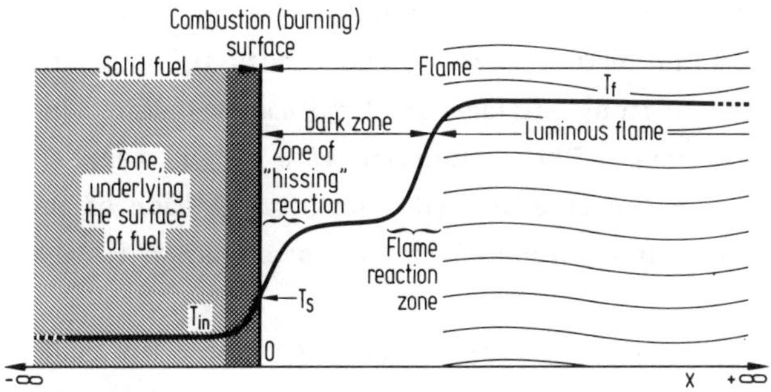

Fig. 2.1 Diagram of the structure of a solid fuel combustion wave

The first zone is sometimes called the cool-flame zone, or pre-flame zone. In cases where further combustion stages are suppressed, by whatever means, the combustion regime that corresponds to the first stage is called the cool-flame stage.

The dispersion of the condensed phase plays a significant part in vapour-smoke gaseous zone formation.

The reaction zone inside the flame is characterised by maximum temperature and light emission. The rate of combustion is determined by either of the stages, depending on the nature of the condensed fuel and the oxidising agent. This zone is called the leading zone.

The diffusion combustion of polymer materials is being studied in the same way as in the above model for solid fuel combustion; in this case five zones are distinguished in the combustion wave: heat transport and reaction zones in the condensed phase, preflame and reaction zones inside the flame, and a combustion products zone [4].

In an experimental study of the combustion of a polymer material it is especially important to investigate all the stages in which phase and chemical transformations of the substance are taking place. The complete scheme of the combustion process could be deduced from the study of such characteristics as the morphological structure of the burning surface, the temperature distribution inside the combustion wave, the structure and concentration profiles of each individual-stage product, the heat liberation and heat losses, the structural properties of the flame and the limitations on combustion stability.

The combustion rate is one of the basic quantitative characteristics of the combustion of polymer materials. It can be expressed in terms of mass and linear rates: combustion mass rate $\dot{m}$ (kg/m$^2\cdot$sec) is equal to the mass of polymer fuel exhausted per time unit from a unit of surface area. The linear, or normal rate of combustion (v, m/sec), represents the rate at which the reaction front (the burning surface or flame mantle) shifts in relation to the unburned products. The combustion rate of polymer-containing materials depends on the rate of heat liberation at each of the combustion process stages and on the thermal and mass-exchange rates between the stages.

The major task in combustion theory consists in determining the relationship between the combustion rate of a substance and the physical and chemical properties of the substance - the process that takes place at each combustion stage. The relationship between the combustion rate, pressure and initial temperature can be expressed by the following coefficients:

$\nu = d(\ln v)/d(\ln P)$ and $\beta = d(\ln v)/dT_0$.

These two correlations and the relationship between the rate of combustion and the concentration of oxidiser in the environment represent a particular idea. The study of the relationship between the combustion rate and the initial temperature of a substance and between the combustion rate and the environmental pressure permits conclusions to be drawn as to the importance of either one of the combustion stages.

An understanding of the general principles underlying the physico-chemical conversion process of an original polymer substance into final combustion products allows a solution to be sought to the problem of the combustion of polymer materials and to a simplification of the problem. It must, however, be mentioned that at present systematic studies of the diffusion combustion of polymer materials are merely at the outset of their development; quantitative information has been derived for a limited number of polymer materials.

The stage of conversion reactions in the condensed phase of a polymer material attracts chief attention of research because it is one of the most important stages of a combustion process. This particular stage involves the generation of fuel gases. The kinetics and mechanism of gasification and the degradation processes of polymers under combustion conditions have not yet been determined for a large number of cases.

The question of whether oxygen participates in the degradation of polymer materials during diffusion combustion is still under discussion. According to F.J. Martin [5], oxygen coming from the surrounding atmosphere is completely consumed, and it is for this reason that the pyrolysis of polymers takes place without the participation of oxygen. At the same time, S.J. Burge and C.F.H. Tipper [6] discovered oxygen in all the flame zones while studying the flames of polyethylene and polymethylmethacrylate. They believe that oxygen is undoubtedly involved in the degradation of polymers. D.E. Stuetz and coauthors found [7], through neutron activation analysis of the superficial layers of burning polypropylene, that oxygen from the surrounding atmosphere diffuses towards the burning surface and reacts with the polymer.

Fuel gas formation is determined by the simultaneous occurrence of thermal and thermo-oxidation degradation of polymers during combustion. A reaction takes place between the surface layers and the active particles that are products of the ignition reaction, which diffuse from the flame towards the surface of the polymeric material.

It is assumed that this reaction also participates to some extent in the reactions of condensed phases during the diffusion combustion of polymers [8]. In the preflame zone, the gaseous phase which results from the gaseous fuel products of polymer degradation is heated and undergoes additional cracking and partial oxidation. Formation of soot is often observed. Depending on the chemical nature of the polymer substance, the heat liberation and heating inside this zone due to an exothermal oxidising reaction can be very considerable. The maximum temperature inside a polymer flame is reached in the flame front.

In theoretical studies of combustion it is assumed that the fuel gas and oxygen flow within this zone are in a state of stoichiometric correlation and that their concentrations in the front of flame are zero. When there are no heat losses, the temperature in the front of flame is equal to a maximum adiabatic temperature. Under real conditions, it has been discovered that there are not only soot particles but also hydrocarbon fragments (the products of partial oxidation) amongst the polymer combustion products. This is a result of different types of heat loss and a fall in temperature in the flame reaction zone, or is due to the fact that the fuel and oxidiser remain in this zone for a sufficient amount of time to ensure that chemical oxidising reactions are not completed.

The formation of carbon black significantly complicates the structure of a diffusion flame. The best example of such diffusion is the candle flame (Fig. 2.2). In this particular case, the accumulation of flame heat leads to paraffin wax vaporisation.

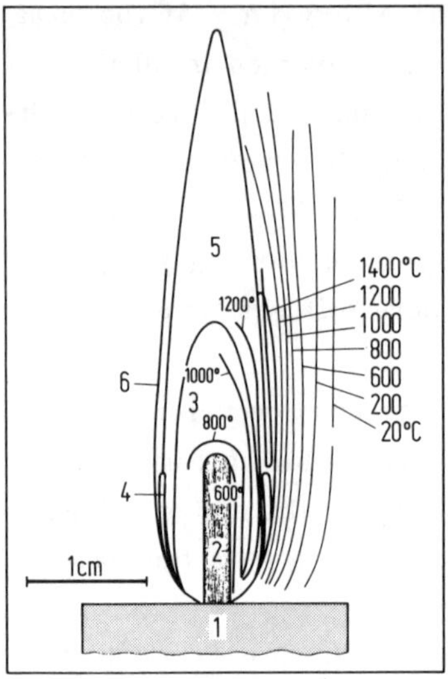

Fig. 2.2 Candle flame. (1) Candle, (2) wick, (3) dark zone (4) zone of radicals: C_2, CH, (5) luminous zone, (6) zone of major reaction.

Cracking of fuel fumes and formation of soot particles occurs in a dark preheating zone (3, Fig. 2.2). These particles are transported further and glow in a luminous flame zone (5). The major oxidation reaction of the fuel is maintained in the diffusion flame zones (4 and 6) where the temperature reaches its maximum magnitude (1400°C).

During the laminar diffusion combustion of polymer materials, when a combustion wave spreads downwards on a sample, a similar candle-like flame is observed. Usually a photograph of these flames clearly shows a dark zone near the polymer surface, a luminous zone inside the flame, a luminous reacting layer on the flame surface and an aura around the flame (Fig. 2.3).

It should be noted from the shape and size of a polymer flame that the distribution of the gaseous zones depends on a number of factors: the direction and rate of flow of the oxidant, the flow of gaseous fuel degradation products, the size of the combustion source and the nature of the polymer substance. Regardless of all these factors, however, the multistage nature of fuel substance combustion product formation still survives. The theory of condensed substance combustion is based on an examination of the physical and chemical processes that occur in a combustion wave and on determination of a leading stage, which limits the combustion rate.

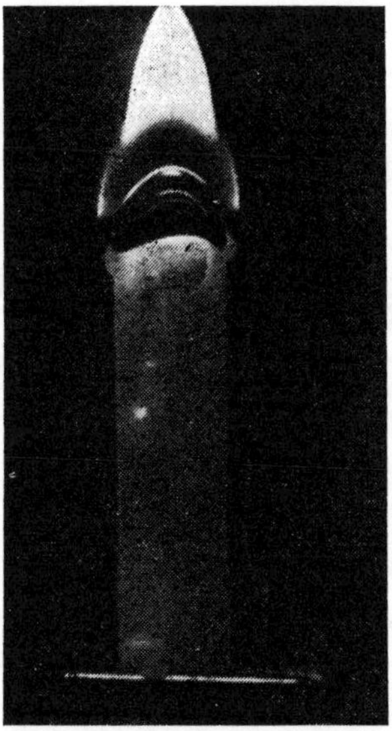

Fig. 2.3 Diffusion flame of a polyethylene sample in an oxygen-nitrogen atmosphere. P = 1 atm, $Y_{ox.}$ = 0.212.

In order to elucidate the complex mechanism of the combustion of condensed substances, the concept of an elementary combustion model was introduced [9]. This elementary combustion model describes the simplest single-stage combustion regimes. The processes that occur in the leading zone are studied separately from the other processes.

On the basis of the simple elementary combustion model in the gaseous phase, there is assumed to be initial gasification of a substance with further combustion in the gaseous phase. In the dispersed phase, the elementary combustion model takes into account the thermal degradation of the condensed substance under the conditions of a considerable reduction in its effective density (due to a volume increase as a result of dispersion and conversion into a foam or aerosol state). The fuel conversion of the reagents into the final products determines the rate of combustion.

Polyvinyl nitrate is an example of a polymer substance with a combustion rate that is determined by a reacting stage in the dispersion phase and calculated on the basis of this elementary combustion model [10]. There is something of interest about the elementary combustion models, which take into account the influence of the partial conversion of a substance in the leading zone (due to the removal of reagents in the dispersed state) on the rate of combustion.

A real combustion process is characterised by a complex interaction of different elemental mechanisms. The theoretical analysis of combustion should be directed towards clarification of how the interaction of the leading and inferior stages affects the combustion process.

Mass Transfer in the Combustion of Polymer Materials

The mass and heat transport in a reacting system are the basic processes of diffusion combustion. D.S. Spolding [11] introduced a dimensionless coefficient, B, into the diffusion combustion theory. This coefficient characterises the intensity of mass transfer within the system.

In the case of diffusion combustion of polymers, this parameter B (Spolding number) defines the driving force of mass transfer from a polymer substance towards the gaseous phase and could serve as a measure of combustion intensity [12, 13]. This parameter depends on the nature of the material, which is why the compilation of B values is so important for the relative determination of a material's combustibility.

Only in real cases does mass transfer occur in the absence of heat transfer. The mass transfer parameter B can be obtained from an examination of.either a material or the thermal balance on the boundary line between phases.

When molecular and convectional heat transfer and mass transfer occur in a reaction system, it is possible to express the B number, which has been determined from a material balance, by the equation

$$B_m = \frac{Y_f + (Y_{ox}/r)}{1 - Y_f}, \qquad (2.1)$$

where

Y_f relative mass concentration of the fuel on the surface

Y_{ox}. relative mass concentration of the oxidiser in the surrounding atmosphere

r stoichiometric relationship between oxygen and fuel

More often, however, the B number derived from an energy balance is used:

$$B_{en} = \frac{(Y_{ox}/r)\,\Delta H_{com} - c_p\,(T_S - T_0)}{\Delta H_e}, \qquad (2.2)$$

where

$\Delta H_{com.}$ heat of combustion of the polymer material

c_p specific heat of the polymer material

T_s surface temperature of a condensed phase

T_0 temperature of the environment

ΔH_e effective heat of vaporisation of the polymer material

The B_m and $B_{en.}$ values correlate as follows:

$$D^{2/3} \ln (1 + B_m) = a^{2/3} \ln (1 + B_{en}).$$

If the heat transfer and mass transfer coefficients are equal (Lewis number Le = D/a = 1), then $B_m = B_{en.}$

Equation 2.2 can be introduced in the form:

$$B = (Y_{ox}/r) B_{com} - b, \tag{2.3}$$

where

$B_{com.} = \Delta H_{com.}/\Delta H_e$

$b = c_p(T_s - T_0)/\Delta H_e$

It is obvious from the above that the relationship between the heat of combustion of a polymer substance and the effective heat of vaporisation is a determining factor in the mass transfer during diffusion combustion of polymer materials. The effective polymeric heat of vaporisation ΔH_e is determined by directly decomposing the substance in a calorimetric bomb [14] or by calculation from equation 2.2 if the $B_{com.}$ number has already been determined by another method [15].

There is a further method by which ΔH_e can be determined for different substances. This is based on a comparion of how the model fuel composition and a standard substance, PMMA, affect the combustion rate [16].

On the whole, the effective heat of vaporisation, ΔH_e, consists of the heat energy consumed when the polymer is heated to its vaporisation temperature [$c_p(T_s - T_0)$], the pyrolysis enthalpy ($\Delta H_{pyr.}$), the evaporation enthalpy ($\Delta H_{evap.}$), the fusion enthalpy ($\Delta H_{fus.}$) and other phase transitions.

For polymers that have a destructive depolymerisation mechanism with further formation of a gaseous monomer ΔH_e is equal to the sum of the heat energy consumed in polymerisation, in liquefying the polymer and in evaporating the polymer [17].

The enthalpies of a number of polymerisations of monomer to hypothetical gaseous polymers, monomer to amorphous polymers and monomer to crystalline polymers have been reported [18-19].

The vaporisation enthalpy for linear aliphatic hydrocarbons and degradation products of hydrocarbon polymers can be calculated according to an empirical formula [18]:

$$\Delta H_{evap.} = 3{,}21 n^{2/3} - 0{,}0193 T + 2{,}92 \text{ kcal/mole}$$

where

n number of atoms in the molecule

T average temperature of the temperature interval in question

This formula was also found to be accurate for unsaturated oligoethers derived from alkylene glycols and hydroxyalkylene glycols [20]. For nonpolar liquids that do not form associated molecules in the gaseous phase, there is an approximation: Truton's law. According to this law $\Delta H_{evap.}/T_{boil.}$ = 21 cal/mole·°K, where $T_{boil.}$ is the boiling temperature of a substance. In equations 2.2. and 2.3 $\Delta H_{com.}$ is assumed to be the complete mass combustion heat, and the physical state of the combustion products is studied at room temperature. In fact, in this case the largest enthalpies of a substance are used. These take into account the heat of water to steam condensation at 298°K (10.52 kcal/mole). The magnitude of $\Delta H_{com.}$ is determined either experimentally by burning a substance inside a calórimetric bomb or by calculation, such as by using known values of the heat of generation of the initial substance and the heat of combustion of the products.

Table 2.1 shows $\Delta H_{com.}$ and ΔH_e values for different polymer materials under real conditions; as in the combustion of any other condensed substance, incomplete combustion is frequently observed. The formation of soot particles or fumes is one of the indications of incomplete combustion. The causes of incomplete combustion differ and depend both on the nature of a substance and on the combustion conditions (lack of oxidant in the flame reaction zone, insufficient residence time of a fuel component in this zone, large heat losses into the surrounding atmosphere, difficult formation of carbonised combustible layers on the surface etc.). Although the causes of incomplete combustion will not be considered in detail, it should be mentioned that one cause is the heat loss into the surrounding atmosphere, due to heat dissipation by the flame, which leads to a further

Table 2.1 Heat of combustion and gasification of polymer materials.

Polymer material	$\Delta H_{com.}$, kcal/g	ΔH_e, cal/g
Polymethylmethacrylate	6.3 [23]; 6.0 [21]	380 [14]; 232 [18]*
Polyethylene	11.0 [23]; 10.0 [21]	924; 683 [18]; 500-565 [16]; 552 [24]
Polypropylene	10.5 [23]; 11.0 [21]	
Polystyrene	9.5 [23]; 9.3 [21] 9.9 [24]	241 [18]; 430-450 [16]; 502 [24]
Polymer ABS	8.4 [23]; 8.5 [21]	
Polyformaldehyde	4.05 [21]	
Polyvinyl chloride	4.3 [23]	
Polycarbonate	7.3 [23]; 7.36 [12]	
Polyamide (nylon 6,6)	7.4 [22]	
Polytetrafluoroethylene	1.10 [23]	415 [18]
Polyamide (Nomex)	6.46 [22]	
Poly-2,6-dimethyl-1,4-phenylene oxide	7.5 [21]	
Natural rubber	9.5 [21]	
Polyisobutylene	11.18 [21]	
Cotton	4.01 [22]	
Cellulose acetate	4.4 [22]	
Viscose rayon	3.84 [22]	
Polyacrylonitrile (Acrylan)	7.32 [22]	
Polyethylene terephthalate	5.47 [22]	
Polyvinyl carbazole (n = 10)		510 [16]
Poly-α-methylstyrene		158 [18]

*In Ref. [18] ΔH_e calculated at 25°C.

decrease in flame temperature and hence prevents the flame from reaching the maximum possible adiabatic temperature.

For luminous flames in which intensive soot formation is observed, the emissivity increases. Residues and heat losses to the surrounding atmosphere increase together with the increase in the thermal flow due to convection. This flow travels along the burning surface of the polymer material. An estimation of the influence of all these factors on mass transfer during the combustion of polymer materials leads to the following expression for mass transfer parameter B:

$$B = \frac{(Y_{ox}/r)\,\Delta H_{com.}(1-\gamma) - c_p\,(T_S - T_0)}{\Delta H_e\,(1-\psi)}. \tag{2.4}$$

where

γ ratio of radiant flame energy to heat of combustion

ψ component of radiation flow which participates in gasification of polymer

If the flame of a burning polymer is characterised by a high radiant power, due to soot formation, then γ flames vary in an 18-25% range, while for light blue flames with low radiant powers, γ = 4-8%.

If it is assumed that ψ varies in proportion to γ for different polymer materials and if the experimental values of heat losses due to radiation, relative to a known ΔH_e, are used (for PMMA ΔH_e = 380 cal/g) [14], then, as established by A. M. Kanury, it is possible to determine the ΔH_e of combustion in air for different polymers (Table 2.2).

It is evident from equation 2.4 that the B number is a function not only of the characteristics of the polymer substance itself but also of the concentration of oxygen in the surrounding atmosphere and the surface temperature of a substance (temperature of gasification). For liquid fuels, the temperature of gasification is close to the boiling point, which means high-boiling substances have lower B values than low-molecular, highly volatile substances with lower boiling points.

A low B number value is assumed for heat-resistant organic polymer compounds. The B number value for the combustion of carbon is taken as a threshold value (Fig. 2.4). The B_{air} for hydrocarbon fuels (Fig. 2.4) was calculated using the following average values [11]: $\Delta H_{com.}$ = 10 300 cal/g, r = 3.48, ΔH_e = 74.1 + 0.479 T_S, T_0 = 15°C; $Y_{ox.}$ = 0.232; B_{air} = [(0.232/3.48)·10 300 - 0.24 (T_S - 15)]/(74.1 + 0.479 T_S).

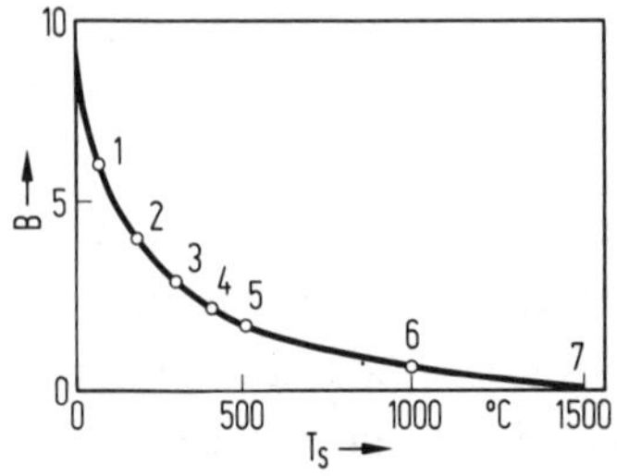

Fig. 2.4 Relationship between transfer parameter B (for liquid and solid fuels burning in an air atmosphere) and the temperature of the burning surface, T_S. (1) Gasoline, (2) kerosene, (3) gas oil, (4) light fuel, (5) heavy fuel, (6) coal, (7) carbon.

Table 2.2 Characteristics of the combustion of polymer materials in an oxygen atmosphere

Polymer material	Stoichiometric fuel/oxygen ratio $f = 1/r$	$\Delta H_{com.}$ kcal/g	B_{air}		γ	Ψ	ΔH_e cal/g
			with radiation	without radiation			
Polymethylmethacrylate	0.523	6.03	1.412	1.32	0.177	0.070	385 [14]
Polyamide (nylon-6,6)	0.191	7.17	0.818	0.81	0.038	0.015	220
Polycarbonate (Lexan)	0.440	7.36	1.206	1.16	0.102	0.040	455
Polypropylene	0.292	11.0	0.940	0.87	0.213	0.084	534
Polyethylene	0.276	9.84	0.640	0.59	0.215	0.085	622
Polyoxymethylene (Delryn 500)	0.938	3.47	0.813	0.79	0.079	0.031	720
Phenolic polymer material	0.414	7.16	0.448	0.42	0.191	0.076	750
Cellulose	0.938	4.00	0.572	0.54	0.169	0.067	710

The mass transfer parameter for the combustion process in an atmosphere of air is related to $B_{com.} = \Delta H_{com.}/\Delta H_e$ by the expression:

$$B_{air} = (B_{com.} - 1)(1 + r/Y_{ox}). \tag{2.5}$$

Table 2.3 gives values of $B_{com.}$ and B_{air} for different polymer materials [12].

There is another important result from equation 2.4: the magnitude of γ is a quantitative index of the susceptibility of a polymer material towards formation of soot or smoke. It should be mentioned that the B and γ variations take on a distinct order for the different polymers (see Table

Table 2.3 Mass transfer parameters in the combustion of polymer materials

Polymer material	B_{air}	$B_{com.}$
Polymethylmethacrylate	1.38	13.6
Polyoxymethylene (Delryn 500)	1.27	8.2
Polyamide (nylon-6,6)	1.2	14
Polyvinyl chloride	0.9	7
Polyethylene, high-density	0.73	12.5
Polyethylene, low-density	0.55	9.7
Polypropylene	0.67	11.6
Polystyrene	0.6	9.4
Cellulose	0.16	2.1
Phenolic polymer material	0.14	2.2

2.2). As mentioned earlier, the mass transfer parameter B (B number) is a measure of the combustion rate for polymer materials: combustion rate $v \sim \ln(1 + B)$.

Heat Transfer during Polymer Combustion

Flame Temperature of Polymers

Combustion and motion are related, which is why the hydrodynamic picture of an environmental flow is usually investigated during the study of combustion. Chemical reactions and transfer phenomena are largely responsible for the process. Combustion can influence the kinetic behaviour in a system, and motion in turn affects the development of the combustion process.

We have already shown an example of the conversion from a laminar to a turbulent regime in a flame with enlargement of the combustion source.

The combustion theory is based on the fundamental laws of the conservation of matter, conservation of energy and conservation of momentum. During the combustion process, a transformation of one type of energy into another takes place: the chemical energy concentrated in a substance is released during a reaction in the form of heat and radiant energy. The radiant energy can be converted into heat energy and back into radiant energy.

Heat transfer in a combustion process is accomplished through heat conductivity, convection, and radiation. These basic forms of heat transfer are not separated in a real situation. There is a contribution from each of them to the total heat exchange between a solid substance and a gas, through heat conduction and convection simultaneously. This kind of heat exchange is called convectional heat exchange or heat emission.

A convectional heat exchange process is similar to a convectional mass exchange process [25]. Radiant heat transfer has no analogy with the mass transfer process of a substance. The mechanism of heat transfer through heat conductivity consists in the transmission of energy as a result of motion and the interaction of molecules, atoms and other particles of a substance. Convectional heat transfer occurs either as a result of environmental movement due to a density difference between the heated and cold layers, in a gravitational field (free or natural convection), or under the effect of external influences (forced convection). The rate of a convectional heat exchange is characterised by a heat transfer coefficient α, which is determined from the Newton formula

$$q = \alpha S \Delta T$$

where

S surface of the heat exchange

ΔT temperature difference

Coefficient α depends on a number of factors and is usually either derived experimentally or calculated by the similarity theory method [25, 26].

Electromagnetic oscillations of wavelength λ (or frequency ν) are the carriers of radiant energy released as a result of electronic transition in atoms, molecules and other particles. This energy is transmitted by photons and is proportional to $h\nu$, where $h = 6.62 \cdot 10^{34}$ J·sec (J = joules).

Heat emission or radiation includes rays of visible and infrared light in the 0.1-100 μm band. It is natural for different substances to have emission and absorption properties that depend on temperature and optical properties. According to the Stefan-Boltzmann law, the radiant energy emitted by a heated body is proportional to the fourth power of the absolute temperature:

$$E = \varepsilon \cdot \sigma \cdot T^4$$

where

ε radiant power (degree of blackness) of a substance

σ radiant constant of an ideal black body

For the latter $\varepsilon = 1$, $\sigma = 5.67 \cdot 10^{-8}$ W/m$^2 \cdot$°K^4).

The flow of radiant energy can be absorbed, dissipated, reflected and transmitted through the medium layer. The portion of a radiant flow that is derived through each of the paths is characterised by specific coefficients (for instance, by specific absorptive power, specific reflective power etc.).

On the basis of Kirchhoff's law, when thermodynamic equilibrium is reached, the absorptive and reflective powers of the substances are numerically equal.

Absorptive and reflective power can be a function of one specific wavelength or an integral characteristic of a specific wavelength range. Values of the average absorptive power, $\bar{\alpha}$, of a polymer substance are given in Table 2.4 [27]. These have been calculated from the absorption spectra of polymer substances in terms of the emission of an ideal black body in the 1000-3500°K range (Table 2.4).

The ability of a polymer to absorb radiation diminishes as the temperature of the material rises. The optical properties (absorption, emission, energy dispersion) of diffusion flames of polymer materials have scarcely been studied.

Table 2.4 Absorptive power of polymer materials

Polymer Material	$\bar{\alpha}$					
	1000°K	1500°K	2000°K	2500°K	3000°K	3500°K
Natural rubber	0.88	0.82	0.76	0.72	0.69	0.68
Polystyrene (white)	0.86	0.75	0.63	0.53	0.45	0.40
Polystyrene (transparent)	0.75	0.60	0.46	0.35	0.28	0.22
Polymethylmethacrylate (white)	0.91	0.86	0.78	0.70	0.62	0.56
PMMA, black (with soot)	0.94	0.94	0.95	0.95	0.95	0.95
PMMA, transparent	0.85	0.69	0.54	0.41	0.31	0.25
Polyethylene, low-density	0.92	0.88	0.82	0.77	0.72	0.68
Polypropylene	0.87	0.83	0.78	0.74	0.70	0.68
Polyvinyl chloride	0.81	0.65	0.49	0.38	0.30	0.24
Organosilicon rubber	0.79	0.66	0.58	0.54	0.52	0.53
Polyamide (nylon-6,6)	0.93	0.90	0.86	0.82	0.75	0.71
Cellulose acetobutyrate	0.84	0.71	0.56	0.43	0.34	0.27
Phenol-formaldehyde polymer	0.90	0.86	0.81	0.77	0.75	0.75

The determination of a heat transfer mechanism during the combustion process and the quantitative evaluation of the contribution of each type of energy transfer to the total thermal balance of the reaction system represent major tasks in polymer combustion research. There is a strong functional connection between the thermal flow, the radiation from a flame to a polymer surface and the rate of combustion.

As already mentioned, stationary combustion is attained when the amount of incoming heat is equal to the amount of heat consumed (in preparation of the next portion for combustion).

The study of temperature distribution in a reacting system, the determination of the so-called temperature field in a combustion wave zone and the temperature gradients are among the most important components of the analysis of the combustion process of polymer materials. Temperature is usually measured during a steady regime using a thermal probe and a pyrometric method. This problem will be discussed in detail in a later section.

The temperature field could be described as a function of one, two or three spatial coordinates. An example is shown in Fig. 2.5, with temperature profiles for gaseous ($y > 0$) and condensed ($y < 0$) phases in a preflame zone ($x \leq 0$). The flame is spreading down the surface of a vertical sheet of polymethylmethacrylate. The half-thickness of the sheet is $L = 0.27$ cm, and the surrounding atmosphere is air [28].

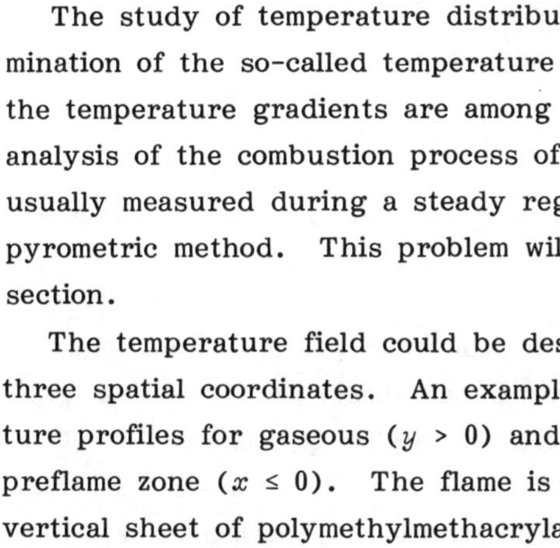

Fig. 2.5 Temperature profiles in a preflame zone ($x < 0$) spreading down the surface of polymethylmethacrylate sheets. (1, 1') x/L = 0.0, (2, 2') (-0.11), (3, 3') (-0.26), (4, 4') (-0.37), (5) (-0.74), (6) (-1.10), (7) x/L = -1.85, L = 0.27 cm.

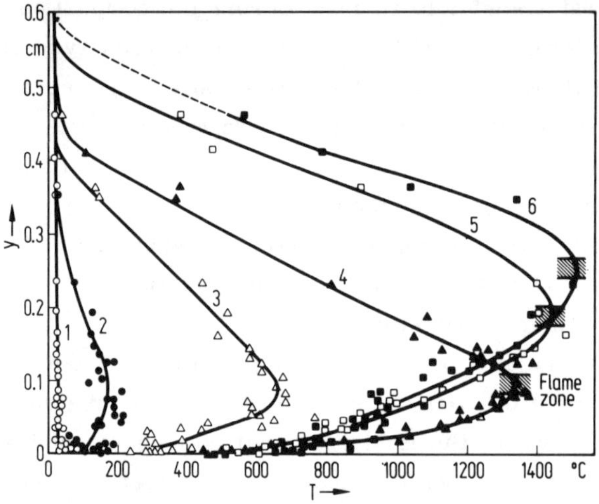

Fig. 2.6 Temperature profiles in the gaseous phase during flame spread down a paper sheet. (1) x = -0.3 cm, (2) (-0.02), (3) (-0.1) (4) 0.0, (5) (+0.2), (6) x = +0.4 cm.

Figure 2.6 shows similar temperature profiles in preflame zones and inside the flame ($x > 0$) spreading down paper sheets [29]. On the ignition boundary time ($x = 0$) the flame temperature is approximately 1620°K. The maximum temperature zone is located at a distance of $y = 0.1$ cm from the surface of the polymer material. The maximum temperature for a blue flame zone can be as high as ~1770°K ($x = 0.4$ cm, $y > 0.3$ cm).

An experimentally found maximum temperature in the flaming reaction zone differs from the theoretical one for the adiabatic flame. The theoretical temperature was calculated for a stoichiometric fuel to oxidiser ratio with a specific composition for the oxidising gaseous atmosphere and pressure and an equally distributed composition of combustion product.

The difference between the experimental and calculated values of the maximum flame temperature of polymer materials is dependent both on incomplete fuel combustion and on heat losses to the surrounding atmosphere through radiation and convectional heat exchange. The maximum temperature of the flame usually increases with an increase in the oxygen concentration in the surrounding atmosphere [20] (Fig. 2.7).

The surface temperature of a burning polymer sample fluctuates in a similar manner at the same time. Table 2.5 shows the values of calculated adiabatic temperatures and measured maximum temperatures for diffusion flames in a number of polymer materials.

Table 2.5 Maximum temperature of diffusion flames of polymer materials [5]

Polymer material	Surrounding atmosphere		$T_{max.}$, °C	
	Inert Component	$Y_{ox.}$	calculated	measured
Polyoxymethylene (Delryn 500)	N_2	0.200	2030	1325
Polyoxymethylene	Ar	0.138	1895	1400
	Ar	0.136	1720	1400
	CO_2	0.277	1635	1400
Polyethylene	Ar	0.153	2030	1680
	N_2	0.212	1855	
	CO_2	0.339	1830	
	N_2 (air)	0.210		1287 [7]
Polymethylmethacrylate	Ar	0.156	2025	1650
	N_2	0.208	1815	
Polyethylene, chlorinated	Ar	0.203	2390	1550
	N_2	0.266	2115	
	CO_2	0.381	1955	
Polyethylene, chlorinated	Ar	0.242	2625	
with 4% Sb_2O_3	N_2	0.346	2445	1700
	CO_2	0.462	2186	
Polytetrafluoroethylene	Ar	13.5	2558*	1905
	N_2	19.0	2554	
Polypropylene	N_2 (air)	0.210		1280 [7]
Poly-(trioxyethylene dimethacrylate)(TGM-3)	N_2	0.167	2090	825 [31]
Poly-(propylene-bis-(meth acryloxyethylene carbonate) (OKM-8)	N_2	0.182	2160	
	N_2	0.300	2540	

* Calculation was made for stoichiometric fuel to oxidiser ratio in the reaction: $1.2(-C_2F_4-)+O_2 \rightarrow 0.4CO + 0.4\ CO_2 + 0.8COF_2 + 0.8CF_4$.

According to the data [8, 30], during candle-type combustion of polyoxymethylene, the surface temperature is $T_S \simeq 750°K$, the gradient on the gaseous phase side is approximately 8000°K/cm and the T_{max} of the flame is approximately 1800°K. The combustion was studied in an atmosphere of argon (79.7%) and oxygen (20.5%) at a reduced pressure (50 mm Hg). S.J. Burge and C.F.H. Tipper [6] discovered that during candle-type

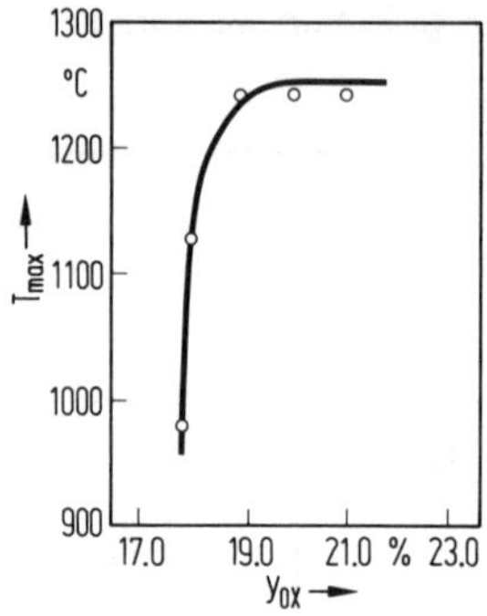

Fig. 2.7 Relationship between concentration of oxygen in surrounding atmosphere ($Y_{ox.}$) and maximum temperature of polypropylene flame ($T_{max.}$).

combustion of polyethylene and polymethylmethacrylate samples in an atmosphere of air, the maximum temperature of the flame did not exceed 1013°K.

A large scattering was observed in the experimental values of the burning surface temperature, the flame temperature and the temperature gradients. These measurements are affected by a difference in the concentration of oxygen in the surrounding atmosphere, the size of the sample and the directions and rates of reagent flow. It is possible that this scattering of the experimental data is due to method-related factors in temperature determination (measurement method, characteristics of the thermocouple bores).

It is evident from Table 2.5 that the calculated temperature of the flame increases when an inert gas has been introduced, in the order: $CO_2 < N_2 < Ar$ and when the concentration of oxygen increases. Thus the temperature characteristics of the combustion of polymer materials substantially depend on the conditions that sustain the process. The information on temperature fields during the combustion of polymer materials allows the temperature gradients to be determined. By using these results it has been possible to calculate the thermophysical and thermochemical characteristics of the initial substance and the combustion and degradation products together with the experimental values of the process rate and thermal flows for a stationary, steady combustion regime. It should be mentioned that another type of heat transfer is also characteristic of polymers. This type of heat transfer is initiated by a purely mechanical motion process in the hot polymer melt [32]. This form of heat transfer is inherent in thermoplastic polymer materials that melt during the heating process.

During the combustion process these polymer materials form droplets and layers, which flow off the surface or spatter around. The type of heat transfer mechanism has not been analysed theoretically. In practice, the heat transfer is (as a rule) unstable but it can be taken into account by averaging the losses of polymer mass per unit time which are due to the removal of the molten mass. Reference [7] is an example of this type of analysis of the combustion of polymer material. A.M. Grishin and A.S. Yakimov [33] attempted to theoretically evaluate the effect of the spread of a molten polymer (with gaseous monomer bubbles included) on the ignition and combustion of PMMA in a heated stream of oxidising gas.

In a stationary combustion regime, heat input is equal to heat removal within the system. The thermal balance for the stationary combustion of polypropylene and polyethylene in an atmosphere of air is given in Table 2.6. Allowance is made for the thermal energy that flows during heating and pyrolysis of a polymer material, during heating of the monomer from the surface temperature to the flame temperature and during heating of the air and the production of a stoichiometric reaction of the monomer's combustion (combustion products CO_2 and H_2O). All the heat losses must be compensated by a heat flow generated during the combustion of the fuel in the front zone of the flame.

The calculation of a thermal radiation flow coming from the polyethylene and polypropylene flames at a flame temperature of 1550°K gives radiant power values of 0.03 and 0.09 respectively. (The flow is being lost in the surrounding atmosphere.) These values are typical for hydrocarbon flames that do not form soot.

Changes in Heat Transfer Parameters and in Hydrodynamics during Polymer Combustion

A burning material is a system in which several parameters are strongly related to one another. The analogy is apparent when the critical conditions of stationary diffusion combustion of polymer materials and the effect of the scale of the combustion source are examined. A change in the characteristics of a combustion process as a result of different conditions

Table 2.6 Thermal balance in the stationary combustion of polymer materials

Type of thermal flow	q cal/min	
	polyethylene	polypropylene
Heat lost:		
losses by heat conductivity through polymer material	38	22.5
heating up of polymer material from T_{in} to T_s	31	14.1
pyrolysis of polymer material at T_s	36	31.5
heating of air from T_{in} to $T_{fl.}$	212.6	297.3
heating of monomer from T_s to $T_{fl.}$	27.1	43.6
heating of CO_2 + H_2O from T_s to $T_{fl.}$	115.2	161.2
Total (endothermal) consumption	459.9	570.2
Heat gained:		
heat of combustion (25°C)	-478.0	-633.0
Difference (losses due to radiation)	- 18.1	- 62.8

is associated primarily with changes in heat transfer parameters and changes in the hydrodynamics of the reacting system.

In small-scale laboratory tests, when the characteristic size of the fuel is less than 0.1-0.2 m, heat conductivity and convection are the predominant types of heat transfer. Flaming diffusion combustion of small-size samples is usually of a laminar nature. As the characteristic size of the fuel increases, however, the process becomes turbulent, which is why the turbulent convective heat transfer from such flames is substantial.

In the case of combustion of medium-sized samples (larger than 0.3 m), the heat transfer through radiation exceeds the free convective heat transfer from the flame, and the former factor cannot be ignored. Finally, the combustion of medium-sized samples is characterised by the fact that heat coming from the flame towards the surface is transferred primarily by radiative processes [34]. Heat transfer in the gaseous phase becomes dominant. Heat transfer by convection, heat conductivity and radiation follows according to inherently different physical laws. It is hence not surprising that the results of small-scale, medium-scale and large-scale tests do not correlate with one another very well.

In recent studies on the combustion process an additional ignition source has been used, and the intensity of the heat flow varied [35-38]. These methods are designed both to determine how heat radiation affects the development of combustion and to determine how the results of laboratory tests can be used to predict the behaviour of polymer materials in actual fire situations.

The increase in the rate of flame diffusion over the surface of polymer materials when an additional ignition source is used is related to an increase in the temperature of the surface in front of the moving flame. However, the latter is not the only cause of the change in combustion rate. It was discovered [35] that when a flame spreads downwards along the vertical surface of thick PMMA sheets under the effect of a source of internal radiation, the rate of combustion varies with time, as a function of the power and duration of exposure.

An additional warming of the surface through a radiant heat flow causes a reduction in the difference $T_{ign.} - T_0$ according to the law $v_{fl.dif.} \sim (T_{ign.} - T_0)^{-2}$. This law was predicted for thermally thick materials. Additional warming causes an acceleration of the combustion process. The increase in the initial temperature T_0 of the polymer surface in front of the edge of flame simultaneously strengthens convection in the border heat layer and reinforces the radiation of heat from the surface, causing partial cooling [35]. As a result, if the intensity of an external radiant flow is low, the rate of flame diffusion over the surface of PMMA will increase with time but will reach a constant value when I_0 is large. The cooling effect as a result of the above-mentioned factors is relatively small, and the rate of flame diffusion must increase exponentially with time. In the latter case, gasification of the polymer occurs along the entire surface, which has also been heated. Diffusion of the flame has a pulsating, unstable character.

The use of external radiation leads to an increase in the heat transfer from the flame to the material through the gaseous phase on account of the increase in the rate and gradient of oxidiser inflow [37, 38]. The effect of a change in the heat transfer mechanism in the case of additional radiation warming of polymer materials is similar to that observed with the variation of oxygen concentration in the surrounding atmosphere. Studies of the diffusion of a laminar flame over the horizontal surface of thermally thin PMMA samples showed [39] that an increase in the concentration of oxygen in the flow of oxidiser causes a change in the nature of the heat transfer and also leads to a shifting of the reaction site from inside the flame to a point closer to the surface of the polymer. When the concentration of oxygen in the flow of oxidiser is $Y_{ox.} > 0.4$, the heat flow transfer to the surface of the polymer (in the zone adjacent to the front edge of the flame) occurs essentially through the gaseous phase. The heat transfer by heat conductivity through the condensed phase drastically decreases (Fig. 2.8).

The way in which the concentration of oxygen affects the various characteristics of polymer combustion when the scale factor changes has been analysed in [40]. In the case of the polymers studied, for those that burn easily in an atmosphere of air when the concentration of oxygen substantially exceeds the atmospheric level in air $Y_{ox.} > 0.233$, all measured characteristics assume their asymptotic values. It was discovered that in small-scale tests with high concentrations of oxygen, a predominant heat transfer from the flame could also occur by means of radiation. This fact, together with the influence that an additional heat flow

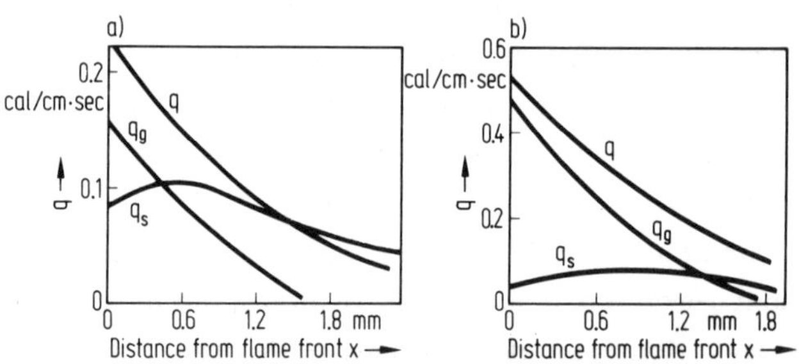

Fig. 2.8 Effect of concentration of oxygen upon the total heat flow (q) and upon the individual contributions to the heat transfer from the flame to the surface of PMMA through the gaseous (q_g) and condensed (q_s) phases. $Y_{ox.}$ a - 2.25, b - 0.40.

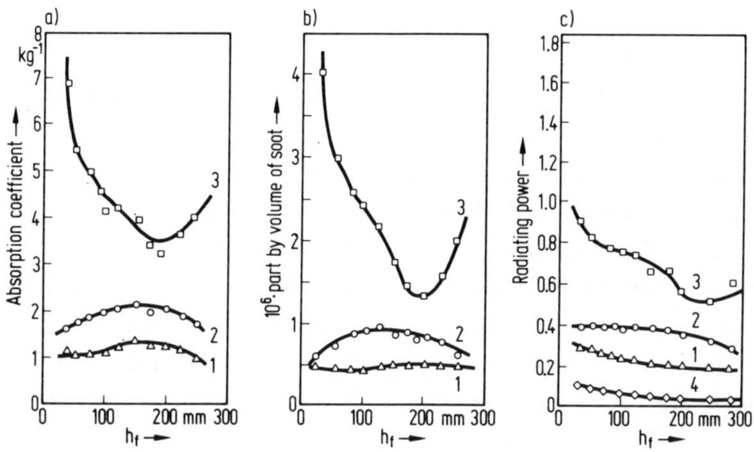

Fig. 2.9 Change in optical properties along the height of the diffusion flames of polymer materials. 1 - PMMA, 2 - PP, 3 - PS, 4 - POM.

exerts on the burning system, suggests an approach to the modelling of fires and the analysis of flammability properties of polymer materials based on scientifically sound programmes of small-scale tests. For a real fire to grow and spread, heat transfer by radiation is essential, and it is for this reason that knowledge of the radiative and optical properties of the diffusion flames of polymers is important.

Markstein [41] developed a simplified method for the evaluation of different optical characteristics of polymer diffusion flames. He discovered a direct relationship between radiating power, coefficient of transmission, absorptive power, other flame characteristics and the formation of soot during the combustion of polymer materials (Fig. 2.9).

The rate of flame radiation for large-sized polymer samples increases in an atmosphere of air in the following sequence: polyoxymethylene (POM) < foam polyurethane (FPU) < polymethylmethacrylate (PMMA) < polypropylene (PP) < polystyrene (PS). The combustion rate changes as follows: POM < FPU < PP < PMMA < PS. The inversion of the order of PMMA and PP can be attributed to a larger heat of vaporisation compared with that for PMMA (~26% more). It should be mentioned that the optical properties of fires depend on their geometry and scale. All flames of large-sized polymers have a grey spectral colour.

Experimental studies of combustion wave structures and the characteristics of polymer combustion in relation to scale and other factors help to determine and elucidate the mechanisms of the process and permit the development of more complete theories of polymer combustion.

Petrella [42] studied the effect of an external heat source on the rate of polymer combustion. From the condition of the conservation of energy, it follows that $\dot{q}_s + \dot{q}_{ex} = \dot{q}_v + \dot{q}_l$. In other words, the heat inflow on to the surface from the flame $\dot{q}_s$ and from the external source $\dot{q}_{ex}$ are balanced by the heat flow used for vaporisation of the polymer, for delivery of fuel into the flaming zone ($\dot{q}_v$) and in loss through the polymer ($\dot{q}_l$). Considering $\dot{q}_v = \dot{m} \cdot \Delta H_e$ ($\dot{m}$ is the mass rate of combustion, ΔH_e the enthalpy of polymer vaporisation, T_0 the initial temperature), and $\dot{q}_s = \xi \cdot Y_{ox.}{}^n$ (ξ and n are constants, depending on the nature of the polymer), we have

$$\dot{m} = (\xi/\Delta H_e) \cdot Y_{ox.}{}^n + (\dot{q}_{ex} - \dot{q}_1)/\Delta H_e$$

When $n = 1$, this equation represents a straight line. From the relationship between mass combustion rate and the concentration of oxygen inside the oxidiser flow at $\dot{q}_{ex}$ = const., it is a simple matter to determine the value of $\xi/\Delta H_e$ from the gradient of the straight line. The segment intercepted by the ordinate axis $Y_{ox.} = 0$ gives $(\dot{q}_{ex} - \dot{q}_l)/\Delta H_e$ (Fig. 2.10a).

In the same way, the gradient of the straight line $\dot{m} - \dot{q}_{ex}$ at $Y_{ox.}$ = const. gives values of $1/\Delta H_e$ (Fig. 2.10b). Thus from the experimental data we can determine $\xi/\Delta H_e$, $\dot{q}_l/\Delta H_e$, $\dot{q}_s$, $\dot{q}_l$, ξ. In the case of polystyrene, for instance, it has been discovered that $n = 1$, $\Delta H_e = 402$ cal/g,

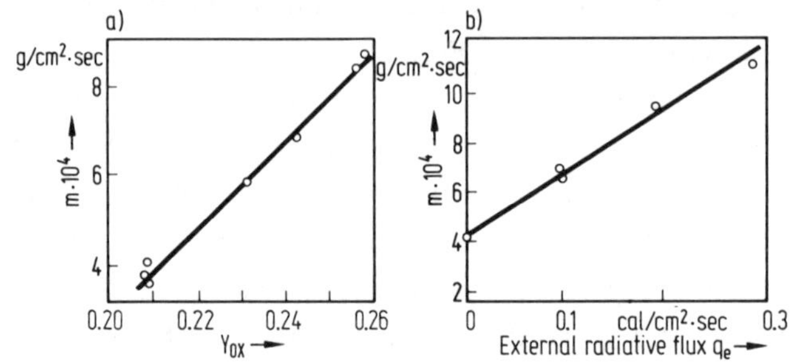

Fig. 2.10a Relationship between mass rate of thorough combustion of polystyrene and concentration of oxygen inside the O_2/N_2 flow at $q_{ex} = 0$

Fig. 2.10b Relationship between mass rate of thorough combustion of polystyrene and the intensity of the external heat flow at $Y_{ox.}$ = const.

$\xi = 3.83$ cal/cm^{-2}·sec^{-1}, $\dot{q}_1/\Delta H_e = 16.2 \times 10^{-4}$ g/cm^2·sec, $\dot{q}_s = 0.8$ cal/cm^2·sec.

With these parameters it is possible to determine the mass rate of polymer combustion for any concentration of oxygen including combustion in air ($Y_{ox.} = 0.21$) and for various external heat flows.

Petrella supposes that if we assume the situation where $\dot{q}_{ex} = \dot{q}_1$ and correspondingly $\dot{m}_{ideal} = (\xi/\Delta H_e)\cdot Y_{ox.}{}^n$ as an "ideal" situation, then it is possible to distinguish three systems of polymer combustion:

1. Small-scale tests: $\dot{q}_{ex} < \dot{q}_1$; $\dot{m} < \dot{m}_{ideal}$
2. Large-scale tests: $\dot{q}_{ex} > \dot{q}_1$; $\dot{m} > \dot{m}_{ideal}$
3. Situations where $\dot{q}_{ex} = \dot{q}_1$ corresponds to the state of fire development.

In Petrella's work the general mass rate of thorough polymer combustion (i.e. for complete or almost complete polymer combustion) for candle-type sample combustion was reported but the case of flame diffusion over the surface was not studied. In order to determine the mass rate of thorough combustion for a flame spreading over the surface of polymer materials, an indirect method was proposed: measuring the profile of the burnout surface with the aid of a profilometer [43].

It has been found that where there is a relatively low concentration of oxygen in the incoming flow ($0.2 \leq Y_{ox.} \leq 0.4$), the rate of thorough combustion for PMMA does not change at all in a large zone beneath the flame. However, if the concentration of oxygen in the oxidiser flow increases, the mass rate of thorough combustion under the edge of the flame is $m_1 \sim Y_{ox.}^4$, and off the edge it stays constant, regardless of $Y_{ox.}$.

By determining the mass rate of thorough combustion for polymer materials under different combustion conditions it becomes possible to establish how different factors affect the mechanism of this complex process. In particular, it allows the effective kinetic parameters of polymer pyrolysis during combustion to be determined, although this does require knowledge of the actual temperatures of the burning surface.

The study of how natural convection affects the combustion of PMMA spheres led investigators [43] to the conclusion that the law of Sreznevsky (d^2) is violated in the case of large Grashof numbers. Thorough combustion of polymer spheres should occur in this case according to the $d^{5/4}$ law.

Going on to develop the analysis of the model for flame diffusion over the horizontal surface of condensed fuel inside the gaseous oxidiser, investigators [44] derived an analytic solution for the two-dimensional problem. In the analysis they derived ratios that are used to determine the location of the chemical reaction zones, the flow of fuel inside the zone and the temperature of the reaction zone. These ratios show that the distance from the surface of the fuel to the reaction zone becomes larger farther away from the front edge of the flame. At such points, the flow of fuel and oxidiser into the reaction zone decreases while the temperature of the zone increases and attains its threshold values, corresponding to adiabatic combustion. Thus, the temperature in the front portion of the flame may be substantially lower than an adiabatic combustion temperature (for instance, on account of the heat transfer to the fuel material). A criterion has been discovered which determines how the rate of chemical reaction in the gas phase affects the structure of the combustion wave and the rate and extent of flame diffusion. When the rate of the gas phase reaction is high, the flame touches the surface. For this case, de Ris' theory of flame diffusion is correct. This theory considers only heat and mass transfer. If the rate of the gas phase reaction is low, the reaction starts a certain distance away from the surface of the fuel material, e.g. a rim of flame exists. It is necessary to consider the kinetics of the chemical reaction in the gas phase to calculate the rate of flame spread.

This model deals with heat transfer as a result of heat conductivity and convection through the gas phase and also as a result of heat conductivity through the polymer material. An increase in the rate of inflow, a fall in the pressure and concentration of oxidiser within the flow and a fall in the value of the pre-exponential factors and the activation energy cause a drop in the heat flow into the fuel material and a reduction in the flame rate. It was predicted that there could be limits imposed by the rate of the incoming flow of oxidiser and the oxidiser concentration inside the flow. Complicated ratios and the lack of quantitative data on different parameter values in the derived equations restrict the use of these equations to the qualitative evaluation of flame diffusion length.

There are a limited number of experimental studies that deal with the structure of the temperature field inside the front edge of the flame, the change of structure due to the change in oxygen concentration in the atmosphere and the rate of the incoming flow. Lalayan [39, 46] showed that the maximum temperature of a laminar diffusion flame that spreads over the horizontal surface of thermally thick samples of PMMA is indeed attained far away from the front of the flame. When the concentration of oxygen $Y_{ox.}$ inside the O_2/N_2 flow changes from 0.2 to 0.4, the maximum temperature of the flame increases from 1720 to 2000°K (the adiabatic temperature of the flame changes from 2230 to 2600°K under such conditions). The temperature at the front edge of the flame, which determines the quantity of heat that flows towards the surface of the polymer material, is substantially lower (by a few hundred degrees) than the maximum temperature.

The early theories for the combustion of polymer materials for both indefinite and terminal rates of flaming gas phase reactions contained no essential mechanism for determining the quenching of the combustion process.

The equations that describe the process for the stationary combustion of polymer materials are also correct for the state of flame quenching.

Frey and Tien [47] theoretically estimated the limitations for flame quenching in the case of combustion of a thermally thin fuel material. They conducted the evaluation in terms of the concentration of oxygen and the pressure and rate of flow of the oxidising agents. They solved numerically the equations that describe the two-dimensional diffusion of the flame by assuming specific values for the kinetic parameters of the solid phase and gas phase reaction and also for the parameters describing the physical and chemical properties of those phases. A state of quenching corresponded to a critical value of the Damkeler number. As was illustrated for PMMA combustion, a change in the oxygen concentration in the surrounding atmosphere affects the temperature of the flame, which in turn affects other parameters. The results were qualitatively correlated with experimental data [47b].

Margolyn and Krupkin [48] summarised the data for different limits of polymer material combustion in respect of how different factors affect the cooling of the gas phase reaction zone. At the end of the diffusion combustion process the ratio between heat expended from the leading combustion zone and the total heat emission resulting from the chemical combustion reaction should be proportional to RT_f^*/E, where T_f^* is the flame temperature at the end of quenching and E the activation energy of the gas phase reaction:

$$\Sigma \dot{q}_1 / \dot{q} \sim RT_f^*/E$$

Each factor contributes towards the total heat loss. Under certain conditions either factor may play a principal role. It is assumed that the maximum rate of thorough combustion of polymer material inside the edge of the flame at the time of quenching corresponds to the rate of combustion of a stoichiometric mixture of degradation and gasification products of polymer and oxidising agent. The latter allows one to calculate the maximum adiabatic temperature of combustion for a particular value of the polymer's standard oxygen index (at normal pressure and acceleration of gravity I_g). By assuming a reasonable value for the effective energy of the gaseous phase (E = 20-40 kcal/mole) and by determining the mass rate of thorough combustion of the polymer (with an additional experiment), such as from the relationship between the limiting oxygen concentration and strain, it is possible to determine the pre-exponential factor for a given polymer from the Arrhenius equation for the gas phase combustion reaction.

Since all factors that affect the cooling of the flaming zone ultimately reflect the change in the limiting concentration of oxygen, it is more convenient to use the relationship between $Y_{ox.}^*$ and either of the factors as a test for determining the nature of heat losses. An equation connects $Y_{ox.}$ with the effective kinetic parameters of a gaseous reaction in the event of a transition to new conditions of combustion differing from the standard ones (P = 1 atm, I_g, no forced flows of oxidiser). Having the parameters is enough for a complete determination of the limitations on the stable diffusion combustion of polymers. The inertial constraints, for instance, increase the convective heat losses on account of the rise in buoyant force and the increase in the rate of convective lift of the hot combustion products. In the case of severe overloading - approximately 300 g, the convective heat losses are the major ones. The ultimate concentration of oxygen in this case is proportional to the cube root of the overloading value $\bar{a}$, which is a weak function of pressure and does not depend on the size of the sample [49].

In the case of zero gravity there is no natural convection, and a forced flow of oxidiser is needed in order to contain the combustion.

Thus the minimum acceleration at which the combustion of PMMA can still be observed in a quiescent medium at $Y_{ox.}$ = 17% is 2 x 10^{-2} [50]. For low accelerations the heat losses are caused by convection and radiation. The relationship between $Y_{ox.}^*$ and pressure shows the effect of radiation on the quenching of flaming combustion when polymers are quenched on account of radiation; $Y_{ox.}{}^{*} \sim 1/P^{k}$ $(k = 0.5\text{-}1.0)$.

The analysis of flame quenching can be extended to include heat losses not only through the gaseous phase but also through the condensed phase of the polymer. The heat losses through the condensed phase intensify the effect of various factors on quenching.

In order to determine the limits of quenching of polymer materials it is necessary to know the kinetic parameters for solid-gas-phase reactions during combustion. In the overwhelming majority of cases, the parameters of flame reactions are unknown, which is why the reverse problem of their determination is of scientific and practical importance. The theory of quenching for diffusion flames of fuel liquids with a flat flame geometry and a counterflow of oxidiser has been used for the flames of polymer materials with similar configurations [51, 52]. The major type of heat loss in this analysis is heat loss by radiation from the surface of either the polymer or the monomer. Figure 2.11 shows the relationship between the rate of flow of oxidiser at the end of quenching and the temperature of the flame at the end. The values of the effective energies of activation have been derived for PMMA and MMA, and the pre-exponential factors calculated from the gradients of these straight lines. The parameters of the gas phase reaction for other polymers [53] have likewise been determined.

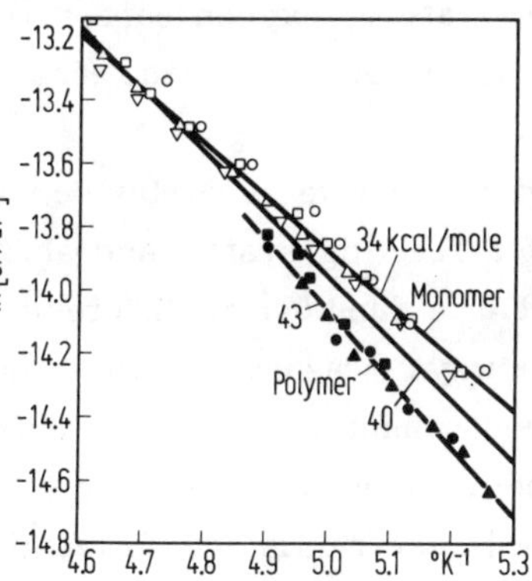

Symbol	App.	Fuel	L in cm	Date
○	Old	Monomer	1.0	Oct 10, 74
□	New	Monomer	1.0	Aug 10, 76
△	New	Monomer	0.75 + (0.45 to 0.9)	Aug 13, 76
▽	New	Monomer	1.0	Aug 16, 76
●	Old	Polymer	1.2	Nov 11, 74
▲	New	Polymer	1.17	Aug 9, 76
■	New	Polymer	0.814 → 0.1	Aug 13, 76

Fig. 2.11 Arrhenius plot showing the best data obtained for extinction with the polymer and the monomer. (Relationship between the function of the limiting velocity of an oncoming flow of oxidiser and the flame temperature at the point of extinction.

The asymptotic method has been proposed for the analysis of diffusion flames (considering heat losses by radiation from the gas phase) [53]. However, similar computations conducted [54] for the quenching of PMMA flames (flames spreading over a vertical surface) showed that the heat loss effects through radiation from laminar diffusion flames are not very important in comparison with the radiation from the surface of a burning polymer.

The Rate of Combustion of Polymer Materials

The rate of combustion is one of the most important quantitative properties of combustion. It is a function of the physical and chemical properties of the reagents, the kinetic parameters of the combustion reaction and the conditions under which the reaction is taking place. It serves as a quantitative measure of the relative combustibility of the substances if the combustion processes are carried out under identical conditions.

The chief problem in a theoretical or experimental study of the combustion of polymer materials is the determination of the rate of combustion. This is a function of the thermophysical properties of the condensed and gaseous phases, the kinetic parameters of the pyrolysis reaction and flame reactions in the gaseous state, the diffusion coefficient of the oxidising agent and the degradation products of the polymer material, the properties of the surrounding atmosphere (pressure, temperature, oxygen concentration, presence of forced convection and external radiant flow), the size of the polymer sample and other factors.

Theoretically, it should be possible to predict the rate of combustion as a function of a large number of parameters of a reacting system and also to predict the limits of steady combustion that is a function of different variables. However, these variables are not always measurable, which is why, when the theory is applied, only the predominant factors determining the combustion rate are taken into consideration. The most productive approach is to separate out the leading stages in the multistage combustion of polymer materials and the development of heat exchange and mass-exchange mechanisms that determine the combustion rate for preset conditions.

The rate of combustion of polymer materials may be expressed in units of mass or linear velocity. In addition, the unidimensional movement of the reaction front vis-à-vis the unburned reagent is taken into consideration. For instance, the normal combustion of a polymer rod from one end or the thorough combustion of polymer spheres is characterised in the same way.

In the case of flame spreading on the surface of polymer materials, the combustion front can shift in two directions: a shift in the flame front along the surface, and normal, complete combustion of the condensed phase in a direction perpendicular to the phase contact area. This situation makes the analysis more complicated, so that, as a rule, the theory considers the unidimensional spreading combustion reaction for purposes of simplifying the analysis.

The Spreading of a Flame over the Surface of a Polymer Material

The spreading of a flame over the surface of a combustible material in a gaseous oxidising atmosphere is the combustion process most frequently observed. A theory for this process is not sufficiently developed as yet. In all the analytical models of the spreading of a flame over the surface of a polymer material, the process is considered as a continuous diffusion ignition process for the gaseous state of the polymer's degradation products. Heat from a flame, radiating on the surface edge of a material, warms up a layer of polymer material to the temperature at which gasification starts. The gaseous fuel products then diffuse from the surface into the oxidising atmosphere. The self-accelerating exothermal oxidation of the fuel is generated in the gaseous phase. In this way a continuous spread of flame is provided.

Different directions of flame spread over the surface of polymer materials have been studied [32]. The downward and upward directions of combustion development along the vertical surfaces have the most distinct characteristics. The other directions occupy an intermediate position, depending on which side of the material surface (top or bottom) the combustion reaction occurs [55].

In theoretical models of flame diffusion over the surface of a polymer material, primary attention has been focussed on the problem of how the transfer of thermal energy takes place from the flame areas that are still not ignited. In general, according to the law of conservation of energy, the rate of flame spread could be expressed by the thermal flow transferred from the flame to the ignitable surface:

$$\rho \cdot v \Delta h = q, \tag{2.6}$$

where

ρ density of fuel material

$\rho \cdot v = \dot{m}$ mass rate of combustion

Δh difference in the enthalpy of the fuel material at its chemical reaction temperature and in the initial state

Either the ignition temperature $T_{ign.}$, or the scale temperature of a chemical reaction T_* (see Chap. 1), are used in the calculation of Δh. The thermal flow q is the specific component of the heat liberated as a result of combustion of the fuel in the flame zone [56].

In order to determine a flame diffusion rate it is thus necessary to know the values of q and Δh. The two boundary conditions that have generally been studied when determining heat accumulation in the condensed phase are:

1. A "thermally thin" layer of polymer material
2. A "thermally thick" layer of polymer material

Flame diffusion over a horizontal surface is presented in diagram form in Fig. 2.12. The direction of the oxidising gaseous flow is opposite to the direction of flame diffusion. A similar picture was used for the examination of mathematical models of flame diffusion under stationary combustion.

J. de Ris [57] analysed a model of the spread of a laminar diffusion flame. According to the model, when there is no forced convection (air

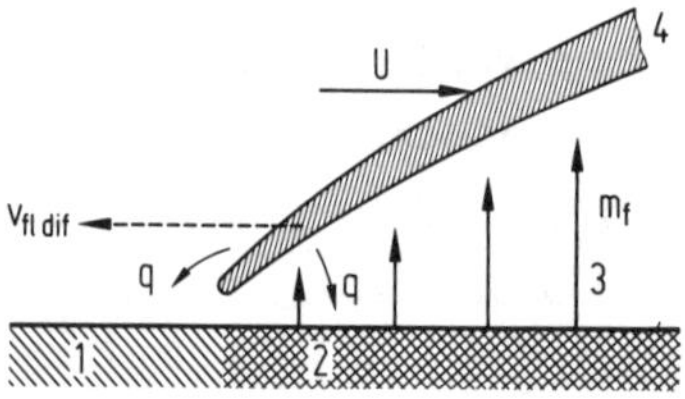

Fig. 2.12 Diagram of flame diffusion over the horizontal surface of a polymer material. (1) zone of not yet ignited fuel, (2) zone of pyrolysis, (3) dark zone of fuel's vapour, (4) flame chemical reaction zone. Oxidising gas flow (U), fuel vapour ($\dot{m}_f$), heat flow coming from flame (q), combustion heat flow ($v_{fl.dif.}$).

velocity = 0) the heat transfer towards the surface of unburned fuel occurs by molecular heat conductivity through the gaseous state; this is true for the thermally thin layers. In the case of indefinitely thick layers, heat transfer occurs by heat conductivity through the gaseous and condensed phases and also by radiation.

When forced convection is present its effect on the combustion process has to be taken into consideration. It is assumed that the rate of flame diffusion is limited by heating and by the diffusion transfer of reagents in the zone of chemical reaction but not by the kinetics of the chemical reaction inside the flame; that is, the time of heat-mass transfer t_m is much larger than the time of chemical reaction t_{ch}. ($t_m >> t_{ch}$). For ignition to be achieved the ignition temperature of the material should be equal to the gasification temperature. The temperature of the flame reaction zone is assumed to be equal to an adiabatic temperature, which is functionally associated with Spolding's mass transfer number B. In this way, the influence of individual characteristics of the material and the concentration of oxidising agent in the surrounding atmosphere becomes apparent.

For thermally thin material layers the rate of flame spread is inversely proportional to the thickness of the layer and does not depend on the rate of oxidising medium:

$$v \simeq \frac{\sqrt{2}\,\lambda_g (T_{fl.} - T_S)}{\rho_c c_c L (T_S - T_0)}, \tag{2.7}$$

where

L thickness of the material

T_s surface temperature, equal to T_{ign}.

For indefinitely thick layers (the thickness of the material is much greater than the depth of concentration of the heat wave) the rate of flame diffusion does not depend on the thickness of the material and is proportional to the velocity of air U.

A cumbersome formula has been derived which does not clearly express the connection between rate v and such factors as the pressure and the concentration of oxygen in the surrounding atmosphere. In the model described [58] an attempt has been made to take into account the gasification of the fuel and the kinetics of the chemical reaction for thermally thick layers of material. Criticism is necessary here because the suggested models do not demonstrate any evident form of relationship between v, the pressure and the concentration of the oxidising agent. They also do not allow one to check whether the relationships correspond to the experiment [59].

In fact, reference [57] disregards the ignition zone near the edge of the flame. This zone is significant because the ratio of mass transfer time to chemical reaction time (Damkeler number) is small and the influence of chemical kinetics essential.

An evaluation was made of the thickness of the heating zone along the surface of the material (δ_{cx}) and its gasification zone ($\delta_{cx,pyr.}$), necessary for maintenance of combustion, on the basis of a study of the thermal balance at the border line of the ignition zone (at the edge of the flame) [60].

The analysed model of a flame is shown in Fig. 2.13. This model considers convectional heat exchange in a gaseous phase at the edge of the flame front and also heat transfer through heat conductivity in the condensed phase. The heat transfer by radiation is not taken into consideration (this is insignificant at the edge of the flame).

It has been discovered for the thermally thin layer that

$$\delta_{cx} \simeq a_c/v, \qquad \delta_{cx,pyr.}/\delta_{g,\min} < RT_*/E,$$

where

a_c thermal conductivity coefficient in the condensed phase

v rate of flame diffusion over the surface

$\delta_{g,min.}$ minimal thickness of the gaseous thermal layer at the edge of the flame, which is determined from the formula:

$$\delta_{g,\min} \simeq \left[\frac{2E\,(T_* - T_0)^2\,\lambda}{RT_*Q\omega_{\max}}\right]^{1/2}.$$

where

Q heat effect

E activation energy

$\omega_{max.}$ maximum rate of chemical reaction in the gaseous phase

T_0 average temperature on the border of the heat layer

The latter formula means that the heating of the combustible material and the initiation of its gasification occur at a distance of less than the

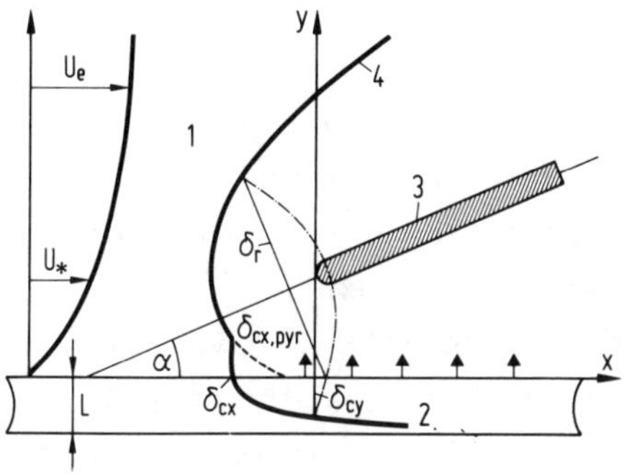

Fig. 2.13 Diagram of a model of diffusion flame spread along the surface of a combustible material. (1) Oxidising medium, (2) combustible material, (zone of chemical reaction), (4) boundary of thermal layer.

size of the gaseous thermal layer, $\delta_{g,min.}$. When the rate of flame diffusion is low, $\delta_{cx,pyr.}$ is small, even in comparison with δ_{cx}.

It was found that the rate of flame diffusion over the surface of a relatively combustible material in the presence of an additional external heat flow $q_{add.}$ and with the rate of oxidising gas motion U is expressed by the equation:

$$v = \frac{v_{max}(1+\Omega)\cos\alpha - U_*}{1+\theta\delta_{c.y}\cos\alpha/\delta_{g,\,min}}, \tag{2.8}$$

where

$$\Omega = \frac{4q_{add.}}{\pi Q\omega_{max}\delta_{R2min}}; \qquad \theta = \frac{\rho_c c_c (T_S - T_0)}{\rho_g c_g (T_* - h_0/c_g)};$$

v_{max} maximum possible rate of flame diffusion over the surface of a solid fuel in the absence of an external heat flow and forced convection.

$$v_{max} = \frac{\pi Q\omega_{max}\,\delta^2_{RZ\,min}}{2\rho_g c_g \delta_{g,\,min}(T_* - h_0/c_g)}.$$

where

ρ_g density of gas

c specific heat of gas

$\delta_{RZmin.}$ minimum value of the half-width of the reaction zone inside the edge of a flame

$h_0 = c_g \cdot T_0 + Y_f \cdot \Delta H_e$ heat consumed for heating and gasification of fuel

Forced convection of the oxidising gas directed against the fuel diffusion leads to a decrease in the velocity of the oxidiser. Forced ignition directed along the flame spread leads to an increase in its rate.

In equation (2.8) the influence of the kinetic parameters of a reaction in a gaseous phase is indicated by the magnitude ω_{max}, which is proportional to the concentration of the reagents. The quantity and volume of gaseous products of fuel degradation in turn are dependent on the pressure of the surrounding atmosphere. Thus the rate of flame diffusion over the surface will depend on the concentration of oxidising agent in the surrounding atmosphere and on the pressure. In order to give a clear presentation of the dependence of flame diffusion rate on the oxygen concentration in the surrounding atmosphere and on pressure [59] the heat transfer and mass transfer processes that take place in the solid and gaseous phases in the ignition region have been discussed separately.

The authors combined the derived solutions for the conservation of energy equations for each individual phase by equating the heat flow of the gaseous state to the heat flow transferred into the polymer. As in reference [59], one of the assumptions applied in the analysis is that of a negligible heat transfer by thermal conductivity through the gaseous and condensed phases in the direction of the spreading flame. In other words, the heat transfer from the flame occurs only in the direction perpendicular to the surface of the combustible material.

In this case the authors derived the following expression for the fuel diffusion rate over the horizontal surface of a thermally thin material:

$$v \simeq \frac{\lambda_g \Delta H_{com.} Y_{ox} f(P, Y_{ox})}{\rho_c c_c L (T_S - T_0)}, \tag{2.9}$$

and for thermally thick layers:

$$v \approx \frac{[\lambda_g \Delta H_{com.} Y_{ox} f(P, Y_{ox})]^2}{\lambda_c \rho_c c_c c_g^2 (T_S - T_0)^2}. \tag{2.10}$$

where

L thickness of the material

$T_s = T_{ign.}$ temperature of the ignitable surface of the material

$\Delta H_{com.}$ combustion heat

$Y_{ox.}$ relative concentration of oxidising agent in the surrounding atmosphere

$f(P, Y_{ox.})$ a simplified function showing the relationship between the relative concentration of vapour of the fuel material in the ignition region, pressure and concentration of oxidiser.

Mathematically, this function is expressed by a double integral, which considers the change in concentration of the vapours of the fuel in the ignition region, due to the reaction in the gaseous phase inside the flame.

These formulae meet the requirements of the empirical function $v \sim (P, Y^m_{ox.})^F$, discovered by R.F. McAlevy [61] for flame diffusion over a horizontal surface of polymer materials. Later the formulae were confirmed for a wider range of polymer materials [62]. The rate of flame diffusion over the horizontal surface of cellulose layers during combustion in an air atmosphere and with a pressure of one atmosphere does indeed fluctuate in inverse proportion to the thickness of the material in a 0.2-2.00 mm range (Fig. 2.14).

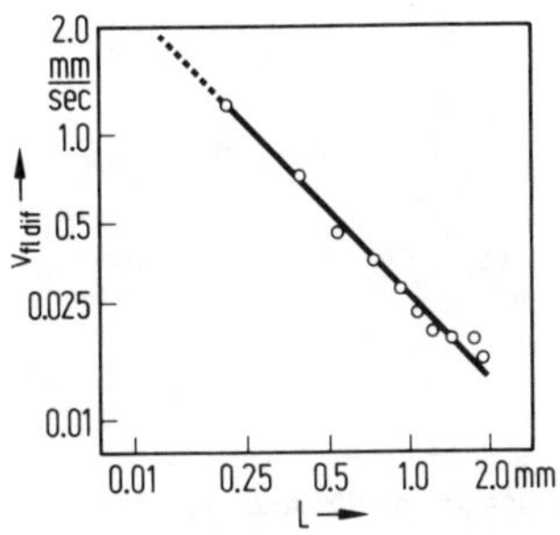

Fig. 2.14 Relationship between flame diffusion rate $v_{fl.dif.}$ and a paper surface of thickness L.

The effect of the oxygen concentration and pressure in the surrounding atmosphere on the rate of flame diffusion over the surface of polystyrene and polymethylmethacrylate samples is shown in Figs. 2.15 and 2.16; the velocity $v_{fl.dif.}$ apparently increases linearly with an increase in pressure and in the concentration of oxygen in the surrounding atmosphere.

It was discovered that, in the presence of forced convection, when the oxidising gas moves in the opposite direction to that of flame spread, the velocity $v_{fl.dif.}$ is proportional to the flow velocity $U^{1/3}$ in the case of thermally thick polymer materials. Lalayan and coauthors [62] studied flame diffusion over the surface of thermally thick samples for a number of polymer materials at atmospheric pressure. They confirmed the accuracy of the ratio $v_{fl.dif.} = kY^n_{ox.}$ when the oxygen concentration is higher than 35%. The constants k and n are a function of the thermophysical prop-

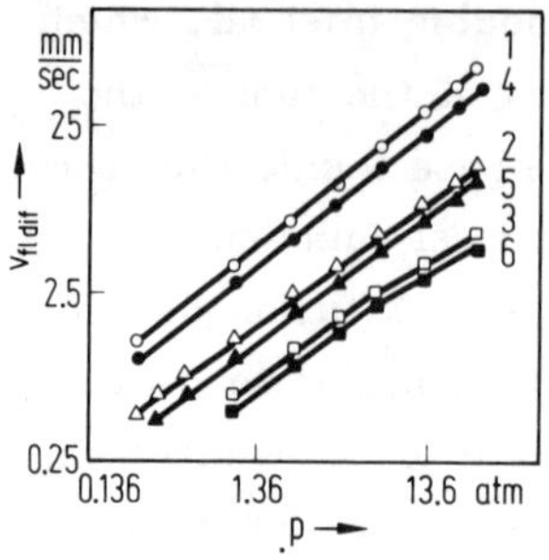

Fig. 2.15 Relationship between flame diffusion rate $v_{fl.dif.}$ and pressure P for polystyrene (1-3) and polymethylmethacrylate (4-6) in an oxygen-nitrogen atmosphere (1, 4) $Y_{ox.}$ = 1.00, (2, 5) 0.62, (3, 6) $Y_{ox.}$ = 0.46.

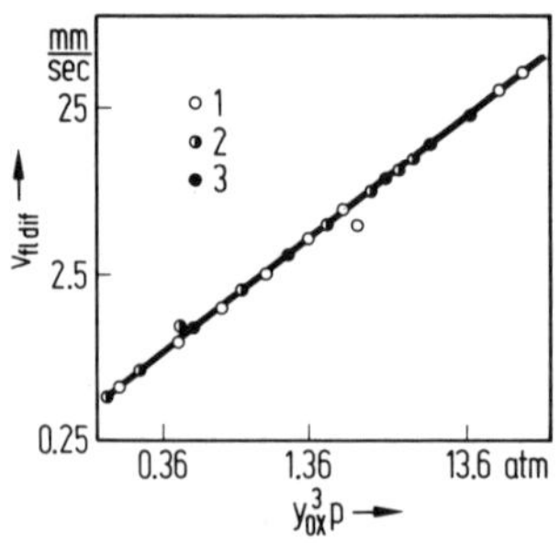

Fig. 2.16 Relationship between $v_{fl.dif.}$ and $Y^3_{ox.}\cdot P$ during combustion of PMMA in an oxygen-nitrogen atmosphere. (1) $Y_{ox.}$ = 0.40, (2) 0.62, (3) $Y_{ox.}$ = 0.1.

erties of the polymer materials. It was determined that the flame diffusion rate over the surface of materials without any impurities and of a size comparable to that of the heated layer could be expressed as an empirical relationship:

$$v_{fl\,dif} = \left(0{,}14 + 1{,}8\cdot 10^{-8}\frac{\Delta H_{com.}}{a}\right) Y_{ox}^{1{,}35+0{,}88\cdot 10^{-4}\Delta H_{com.}} \text{cm/sec}$$

where

$\Delta H_{com.}$ heat of combustion (cal/g)

a thermal conductivity coefficient of the polymer material (cm^2/sec)

Despite the fact that formulae 2.9 and 2.10 were introduced for the case of flame diffusion over a horizontal surface, they are valid for experimental results [63] that have been derived from the study of combustion of vertically positioned polymer samples (in a downwards direction). The results are explained by the fact that flame spread over the horizontal surface has characteristics that very closely resemble those of flame spread downwards over a vertical surface.

The study of different orientations of flame spread along the surface of polymer materials indicated that the flame spreads downwards and along the upper side of the horizontal plane of the samples (the fluctuation range of the angle of orientation is -90° to 0°). Flame diffusion upwards and along the underside of the surface of plane polymer samples (the fluctuation of the angle of orientation is in the +90° to 0° range) is unstable; it accelerates in proportion to the acceleration of the process itself [28, 29, 55, 63, 64].

The difference in the character of a flame spreading over the surface of polymer materials with a different orientation is determined through variation of the hydrodynamics of the flows and through variation of a predominant heat transfer mechanism during combustion.

Photographing the tracks of solid particles (BaO or MgO) introduced into the oxidising gas flow gives a clear picture of the change in character of the gaseous flow for a change in the orientation of flame diffusion over the surface of a polymer material (Fig. 2.17). In this case, the laminar motion during flame diffusion downwards over the surface of the paper is very distinct. In the opposite direction the flow and flame become turbulent.

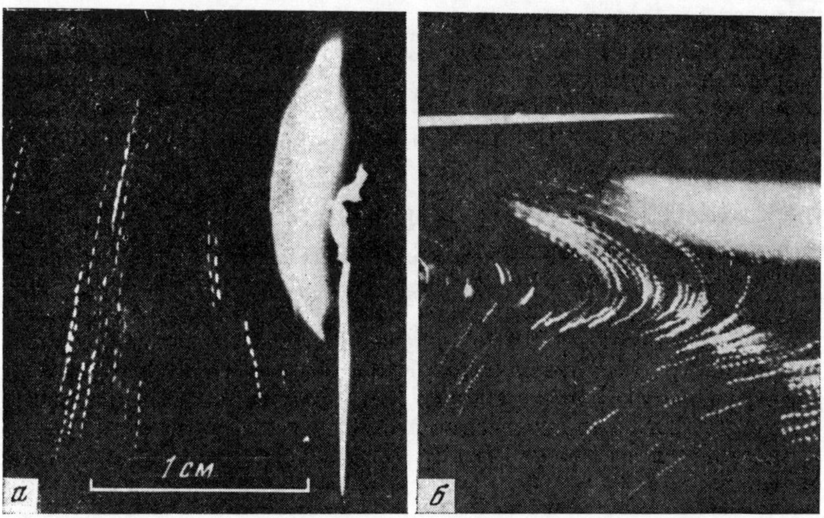

Fig. 2.17 Pictures of tracks of MgO particles inside an air flow near a flame, spreading downwards over paper (a) and along the lower horizontal area (b).

The simplified model of a diffusion flame spreading down the surfaces of polymer materials has been studied [28]. As in references [59] and [60] a zone on a front edge of the flame was examined. Assuming that the Damkeler number De = 1, the performance time of the chemical reaction inside the flame, $t_{ch.}$, is equal to the time of mass transfer, t_m, which occurs in a stationary medium through heat transfer in the gaseous and condensed phases.

For a reaction in the gaseous phase inside the flame

$$t_{ch} = \rho_g \, [Y_f \, Y_{ox} P^n k_{0,\,g} \exp(-E_g/RT_{fl})]^{-1}.$$

The time of mass transfer inside the flame zone is directly proportional to the square of the distance from the surface to this reaction zone (y_g) and inversely proportional to the diffusion coefficient D:

$$t_m = y_g^2/D$$

When the mass flow of fuel inside the flame zone is expressed through the kinetic parameters of the degradation reaction of the condensed phase of a polymer material, it is possible to determine the flame diffusion rate as a function of the thickness of the material L, the pressure of the surrounding atmosphere P, the oxygen concentration $Y_{ox.}$ and the orientation, expressed in the form of acceleration due to gravity g.

By using experimental data it was discovered [28] that the flame diffusion rate over the surface of polymethylmethacrylate sheets in that case can be expressed by the equation

$$v_{fl.dif.} = k\,(1 + 0{,}2/L)^{3/4} g^{-1/4} P^{0{,}8} Y_{ox}^{2{,}5}, \tag{2.11}$$

where k is a constant. Thus the derived result $v = f(P \cdot Y_{ox.})$ is compatible with the data [65].

It was discovered from temperature field measurements in the condensed and gaseous phases during downwards flame diffusion over thermally thick layers of polymethylmethacrylate and from the heat balance calculation inside a control volume that flame diffusion is caused by heat conductivity through the condensed phase of a polymer material.

Heat flow carried through the gaseous phase along the direction of flame spread:

$$q_g = \int_{-\infty}^{x} \lambda_g \left(\frac{\partial T}{\partial y}\right)\bigg|_{x,\,0} dx,$$

plus a heat flow carried through the gaseous phase:

$$q_c = \int_{-L}^{0} \lambda_c \left(\frac{\partial T}{\partial x}\right)\bigg|_{x,\,y} dy,$$

must compensate for the total heat flow consumed on complete combustion of the polymer material:

$$q = \int_{-L}^{0} \rho v \left(\int_{T_0}^{T} c dT\right) dy.$$

The correlation between q, q_c and q_g shows that less than 10% of the energy consumed in heating a polymer substance to the ignition temperature is carried through the gaseous phase by molecular thermal conductivity [28]. A completely different mechanism of heat transfer characterises flame diffusion over the surface of cellulose samples, namely heat transfer by heat conductivity through the gaseous phase. The chief cause of such a change in flame spread mechanism is the higher thermal conductivity of polymethylmethacrylate.

When $L \geq 1$ cm, $v_{fl.dif.}$ is not dependent on the thickness of the flat polymethylmethacrylate sheets during downwards flame spread. Thinner sheets lead to a lessening of the heat transfer contribution through the condensed phase. For $L \leq 0.4$ mm heat transfer through the gaseous phase dominates. Near the edge of a laminar diffusion flame, radiation has no effect. With an increasing distance over the height of the flame, however, the contribution of heat radiation to the total heat flow on the surface of the polymer material increases.

A ratio was derived during a study of a model for the downwards spread of a laminar flame over the surface of thermally thick polymer materials. This ratio shows the dependence of the combustion rate on many more parameters than those given in equation (2.11) [66]:

$$v_{fl.dif.} \approx [k c^{1/4} P^{5/4} Y_{ox}^{7/4} \exp(-E_g/RT_{fl})]/[\rho \lambda^{1/4} g^{1/6} (T_S - T_0)^2],$$

where

k constant that depends on the kinetic parameters in the solid phase

c, ρ, λ thermophysical characteristics of the polymer material

E_g effective activation energy of the reaction in the gaseous phase

T_s surface temperature, equal to $T_{ign.}$

T_0 initial temperature of the surrounding medium

$T_{fl.}$ temperature of the flame

When the flame spreads upwards over the vertical surface of the material, heat radiation becomes the dominant heat transfer factor which is especially characteristic of the combustion of large polymer samples. Thus, according to reference [67], up to 85% of the total heat flow on the surface of polymethylmethacrylate is transferred by radiation during upwards spreading of the flame. In this case the flame is similar to the turbulent diffusion flames of hydrocarbons.

In an approximate examination of the dominant heat transfer mechanism by means of radiation, the rate of flame spread in the upwards direction over the surface of a material can be expressed by the following equation [32]:

$$v_{fl.dif.} = [\varepsilon_{fl.}\sigma T_{fl.}^4/(T_{ign.} - T_0)]^2[h_{fl}/(\rho_c c_c \lambda_c)], \quad (2.12)$$

where

$h_{fl.} = \alpha \dot{m}_f^n$ height of the flame

$\dot{m}_f$ mass flow of fuel

α, n constants

When the flame spreads upwards it washes that part of the surface that has still not ignited and, in the long run, becomes larger in size and turbulent. A theoretical analysis of flame spread upwards over the surfaces of materials of indefinitely large thickness shows that mass flow accelerates as the front of the flame moves upwards. The height of the flame increases approximately in proportion to the square of the time of flame spread [64].

M. Sibulkin and J. Kim [68] disregarded the effect of the acceleration of a flame front on temperature and studied only the convectional heat transfer towards the ignitable surface of the polymer material. They discovered that, in this case, the rate of upwards flame spread over the vertical surface of a polymer material can be described by the relationship:

$$v_{fl.dif.} = \frac{\Delta H_e \Delta H_{com} [\dot{m}_f (x_{pyr.})]^2}{(^4/_3)^j \rho c \lambda (T_{ign.} - T_0)^2},$$

where

F ratio of the rate of heat transfer to the fuel rate of heat release during combustion

$\dot{m}_f(x_{pyr.})$ mass flow of fuel as a function of the length of the pyrolysis zone

j index (for a laminar flame $j = 0$, for a turbulent flame $j = 1$)

When the heat transfer acceleration parameter is larger than the critical one, the flame spread is non-stationary. For PMMA it was disçovered that $F_{crit.}$ = 0.13 in the region of turbulent heat transfer. It is easy to see that when the flame spreads over the surface of thermally thin samples the rate changes in inverse proportion to the difference between the samples' ignition temperature and the initial temperature.

With thermally thick samples the rate of flame spread is inversely proportional to the square of this difference. The spread of flame over the surface of a polymer material is a rather complex phenomenon. As we have seen, the models that have been taken as a basis for the theoretical analysis allow for different aspects of this phenomenon.

At the present time it is still not clear what kind of processes take place in the ignition zone at the front edge of the flame. It is these processes that determine the rate of flame spread over the surface of a polymer material. Outside the edge of the flame the quasi-stationary rate of combustion of the polymer material is practically independent of the chemical kinetics, as has been shown by studying diffusion processes of mixing and the transfer of reagents.

The bulk of polymer material is consumed far away from the edge of the flame. Critical conditions of flame spread over the surface of polymer materials represent critical points of its ignition and extinguishing. This is why an analysis of the critical phenomena of flame spread over the surface of a material is usually tied to an investigation of how the different parameters of the system affect the processes taking place in the ignition zone.

It is important to mention that chemical kinetics and, in particular, the ratio of the values of effective activation energy for the reaction in the flame zone to those in the condensed phase of the polymer material can be regarded as a relationship between the rate of laminar flame spread over the surface of a polymer material and the rate of the oxidising gas flow [28].

Combustion of Spherical Polymer Particles

A spherical flame is the ideal type of unidimensional flame, and the mathematical analysis of the combustion process has been developed for this particular flame type. The characteristic property of a spherical flame is that the cross-gradients of its properties (temperature, concentration, etc.) are equal to zero, regardless of the radius of the sphere.

A study of polymer sphere combustion is useful in a number of respects, since it allows a detailed verification to be conducted of many aspects of combustion theory. However, the diffusion combustion of polymer spheres has hardly been studied because the combustion of this type of polymer material is of no practical use. A utilitarian interest only emerges when there is a problem with the flammability of dust particles from a polymer material (fire hazard).

The diffusion combustion of polymer spheres is similar to the combustion of liquid fuel drops. The burning of spherical particles in an atmosphere of gaseous oxidiser can be described by Sreznevsky's law:

$$d^2 = d_{in}^2 - kt$$

where

d_{in} initial diameter of particle

k linear rate constant of combustion

t time

The theory of diffusion combustion of fuel drops has been described in [69]. The combustion process of polymer spheres can be characterised either by mass rate or by linear rate, on the basis of measurements of the reduction over time in the diameter of the particle or of its mass.

Essenhigh [70] studied the combustion of spherical particles for a significant number of polymer materials. The diameters of the particles vary in a 1-2 mm range. Combustion was carried out in an atmosphere of stationary air. The derived constants of the polymer sphere combustion rates are given in Table 2.7. Thermoplastic polymer materials melted just before combustion. The combustion of the polycarbonate and the phenolic polymer is accompanied by the formation of a carbon layer. It should be mentioned that melting often distorts the spherical shape of the particles. Different values for the Sreznevsky constant have been derived for the combustion of PMMA, as given in reference [71].

In a more detailed study of the combustion of particles of polymethylmethacrylate [72], the authors observed the combustion of a particle in a gaseous atmosphere and assumed that the heat from the flame was trans-

Table 2.7 Combustion rate constants of spherical particles of polymer materials

Polymer material	T_s, °C	k, cm^2/sec
Polyethylene	538	39.8
Polypropylene	528	40.4
Nylon 6,6	542	29.4
Polymethylmethacrylate	416	36.3
Copolymer of styrene and butadiene	-	44.1
Polyvinyl acetate	-	25.4
Cellulose acetate	309	6.1
Polystyrene	494	57.6
Polycarbonate	578	11.6
Soluble phenol-formaldehyde resin	-	80.5

ferred towards the surface of the polymer material through heat conduction [78].

In this case, the extinction rate constant k is expressed by the equation

$$k = \frac{8\lambda_g}{c_c\rho_c(1 - r_c/r_{fl})} \ln\left\{1 + \frac{1}{\Delta H_e}\left[c_c(T_0 - T_S) + \frac{\Delta H_{com.} Y_{ox}}{f}\right]\right\}, \tag{2.13}$$

where

λ_g heat conductivity coefficient of the gas

r_c radius of particle

$r_{fl.}$ radius of the reaction zone inside the flame

ΔH_e heat of gasification

$\Delta H_{com.}$ heat of combustion

f stoichiometric coefficient

The constant of extinction of a polymer sphere is, in fact, proportional to ln (1 + B).

A large number of polymers undergo carbonisation when heated to a high temperature. According to Spolding, the linear rate of extinction of spherical particles in an atmosphere of immobile gaseous oxidiser is inversely proportional to the diameter of the particles and is a function of the mass transfer number B:

$$v = \frac{2D\rho_g \ln(1+B)}{d_{in}\rho_c}, \tag{2.14}$$

where

D diffusion coefficient

ρ_g density of the oxidising gas

ρ_c density of the particles

Combustion of Polymer Materials for a Counterflow of Gaseous Oxidiser

In studying the mechanisms of the diffusion combustion process of a polymer material, it is interesting to consider the downwards combustion of vertical polymer samples under conditions where the fuel products of polymer degradation and the oxidising agent flow in opposite directions to each other. In this case the burning surface of the polymer sample is sustained by using a special device placed at a specific level and distance from the oxidiser supply jet. A diffusion flame is formed near the surface of the polymer sample (distance ≅ 1 mm). The flame is approximately flat in shape. Convectional heat exchange and mass exchange predominate. It is assumed that there are no radial property gradients, and hence a quasi unidimensional diagram of the combustion process is considered.

An analysis of this type of diffusion flame of polymer materials has been made by D.J. Holve and R.F. Sawyer [15]. As in other analytical models for the diffusion combustion of polymers, several assumptions were made: the thermophysical properties and diffusion coefficients of reagents and

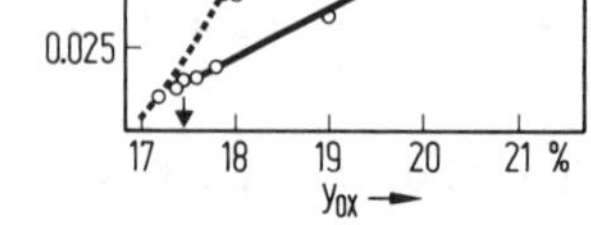

Fig. 2.18 Relationship between the combustion rate $v_{fl.dif.}$ of PMMA and oxygen concentration $Y_{ox.}$ at different rates of oxidiser flow U. (1) U = 71 cm/sec, (2) 95, (3) 118, (4) U = 142 cm/sec. The dotted line sets boundaries for the conditions of flame extinction. The arrow shows the value of an oxygen index, OI, obtained by a standard method.

gaseous products are assumed to be constant. However, the Lewis number Le = $D/a \neq 1$. The linear rate of polymer material extinction under such conditions can be expressed as a function of the concentration of oxidiser in the surrounding atmosphere, the mass transfer number, and other characteristics of the system:

$$v=\left(\frac{2\lambda\,\mathrm{Le}\,\gamma_D}{d}\right)^{1/2}\frac{\dot{m}_{\mathrm{ox.}}^{1/2}}{\rho_c}\ln\left\{1+\frac{2Y_{\mathrm{ox.}}}{\pi^{1/2}r}\left[1+\frac{\tau'(B_{\mathrm{com.}}-1)}{\mathrm{Le}}\right]\right\}, \quad (2.15)$$

where

d diameter of the sample and the oxidiser jet (these being equal)

λ dimensionless parameter of the oxidiser mass flow gradient

$\gamma_D = \rho_g \cdot D$

$\dot{m}_{\mathrm{ox.}}$ mass flow of oxidiser

$B_{\mathrm{com.}} = \Delta H_{\mathrm{com.}}/\Delta H_e$

ρ_c density of polymer material

$\tau' = (T_{fl} - T_s)/(T_{fl} - T_0)$

$Y_{\mathrm{ox.}}$ relative oxygen concentration in the surrounding atmosphere

r stoichiometric oxygen to fuel ratio, identical to that for the combustion of spherical particles, $v \sim \ln(1 + B)$.

The relationship between the linear rate of combustion of a polymer material and the concentration of oxygen in the surrounding atmosphere, and also the flow rate of the oxidiser, has been confirmed experimentally (Fig. 2.18). The author used the method of opposed, bidirectional oxidiser flow to determine the combustion rate under the limiting conditions of flame extinction. However, there is another problem to be tackled: according to the experimental values of the combustion rate under various conditions, other characteristics, such as ignitability and combustibility, that indicate the individual properties of a polymer material, should also be determined. This approach permits a quantitative determination of the temperature coefficient of the polymer's flaming reaction in the gaseous phase, the mass transfer number B and the effectiveness of the addition of flame retarder etc. [13, 73]. The method not only allows the relative ignitability and combustibility of polymer materials to be revealed but also gives insight into the mechanism of the combustion process itself and the influence on the mechanism of either component of the polymer material. A study of the limiting conditions of diffusion combustion is particularly useful in this respect.

Stability Limitations on the Diffusion Flames of Polymer Materials

A qualitative picture of the limits on the stable existence of a diffusion flame of a polymer material can be presented on the basis of the above-mentioned models of polymer ignition in the gaseous phase and the spread of burning combustion. For a stable combustion process to be maintained, it is necessary for the formation of gaseous degradation products from the condensed phase and the flame reaction in the gaseous phase to be balanced.

The reaction between unmixed vapour composed of fuel and oxidiser, in turn, is only possible within a specific range of reagent flow rates towards the flaming reaction zone. According to the theory of unmixed gas and combustion [74], when the flow of reagents takes on less than a minimum value, ($\dot{m}_{crit., min}$) or when it exceeds its maximum value ($\dot{m}_{crit., max}$), the combustion reaction does not occur. This can be explained by the cooling of the flame and the lowering of its temperature.

Many of the parameters of a burning polymer system are closely connected in the sense that if one of them changes it leads to a change in other combustion characteristics. In the long run, since a stable combustion process is established when the heat supply is equal to heat elimination in the system, the condition of heat balance is a critical condition for the stability of the diffusion flame.

It is obvious from a model of polymer ignition in the gaseous phase [75] that there has to be a temperature limit up to which a surface layer of polymer material must be heated in order to provide the diffusion flame with fuel and in order for the degradation reaction to occur. This temperature correlates with the temperature of polymer gasification under certain specific conditions for the surrounding atmosphere. To provide a sufficient amount of reagent coming into the gaseous phase, it is necessary to create a certain temperature gradient near the surface of the polymer material. If a critical limit value of the temperature gradient near the surface is exceeded, then this is equivalent to creating conditions for surpassing the maximum value of fuel flow inside the diffusion flame and for decreasing the flow of oxidising agent inside the reacting zone to below the minimum critical value.

The model of laminar diffusion flame spread over the surface of a combustible material allows further types of limits to be distinguished [60]:

1. Limits due to the thickness of a layer of combustible material. This limit is tied to the minimum thickness of a material layer that is able to supply the diffusion flame with a sufficient amount of fuel
2. Limit due to heat losses into the surrounding atmosphere
3. Limit due to the rate of convection in the oxidising medium, which is associated with a disruption of the flame by a flow opposed to the flame spread

One of the important critical conditions of stability of polymer diffusion flames is the existence of the limit set by the oxygen concentration in the surrounding atmosphere. This limitation was first discovered by C.P. Fenimore and F.J. Martin [76]. They showed that combustion cannot take place in an atmosphere with an oxygen concentration below the critical concentration.

Martin discovered during the analysis of a model of candle-type polymer diffusion combustion that this limiting oxygen concentration is a function of the heat conductivity of an inert diluent and the individual properties of the combustible material. The latter indicates that a stoichiometric ratio exists between the oxidiser and the fuel inside the flame and for the temperature of the flame. Later, this limit, which is set by the oxygen concentration, was named the limiting oxygen index and was used for the relative evaluation of ignitability and combustibility of polymer materials.

According to Martin, the limiting oxygen concentration is expressed in the following way:

$$\frac{n_{ox.}}{n_{int.}} = \frac{T_{fl.} - T_0}{B}\left(-\frac{Q}{T_{fl.} - T_0} + c_{p,int.}\right), \qquad (2.16)$$

where

$n_{ox.}$, $n_{int.}$ molar concentrations of oxygen and inert gas in the surrounding atmosphere

$T_{fl.}$, T_0 temperatures inside the flame reaction zone and in the surrounding atmosphere

B mass transfer number

Q heat losses

$c_{p,int.}$ specific heat of the inert gas

It is obvious that if the temperature of the surrounding atmosphere T_0 is raised, then $n_{ox.}/n_{int.}$ should decrease. Indeed, a relationship was discovered between the oxygen index (OI) of a polymer material and the temperature of the surrounding atmosphere. For materials with OI > 0.21 (which do not ignite in an air atmosphere), when the temperature of the medium (or the temperature of the polymer samples) increases, a decrease is observed in the limiting value of the oxygen concentration at which stable combustion is still possible. The nature of this decrease, which is due to temperature, is determined by the individual properties of the polymer material.

The limiting temperature at which OI = 0.21, i.e. stable burning of the material still occurs in an air atmosphere, can also be used as a measure of the ignitability and combustibility of polymer materials. This limit is called Ratly's temperature index (TI). The effect of temperature on a relative change of the OI value, measured at 25°C, and temperature indexes of polymer materials with different OI are shown in Fig. 2.19. Stability limits on diffusion flames of polymers due to the pressure of the surrounding atmosphere (critical pressures) have been studied [77-79]. The relationship between the rate of flame spread over the surface of a polymer material and the pressure is described by a rate law (see equations 2.9 and 2.10). This law is violated when the flame reaches its extinction limit.

A disruption of the flame during its spread downwards over vertical cellulose samples has been observed when the pressure drops to a critical value (Fig. 2.20).

L. Krishnamurthy [77] studied how a change in the concentration of oxygen in a flow of oxidiser and a change in the surrounding atmospheric pressure in a 0.1-1 atm range affects the diffusion combustion of polymethylmethacrylate. He discovered that combustion of the polymer in an atmosphere of pure oxygen in the above pressure range was stable when the percentage of oxygen was reduced; the pressure dropped to $P_{crit.}$ at which point the flame died out. The relationship between combustion rate, pressure, and oxygen concentration at the extinction point was used to calculate the kinetic parameters of the flaming reaction in the gaseous phase: $E_{ef.} = 21$ kcal/mole, $k_0 = 3 \times 10^{12}$ cm^3/mole·sec. The analysis was conducted with extinction criteria in mind, i.e. attainment of Damkeler's number.

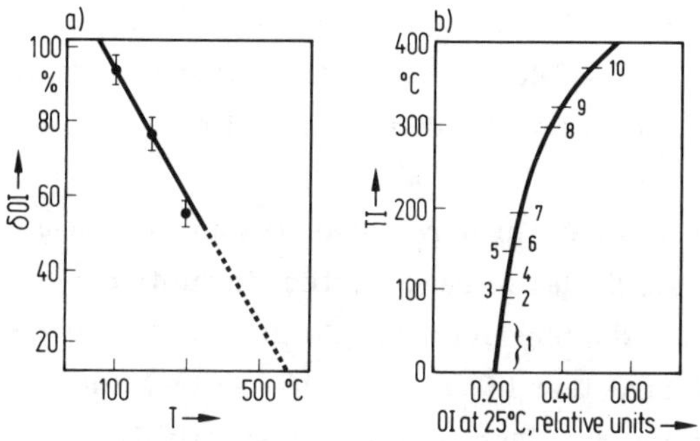

Fig. 2.19 Relationship between relative change (fluctuation) of an oxygen index OI and temperature (a); and temperature indexes TI of polymer materials with a different OI (b). (1) Foam polyurethane (FPU), (2) fire-resistant polystyrene, (3) polyamide, (4) fire-resistant FPU, (5) polycarbonate, (6) wool, (7) polyamide (Nomex), (8,9) neoprene, (10) polyvinyl chloride.

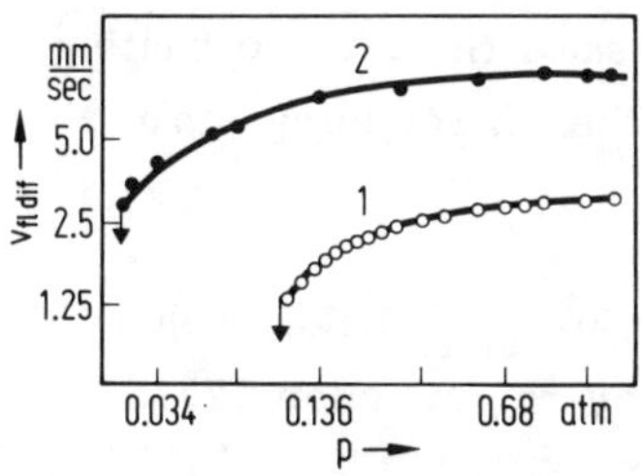

Fig. 2.20 Relationship between $v_{fl.dif.}$ and P during diffusion combustion of cellulose (thickness 0.2 mm, width 20 mm). (1) $Y_{ox.}$ = 0.30, (2) 0.75. Arrows show the limits of flame extinction.

A study of orientational effects during flame spread over polymer materials indicates that gravitational forces influence the combustion process, which means that the presence of critical combustion effects can be expected during a variation of gravitational forces (during overload conditions and possibly at zero gravity). These effects have not been sufficiently studied.

The effect of an overload of up to $300g$ mass on the combustion of a paper layer in an atmosphere of oxygen at pressure P = 1 atm was investigated in reference [79]. It was discovered that at a certain acceleration, a disruption of combustion occurs and, at a critical overload value, the combustion of polymer materials is impossible. This critical overload value is markedly dependent on the direction of flame diffusion and on the thickness of the sample.

In the case of downwards flame spread (in a radial direction out from the centre) over the surface of a single-layer sample, the disruption of combustion occurs at $5g$, and with a double-layer sample at $2.5g$. During upwards flame spread (in a radial direction in towards the centre) the corresponding critical overload values are $100g$ and $70g$ respectively.

The effect of the size of polymer samples on diffusion combustion stability has been the subject of many studies [28, 75, 79, 83]. Different situations were analysed simultaneously: the presence of heat-conducting and insulating coatings or layers and the existence of space between the polymer material and the wall - a "sandwich" type system. It was discovered that there are limiting size values for polymer materials which depend both on the individual properties of the polymer material and on the properties of the layer or the wall. At the same time, these limiting sizes are a function of pressure and oxygen in the surrounding atmosphere.

The critical size of a multilayer cellulose sample $L_{crit.}$ below which combustion is impossible can thus be described by the function [79]:

$$L_{crit} \simeq 20/Y_{ox.}P^{0,75} \text{ cm}.$$

In turn, it was found that $L_{crit.}$ increases sharply when P decreases; at $P < P_{crit.}$ combustion is impossible regardless of how large the sample is. This critical pressure is equal to 120 mm Hg when a single-layer paper sample burns in an air atmosphere and to 250 mm Hg for a double-layer paper sample (Fig. 2.21).

The stability region for the combustion of polymer materials in general should therefore be described by the following functional relationship:

$$f\,(Y_{ox.}, P, g, T, \alpha_i, L) = \text{const},$$

where

α_i parameter characterising the individual properties of the polymer material

g acceleration of gravity

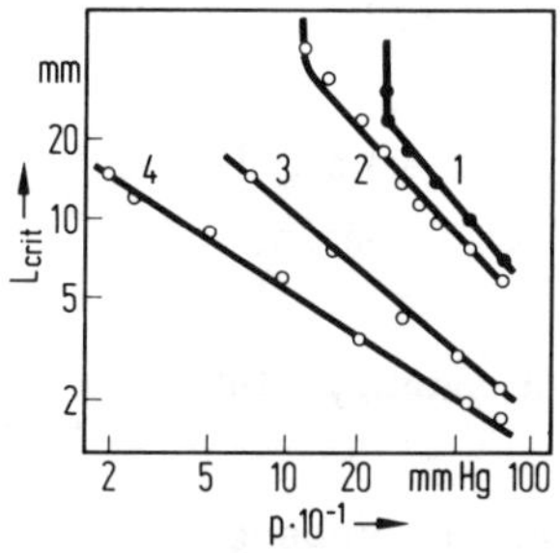

Fig. 2.21 Relationship between critical combustion size of paper sheet $L_{crit.}$ and pressure P of surrounding atmosphere at various oxygen concentrations $Y_{ox.}$ (1, 2) $Y_{ox.}$ = 0.21, (3) 0.60, (4) $Y_{ox.}$ = 1.0. 1 = Two-layer paper sample, 2-4 = single-layer sample.

A change in any parameter which leads to a violation of the thermal balance in the system causes a disturbance in the stability of the diffusion combustion of the polymer material.

References

1. Pokhil P. F., Phyzika vzriva, 1955, no. 2, pp. 181-185.
2. Protessi goreniya. Lewis B., Pierce H. S., Taylor L. eds. Phyzmatgiz, 1961.
3. Novikov S. S., Pokhil P. F., Riazantsev U. S. Phys. goreniya. i vzriva, 1968, vol. 4, no. 4, pp. 469-473.
4. Hoke C. E., SPE Journal, 1973, vol. 29, no. 5, pp. 36-40.
5. Martin F. J., Combust. and Flame, 1968, vol. 12, no. 2, pp. 125-135.
6. Burge S. J., Tipper C. F. H., Combust. and Flame, 1969, vol. 13, no. 5, pp. 495-505.
7. Stuetz D. E., Diedwardo A. H., Zitomer F., Barnes B., J. Polymer Sci., Polymer Chem. Ed., 1975, vol. 13, no. 3, pp. 585-593.
8. Fenimore C. P., Jones G. W., Combust. and Flame, 1966, vol. 10, no. 3, pp. 295-301.
9. Merzhanov A. G., Combust. and Flame, 1969, vol. 13, no. 2, pp. 143-156.

10. Maksimov E. I., Merzhanov A. G., Dokl. AN USSR, 1964, vol. 157, no. 2, pp. 412-416.
11. Spolding D. S., Osnovi teoree goreniya. M.: L.: Gosenergoizdut, 1959.
12. Kanury A. M., in 15th Symposium (Intern.) on Combustion, Pittsburgh: Combust. Inst., 1974, pp. 193-202.
13. Matthews R. D., Sawyer R. F., J. Fire and Flammability, 1976, vol. 7, no. 1, pp. 200-216.
14. Dmitriev B. M., Kochetkov O. A., Ulibin V. B., Shteynberg A. S., Phys. goreniya i vzriva, 1969, vol. 5, no. 1, pp. 26-31.
15. Holve D. J., Sawyer R. F., in 15th Symposium (Intern.) on Combustion. Pittsburgh: Combust. Inst., 1974, pp. 351-361.
16. Braginsky V. I., Bakhman N. N., Phys. goreniya i vzriva, 1971, vol. 7, no. 3, pp. 376-381.
17. Kishore K., Verneker V. P., J. Pol. Sci., Pol. Let. Ed., 1976, vol. 14, no. 12, pp. 761-765.
18. Wall L. A., Fire Res. Abstrs and Revs, 1971, vol. 13, no. 3, pp. 204-219.
19. Berlin A. A., Volfson S. A., Enikolopian N. S., Kinetika polymerizationnikh processov, M: Khimiya, 1978.
20. Aseeva R. M., Zaikov G. E., et al., Visokomolekul. soed., 1980, vol. 22A, no. 9, pp. 2075-2081.
21. Flame retardancy of polymeric materials, Kuryla W. C., Papa A. J., eds., N.Y.: Marcel Dekker, 1973, vol. 2, p. 143.
22. Bostic J. E., Kwan-Nan-Yeh, Barker R. H., J. Appl. Polymer Sci., 1973, vol. 17, no. 2, pp. 471-477.
23. Hilado J., Flammability Handbook for Plastics. Stanford: Technomic Publ. Co., 1969.
24. Magee R. S., Reitz R. D., in 15th Symposium (Intern.) on Combustion, Pittsburgh: Combust. Inst., pp. 337-347.
25. Frank-Kamenetskii D. A., Diffuziya i teploperedacha v khimicheskoy Kinetike, 2nd ed.; rev. and ed. M: Nauka, 1967.
26. Mikheev M. A., Mikheeva I. M., Osnovi teploperedachi. M.: Energiya, 1973.
27. Hallman J., Walker J. R., Slipcevich C. M., SPE J. 1973, vol. 28, no. 1, pp. 43-51.
28. Fernandez-Pello A., Williams F. A., in 15th Symposium (Intern.) on Combustion, Pittsburgh: Combust. Inst., 1974, pp. 217-231.

29. Hirano T., Noreikis S. E., Waterman T. E., Combust. and Flame, 1973, vol. 23, no. 1, pp. 83-88.
30. Fenimore C. P., in Flame Retardant Polymeric Materials, Lewin M., Atlas S. M., Pearce E. M., eds., N.Y.-L.: Plenum Press, 1975, ch. 9, p. 371.
31. Aseeva R. M., Ushkov V. A., Andrianov R. A., Zaikov G. E., in Nehoriavost polymernych materialov. Bratislava: Dom Techn. SVTS, 1976, pp. 3-5.
32. Williams F. A., in 16th Symposium (Intern.) on Combustion. Pittsburgh: Combust. Inst., 1976, pp. 1281-1294.
33. Grishin A. M., Yakimov A. S., Phys. goreniya i vzriva, 1978, vol. 14, no. 2, pp. 39-48.
34. Ris J. de, in 17th Symposium (Intern.) on Combustion, Pittsburgh: Combust. Inst., 1979, p. 1003.
35. Fernandez-Pello À. C., Combust. Sci. and Technol., 1977, vol. 17, no. 1/2, pp. 1-9.
36. Kashiwagi T., Newman D. L., Combust. and Flame, 1976, vol. 26, p. 163.
37. Hirano T., Tazawa K. Combust. and Flame, 1978, vol. 32, p. 95.
38. Kashiwagi T., in 15th Symposium (Intern.) on Combustion, Pittsburgh: Combust. Inst., 1974, pp. 255-265.
39. Lalayan V., Khalturinsky N. A., Berlin A. A., Vysokomol. soed., 1979, vol. 21A pp. 1139-1142.
40. Tewarson A., Lee J. L., Pion R. F., in 18th Symposium (Intern.) on Combustion, Pittsburgh: Combust. Inst. 1981, pp. 563-570.
41. Markstein in 17th Symposium (Intern.) on Combustion, Pittsburgh: Combust. Inst., 1979, p. 1053.
42. Petrella R. V., J. Fire and Flammability, 1980, vol. 11, p. 3.
43. Lalayan V. M., Tovmasyan Y. M., Khalturinsky N. A., Berlin A. A. Vysokomol. soed., 1980, vol. 22B, pp. 150-153.
44. Fernandez-Pello A. C., Law: Combust. and Flame, 1982, vol. 44, pp. 97-112.
45. Rihbarin S. S., Sobolyev S. L., Stesyk L. N., in Khimicheskaya physika protsessov goreniya i vzriva Goreniye condensirovannih i geterogennih system. Tchernogolovka: Ed.-publ: div. of OIKhF AN USSR 1980, pp. 32-38.

46. Lalayan V. M., Experementalnoye Izucheniye zakonomernostey rasprostraneniya laminarnogo plameny po poverkhnosty polymerov. Moscow: IKhF AN USSR, 1980 (Ph.D. dis.).
47. Frey A. F., Tien J. S., (a) Combust. and Flame, 1979, vol. 36, p. 263; (b) ibid, 1978, vol. 33, p. 55.
48. Margolyn A. D., Krupkin V. G., Doklads AN USSR, 1981, vol. 257, p. 1369.
49. Margolyn A. D., Krupkin V. G., Physika goreniya i vzriva, 1980, vol. 16 p. 47.
50. Melikhov A. S., Potiakin V. I., Khimincheskaya physika protsessov goreniya i vzriva. Goreniye geterogennih i gasovih system. Tchernogolovka: OIKhF AN USSR, 1980, p. 48.
51. Sechadri K., Williams F. A., J. Pol. Sci. Pol. Chem. Ed., 1978, vol. 16, pp. 1755-1788
52. Sohrab S. H., Williams F. A., J. Pol. Sci. Pol. Chem. Ed., 1981, vol. 19, p. 2955.
53. Sohrab S. H., Linan A., Williams F. A., Combust. Sci. Technology, 1982, vol. 27, pp. 143-154.
54. Sibulkin M., Kulkarni A. K., Annamalai K., Combust. and Flame, 1982, vol. 44, pp. 187-199.
55. Hirano T., Noreikis S. E., Waterman T. E., Combust. and Flame, 1974, vol. 22, no. 3, pp. 353-363.
56. Rasbach D. T., Combust. and Flame, 1976, vol. 26, no. 3, pp. 411-420.
57. Ris J. de, in 12th Symposium (Intern.) on Combustion. Pittsburgh: Combust. Inst., 1969, pp. 241-255.
58. Tarifa C. S., Notario P. P., Torralbo A. M., in 12th Symposium (Intern) on Combustion. Pittsburgh: Combust. Inst., 1969, pp. 229-240.
59. Lastrina F. N., Magee R. S., McAlevy R. F. III, in 13th Symposium (Intern.) on Cmbustion. Pittsburgh: Combust. Inst., 1971, pp. 935-946.
60. Rihbanin S. S., in Khimicheskaya physika protsessov goreniya i vzriva. Goreniye geterogennih i gazovih system. Tchernogolovka: Ed.-publ. div. of OIKhF AN USSR, 1977, pp. 3-7.
61. McAlevy R. F. III, Magee R. S., in 13th Symposium on Combustion. Pittsburgh: Combust. Inst., 1971, pp. 215-225.

62. Lalayan V. M., Khalturiski N. A., Berlin A. A., Visokomolec. soed., 1979, vol. 21A, no. 4, pp. 11-13.
63. Funt J. M., Magill J. H., J. Appl. Polymer Sci., 1974, vol. 18, no. 4, pp. 1243-1247.
64. Orloff L., Ris J. de, Markstein L., in 15th Symposium (Intern.) on Combustion. Pittsburgh: Combust. Inst., 1974, pp. 183-192.
65. Magee R. S., McAlevy R. F. III., J. Fire and Flammability, 1971, vol. 2, no. 2, pp. 271-278.
66. Fernandez-Pello A. C., Combust. Sci. and Technol., 1977, vol. 17, no. 1/2, pp. 1-9.
67. Orloff L., Modak A. T., Alpert R. L., in 16th Symposium (Intern.) on Combustion. Pittsburgh: Combust. Inst. 1976, pp. 1345-1354.
68. Sibulkin M., Kim J., Combust. Sci. and Technol., 1977, vol. 17, no. 1/2, pp. 39-49.
69. Varshansky G. A., Peshchanskaya L. G., Trudi/Odessa State Univ. I. I. Mechnikov ser. phys. nauk. 1962, vol. 152, no. 8, pp. 5-10.
70. Essenhigh, Drier W. L., Fuel, 1969, vol. 68, no. 4, pp. 330-338.
71. Shteynberg A. S., Ulihbin V. B. Dolgov E. I., Manelis G. B., in Goreniye i vzriv. M.: Nauka, 1972, pp. 124-127.
72. Tumanov V. V., Berlin A. A., Khalturinskyi N. A., Visokomolecul. soed., 1978, vol. 20A, no. 12, pp. 2784-2790.
73. Holve D. T., Sawyer R. W., Polymer Flame Retardant Mechanisms. Report ME-75-2, College of Engineering, Department of Mechanical Engineering, University of Berkeley, 1975.
74. Zeldovich Y. B., ZhTF, 1949, vol. 19, no. 10, pp. 1199-1205.
75. Librovitch V. B., PMTF, 1968, no. 2, pp. 36-41.
76. Fenimore C. P., Martin F. J., Combust. and Flame, 1966, vol. 10, no. 2, pp. 135-140.
77. Krishnamurthy L., Combust. Sci. and Technol., 1975, vol. 10, no. 1-2, pp. 21-26.
78. Frey A. F., Tien I. S., Combust. and Flame, 1976, vol. 26, no. 26, pp. 257-264.
79. Margolyn A. D., Krupkin V. G., Phy. goreniya i vzriva, 1978, vol. 14, no. 2, pp. 56-63.
80. Komamiya K., J. Fire and Flammability, 1973, vol. 4, no. 2, pp. 82-87.

81. Margolyn A. D., Krupkin V. G., Dokl. AN USSR, 1976, vol. 228, no. 6, pp. 1395-1397.
82. Margolyn A. D., Krupkin V. G., in Khimicheskaya physica protsessov goreniya i vzriva. Goreniye geterogennih i gazovih system. Tchernogolovka: Ed.-publ. div. of OIKhF AN USSR, 1977, pp. 11-17.
83. Bakhman N. N., Aldabaev L. I., ibid, pp. 7-11.

Chapter 3

Chemical Processes in the Combustion of Polymeric Materials

Chemical processes are principally responsible for the combustion of polymeric materials. It is impossible to control the combustion process and solve the practical problems related to the flammability of polymeric materials without understanding the general principles that describe the chemical transformation of substances into final combustion products. Naturally, the chemical aspects cannot be isolated from other important aspects of combustion. The interaction of physical and chemical processes is the essence of a complex combustion phenomenon. Nevertheless, precise knowledge of the mechanism and kinetics of the chemical transformation of polymers during the combustion process is the key to predicting how polymer materials behave in critical fire hazard situations.

Flammability of polymer materials, ease of flame spread, combustion products and their toxicity, and retardation and suppression of the flame are properties connected with the chemistry of combustion.

The chemical transformation of a polymeric substance into final combustion products has a multistage character, as noted before. The information on the chemistry of the combustion of a polymeric substance must contain data on the process taking place in the condensed and gaseous phases at the interface. It should be mentioned that information is still limited, and there is no doubt about the necessity for expansion of future research.

It is obvious that the wealth of experimental and theoretical material on gas phase flame reactions of various low-molecular components must be used [1-4]. This information is applicable to the flaming plumes of polymers if it is based on knowledge of what kind of compounds are formed as a result of the degradation and vaporisation of polymers under the conditions of conflagration.

Chemical Processes in the Condensed Phase

The kinetics and mechanism of degradation for the majority of polymers under the conditions of combustion have still not been studied to any notable extent. As a result, the following question arises: to what degree is data derived from experiments involving moderate heating of small samples relevant? These conditions are useful in eliminating diffusion limitations from the study of the chemical kinetics of the process. The second question is then: what is the role of oxygen in the degradation of the condensed phase if a thermal process occurs during combustion or if the degradation of the polymer has a thermal-oxidative character?

Degradation and gasification of polymer substances during stationary combustion in an atmosphere of gaseous oxidiser are characterised by the existence of large temperature gradients near the surface. The degradation process in fact occurs in a narrow zone, the width of which depends on the thermophysical properties of the polymer and on the conditions of combustion. The width of the pyrolysis zone in the combustion of a polymer does not exceed 2-3 mm. According to the data [5], the active pyrolysis zone is (1 to 2) x 10^{-2} mm for polymethylmethacrylate combustion and (1 to 2) x 10^{-3} mm for polystyrene combustion.

According to other research [6], it is impossible to extrapolate thermal analysis data (TGA, DTA) of polymeric materials to the region of high surface temperatures such as are observed during the combustion of polymers.

The ability to obtain information directly on macrokinetic characteristics of the high-temperature degradation of condensed substances led to the creation and development of the linear pyrolysis method. Linear pyrolysis (LP) is a unidimensional spreading of the front of a thermal degradation reaction under the influence of an external heat source. The development of the LP method, as proposed by R.D. Schultz and A.O. Dekker [7], is cited in the literature [8-12]. The pyrolysis of a polymer in this context occurs as a result of the samples being heated with a hotplate. The existence of a gaseous film between the plate and the surface of the sample [9-12] enables the surface temperature to be adjusted to calculate the kinetic parameters of thermal degradation due to the experimentally derived rates of the LP process.

The theory of the LP process of condensed substances, which considers heat processes in the surrounding medium, was developed by A.S. Shteynberg and V.B. Ulihbin [11]. This work demonstrated that the existence of different LP regimes is possible. At a low LP rate the thermal resistance of the gaseous space is low and the surface temperature is practically equal to the temperature of the plate ($T_S = T_{pl.}$ = const.).

The thermal losses through the side surface of a sample can be considerable. The LP rate (v_{LP}) that accounts for this factor can be expressed by an equation:

$$v_{LP} = \frac{L\,(RT_S^2/E)\,k_0 \exp(-E/RT_S)}{(T_S - T_0)(\alpha L/\lambda)^{1/2}}, \quad (3.1)$$

where

L size of sample

α heat emission coefficient

λ heat conductivity coefficient

$\alpha L/\lambda$ Biot criterion

T_S, T_0 surface temperature of sample studied and temperature of surrounding atmosphere respectively.

When the external heat losses are small and LP is taking place, the rate v_{LP} is calculated according to the Merzhanov and Dubovitskyi formula [13].

$$v_{LP} = \sqrt{\frac{aRT_S^2 k_0 \exp(-E/RT_S)}{E\,[(T_S - T_0) + Q/2c_c]}}, \quad (3.2)$$

where

a thermal conductivity factor

Q heat of thermal degradation reaction of a polymer

C_c specific heat of a polymer

The kinetic parameters k_0 and E of the thermal degradation of a polymer can be calculated on the basis of experimental v_{LP}, T_S and Biot data and the thermophysical characteristics of the polymer [10, 12]. The criticism of the LP method in the case of a sample heated by a single hotplate emphasises that:

1. The presence of a plate affects the conditions of polymer pyrolysis to the extent that they deviate widely from the conditions of combustion

2. Released vapours cause erosion of the superficial layers, particularly in the case of fusible polymers

In order to eliminate these disadvantages in the study of LP, a porous plate was used [14]. Later, the study of the dynamics of the degradation and vaporisation processes of the polymer was conducted under the conditions of stationary combustion [6].

It is interesting to compare kinetic parameters of the thermal degradation of polymers that have been derived by regular methods of thermal analysis, by the LP method and also under combustion conditions. Figure 3.1 shows how the conditions of the process affect the rate of vaporisation and the degradation of linear polymethylmethacrylate and a network polymer, obtained by the copolymerisation of methylmethacrylate and triethyleneglygol dimethacrylic ether (TGM-3). At that time three pyrolysis regimes were observed (I-III).

The kinetic parameters derived for the study of the thermal degradation of polymers at lower temperatures and for linear pyrolysis tally very well with each other (Fig. 3.2 and Table 3.1). The extinction rates of the burning polymer spheres in a stationary atmosphere of air correspond to the rates of linear pyrolysis in regimes I and II [15]. In regime III, it was observed that a small rise in temperature produced a sharp increase in the rate at which the condensed phase (C phase) was dispersed, a condition which is due to the dispersion of the polymers. Visually, the dispersion was observed in the form of microtracks of polymer particles in the gaseous phase. It may thus be concluded that the conditions on the surface (in the pre-surface layer) of the polymer determine the character of its vaporisation and degradation.

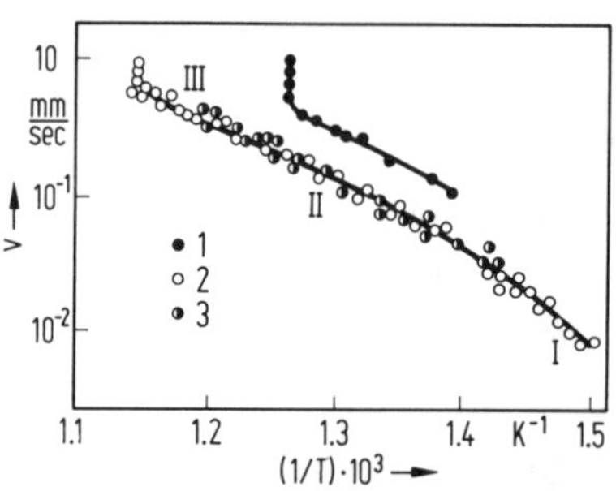

Fig. 3.1 Relationship between rate v of LP process (1, 2) and combustion process (3) of polymers PMMA (1), PMMA + TGM-3 (2, 3) and temperature.

Table 3.1 Kinetic parameters of thermal degradation (TD) and linear pyrolysis (LP) of polymers

Polymer	Temperature, °C		E, kcal/mole		lg, k_0	
	TD	LP	TD	LP	TD	LP
MMA	290-340	450-540	42.6 ± 2	43 ± 3	12.47 ± 0.7	13.32 ± 1
MMA + 2% TGM-3	310-360	400-590	43.5 ± 2	43 ± 3	12.14 ± 0.7	12.51 ± 1
MMA + 10% TGM-3	310-360	400-590	43.8 ± 2	43 ± 3	12.3 ± 0.7	12.51 ± 1

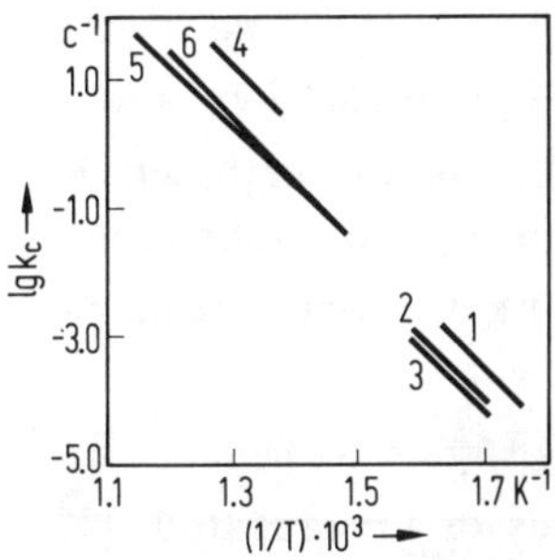

Fig. 3.2 Relationship between $l_g k_0$ and temperature of TD process (1-3), of LP process (4, 5), and of combustion process (6) of PMMA (1, 4), PMMA + TGM-3 (2, 3, 5, 6). (2) = 2% TGM-3; (3) = 10% TGM-3.

It is useful to compare the chemical transformation of the C phase during laminar diffusion combustion of polymers with the degradation of polymers that occurs at lower temperatures. Agreement of the thermal degradation and LP parameters and LP rates, and the disappearance of the C phase during combustion indicates the analogy between the chemical transformation processes of the polymers under the heating conditions studied.

Going by Madorsky's data [16] on the thermal degradation of polymers in a vacuum and data on the linear rate of disappearance of the condensed phase during combustion, L.A. Wall [5] mentioned that the activation

Table 3.2 Activation energies of thermal degradation (TD) and disappearance of the C phase during the combustion of polymer materials

Polymer materials	*E*, kcal/mole	
	TD	Regression of C phase
Polymethylmethacrylate	32-52	42-48
Polyethylene	72	73
Polypropylene	58	40
Polyformaldehyde	≧ 20	24
Polyamide (nylon 6,6)	42	65

energy values for these processes tally sufficiently well (Table 3.2). Certain discrepancies in polypropylene and polyamide *E* values were attributed to a difference in the samples tested. According to McAlevy and coworkers [6], a number of polymers (polyurethane, polyoxymethylene) have an activation energy of C phase regression during the combustion process that is only half the activation energy during thermal volumetric degradation.

S.K. Brauman [17] studied the rate of linear pyrolysis of impact-resistant polystyrene samples. He also used samples of unsaturated polyether with a base of o-phthalic acid, maleic acid and propylene glycol crosslinked by styrene.

The experiment was carried out in an inert and oxidising atmosphere under the influence of a radiant heat flow. A comparison of the derived data with the rate of polymer regression during combustion in a similar oxidising atmosphere (but without an external heat flow) caused the author to conclude that superficial oxidation is not essential during stationary combustion. According to the author, such a situation is only necessary for the combustion of polymer samples of a round bar shape or for polymer tablets with a diameter of approximately 6 mm. The effect of oxygen is such that, in its presence, the conditions of stationary combustion are established more rapidly.

There has been an impression that a pure thermal degradation process makes up the chief contribution to the vaporisation of polymers during

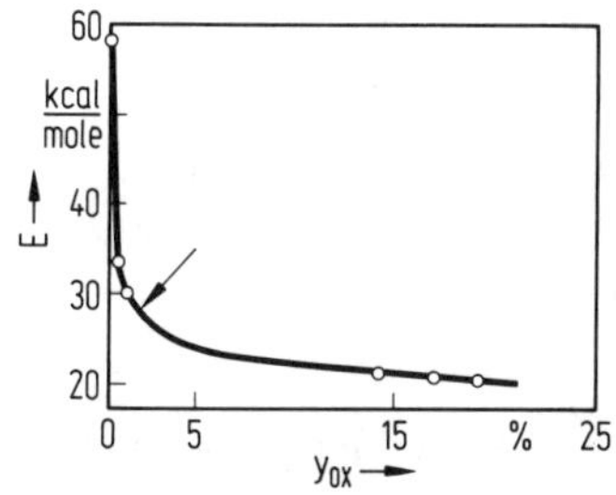

Fig. 3.3 Effect of oxygen concentration in surrounding medium Y_{ox}. on activation energy E of polypropylene destruction. The arrow points to a spot that corresponds to the conditions on a surface during combustion of PP in an air atmosphere.

laminar diffusion combustion. Recently, however, a number of works have been published that contain indirect data on the participation of oxygen in C phase degradation during the combustion of polymers. Small quantities of oxygen have been discovered in the flame zone adjoining the burning surface of a polymer sample [18-19]. The direct detection of oxygen in the superficial layer of burning polypropylene by neutron activation analysis [19] explains the lower E values of C phase regression during the combustion of this polymer (see Table 3.2 and Fig. 3.3).

By analysing the microtome polymer cuttings for oxygen content, D.E. Stuetz and coauthors [19] showed that the concentration of oxygen drops from 26% to 0.3% at a distance of 1.6 mm beneath the boundary line of the polypropylene burning surface. Thus in the condensed phase reaction zone thermo-oxidative degradation is also possible. The authors concluded that a sufficiently large amount of oxygen can diffuse through a high temperature reaction zone inside the flame towards the polymer surface and be absorbed by a superficial layer. This cited work represents the first attempt to directly analyse the possible oxidative transformations of a polymer under combustion conditions. The measurement of the oxygen concentration in the superficial layer, however, does not correlate with the dissolved oxygen but with the chemically bonded oxygen.

It is known [20] that the solubility of oxygen in solid polypropylene depends on the degree of crystallinity and the nature of the superstructure of the polymer sample. At a high temperature a melt is formed on the hot surface but the solubility of the oxygen sharply diminishes. The diffusion of oxygen in the melt is complicated by a chemical reaction under these conditions. The combustion conditions, however, play a determining role in the transfer of oxygen towards the polymer surface, and the contribution of the thermal oxidative process to the total vaporisation rate of the polymer may fluctuate substantially. As a result of the latter, turbulisation of the flame is conducive to a mixing of the fuel products of polymer degradation with a gaseous oxidiser and to a transfer to the condensed phase.

It is possible that microturbulence occurs near the surface at a fast rate with the oxidising gas used in [19]. In view of the specific nature of the polymer (existence of ternary carbon atoms increases the tendency of polypropylene to oxidation) the observed result becomes comprehensible. Indeed, this situation further complicates the picture of the chemical reaction on the surface of the condensed phase. It is impossible to describe the process in detail in such a way that it will be applicable to polymer materials of different structure. Even in the simplest case when pyrolysis of a polymer is taking place as the result of depolymerisation and leads to formation of the sole product, monomer, the process includes many simple reactions. The specific interaction of these simple reactions determines the total rate of polymer pyrolysis transformation and its temperature coefficient.

It is appropriate to discuss the general tendencies in chemical transformation reactions of polymer materials at high temperatures in this chapter. It was decided to limit the range of problems by analysing pure thermal conditions. In spite of the diversity of chemical reactions taking place during pyrolysis, it is possible to divide all organic polymers into two groups on the basis of thermal effect [21, 22].

The first group includes those polymers that undergo destruction with rupture of the bonds of the principal chain and the formation of low-molecular gaseous and liquid volatile products under the influence of heat. Polymers of this type are destroyed virtually completely at high temperatures (300-600°C) and form a very small quantity of nonvolatile residue. Such polymers as polymethylmethacrylate, poly-α-methylstyrene, polyoxymethylene and polytetrafluoroethylene, which depolymerise with almost 100% yield, belong to this group. Hydrogenated polystyrene, polyoxypropylene, polymethylmethacrylate, polypropylene, polyethylene, polyethyleneterephthalate and other polymers which are destroyed almost completely and have a relatively low yield of monomer also come under this group.

The second group consists of polymers with a tendency to intramolecular detachment of specific atoms of groups and to cyclic, condensation, recombination and other types of reactions that lead to the formation of nonvolatile carbonisation products. This group of polymers takes in polyvinyl alcohol and its derivatives, chlorine-containing polymers of the vinyl and diene series, polyacrylonitrile, cellulose and many aromatic and heterocyclic polymeric compounds. A common feature of all the polymers of this group is the formation of a zone with conjugate multiple bonds inside the macromolecules and the conversion from a linear structure to a three-dimensional network structure during the pyrolysis process. Unlike the first group of polymers, the pyrolysis of similar high-molecular compounds is as a rule characterised by an exothermal heat effect; subdivision of polymers into the specified groups is not rigorous because the direction of a chemical reaction can change as a result of external conditions. At the same time it should be considered from the viewpoint of structure. The formation of molecular and submolecular structures occurs during synthesis and reprocessing stages. Formation of heterocycles and defective structures is possible at one and the same time [23]. The heterocycle in the molecular structure of a polymer has a substantial effect on its thermal and other properties. This factor was used by chemists for a long time for purposes of creating new materials with desired properties. It also represents one of the most important trends in the chemical modification of polymeric materials in order to reduce their ignitability and combustibility. When there is a thermal influence acting on a polymer material the type of chemical reaction is determined by the nature and stability of the bonds inside the molecule. (We are not discussing the effect of impurities and admixtures on polymer pyrolysis here.) The heat of polymerisation was selected as a criterion of polymer behaviour during heating [21]. Polymers with low values of polymerisation heat ($\leq$ 10-13 kcal/mole) are the least thermostable and are inclined to form monomers by a reverse reaction.

A similar inclination to depolymerise, however, is observed with polytetrafluoroethylene, which has a polymerisation enthalpy equal to 41.5 kcal/mole. Pyrolysis of the olefin, vinyl chloride or vinyl acetate polymers with $\Delta H_{pyr.}$ in the 17-26 kcal/mole range occurs through totally different mechanisms. The concept of the energy capacity of a middle bond in the polymer $q = \Delta H_{com.}/n$ [24] (n being the number of bonds in the polymer and $\Delta H_{com.}$ the combustion heat), is not useful in this case, since the neutralisation of the bonds blacks out the general picture and does not solve either the problem of real thermal stability or that of the possible direction of the reaction. This property does not indicate the real situation with regard to the flammability of a polymer.

Unfortunately there is practically no data on the thermodynamics of polymer pyrolysis processes. L.A. Wall [25] calculated the changes in enthalpy, entropy and free energy at 25°C for a number of hypothetical directions of pyrolysis for two polymers, polyethylene and polytetrafluoroethylene (Table 3.3). The calculation shows that from the viewpoint of thermodynamics, polymers are non-equilibrium systems. Decomposition with the further formation of carbon during pyrolysis is thermodynamically beneficial. Reactions 5 and 8 (Table 3.3) are exothermal and are characterised by negative values for a change in the thermodynamic potential, Gibb's free energy ΔG. In the case of polytetrafluoroethylene, explosive reaction rates and the formation of carbon could be expected. However, both polymers are stable not only at 25°C but also at higher temperatures. The real reactions, 4 and 6, have positive ΔG values.

Thus a thermodynamic consideration of the possible directions of pyrolysis is not in itself sufficient for the prediction of polymer behaviour under thermal treatment. In order to understand polymer behaviour

Table 3.3 Thermodynamic parameters of pyrolysis for polyethylene and polytetrafluoroethylene

No. of reaction	Direction of reaction	ΔH kcal/mole	ΔS kcal/mole·°K	ΔG kcal/mole
1	$\sim CH_2-CH_2\sim \rightarrow CH_2=CH_2$	22.35	34.27	12.19
2	$\sim CH_2-CH_2\sim \rightarrow 1/3C_6H_6+H_2$	16.46	34.28	6.23
3	$\sim CH_2-CH_2\sim \rightarrow 2C$(graphite)$+2H_2$	9.85	46.76	-4.10
4	$\sim CH_2-CH_2\sim \rightarrow n$-alkene [$1/18CH_3-(CH_2)_{33}-CH=CH_2$]	1.09	2.03	0.49
5	$\sim CH_2-CH_2\sim \rightarrow C$(graphite)$+CH_4$	-8.04	27.49	-16.23
6	$\sim CF_2-CF_2\sim \rightarrow CF_2=CF_2$	46	45	33
7	$\sim CF_2-CF_2\sim \rightarrow 2C$(graphite)$+F_2$	194	71	172
8	$\sim CF_2-CF_2\sim \rightarrow C$(graphite)$+CF_4$	-27	35	-38

vis-à-vis heat treatment it is necessary to conduct a serious study of the mechanism and kinetics of the polymer's pyrolysis reactions. The thermal degradation of first and second group polymers can go through a radical, ionic or molecular mechanism (either chain or non-chain processes). The chain nature of thermal degradation in a large number of high-molecular compounds is strongly connected to the polymer structure. A polymeric structure generally affects the direction and rate of chemical transformation of a substance (manifestation of conformation, configuration and other effects) [26]. The effect of configuration ("neighbour" effect) is especially clear during the pyrolysis of both groups.

To determine the pyrolysis mechanism of a substance, it is important to know the nature of the active centres. Yet it is always possible to determine and clearly show the nature of active centres in the pyrolysis processes of carbo- and heterochain polymers. This is why this problem is still a subject of active research for a number of polymer systems. The kinetics of a chemical reaction in polymers are connected to the kinetics of the molecular movements of a matrix, which, in turn, are determined by the structure and submolecular organisation.

At high temperatures corresponding to the temperature of a burning surface the thermoplastic polymers are in a viscous fluid state. Molecular movements are "defrosted" in this state and it should be expected that the rate of elementary reactions will be equal in view of the order of magnitude of the reaction rate in liquid viscous media. A large number of rigid-chain polymers with a linear structure, however, have a second-order phase transition temperature. Close to degradation temperatures rigid-chain, three-dimensional network polymers do not become soft at all. This puts diffusion limitations on the movement of active centres within the matrix and thereby affects the kinetics of the chemical transformation of the substances (see Table 3.3).

Some common kinetic characteristics of radical solid phase reactions of the thermal and thermo-oxidative degradation that occurs within the matrix are discussed in a survey by V.S. Pudov and A.L. Buchachenko [27]. In systems in which diffusion of large active and inactive molecules is difficult, it is possible to have a relay-race mechanism of active centre migration [28]. Frequently, heterochain polymers undergo heterocyclic degradation under the influence of catalytic admixtures (additions of an ionic nature during heating).

The catalysts substantially reduce the activation energy of the dissociation reaction of the bonds, as opposed to homolytic fracture. The presence of moisture vapours is particularly apparent from the rate of liberation of volatile degradation products.

Pyrolysis of First-Group Polymers (Chain Scission/Volatile Products)

As mentioned previously, polymers with a macromolecular structure that is completely destroyed on pyrolysis (as a result of the fracture of principal chain bonds) belong to the first group. When pyrolysis of members of this group of polymers occurs there are two limiting cases:

1. The process occurs according to a depolymerisation mechanism with the formation of only one product: monomer
2. The process of macromolecular degradation is not followed by formation of a monomer.

Nevertheless, the thermal degradation of a considerable number of polymers of the addition type falls into an intermediate pyrolysis category. The kinetics and mechanism that determine the direction of a chemical reaction during the thermal destruction of this group of polymers are completely determined by the stability of the bonds in the macromolecules, by the reactive ability of the active centres, by the probability of a transfer reaction of the chain and by conditions that bring the process to completion.

It is known, for instance, that few active alkyl radicals, which are formed during the destruction of polymethylmethacrylate, disintegrate with formation of a monomer. More active polypropylene radicals, which occur during inter- and intramolecular chain transfer, undergo isomerisation with bond rupture and formation of fragments with various molecular weights. The most common type of chain-transfer reaction in the polymers is the separation of a mobile hydrogen atom from the valence-saturated molecules by an active centre (macroradical).

The reactive capacity of different molecules and of the radicals formed from those molecules is antibatic* [29]. It is easy to see that polymers that contain mobile atoms, for instance α-hydrogen atoms (polyacrylates, polyolefins), decompose with a low yield of monomer. Polymers with blocked α-hydrogen atoms and radicals of substituents do not participate in the

* i.e. unreactive molecule forms reactive radical, reactive molecule forms unreactive radical

transfer of the kinetic chain; they are characterised by a high yield of monomers (polymethylmethacrylates, poly-α-methylstyrene, polytetrafluoroethylene). At the same time, if there is no chain transfer of the side substituent groups, the nature of the polymer structure is irrelevant: a linear or a three-dimensional network. It was discovered that oligoethermethylacrylate polymers also depolymerise with a virtual 100% yield of initial oligomers [30-32]. The nature of the end bonds in macromolecules and the presence of various weak bonds have a substantial influence on the kinetics of the pyrolysis of polymers which degrade by the depolymerisation mechanism. A fairly large role during degradation can be attributed to the molecular weight and the molecular weight distribution (MWD) $\bar{P}_w/\bar{P}_n$ of the polymers.

It should be mentioned that these factors play a role not only in the case of carbochain polymer destruction but also during destruction of heterochain polymers. When free hydroxyl groups or group fragments of ionic catalysts are attached to the ends of polydimethylsiloxane macromolecules [33] and polyoxymethylene macromolecules [34], for instance, the destruction of these polymers is initiated from the end of the chain by a depolymerisation mechanism. In general, the degradation process of this group of polymers can be represented by the following scheme:

Initiation		
from end of chain	$P_i \xrightarrow{k_{1,a}} R_{i-1} + R_1,$	$k_{1,a} \sum P_i$
according to the law of probability (random scission)	$P_i \xrightarrow{k_{1,b}} R_{i-j} + R_j.$	$k_{1,b} \sum P_i$
	$R_i \xrightarrow{k_2} P_{i-1} + M.$	$k_2 \sum R_i$
Formation of chains		
Reverse polymerisation:	$R_i + M \xrightarrow{k_{2,a}} R_{i+1}.$	$k_{2,a} \left(\sum R_i\right) M$
Chain transfer		
with chain rupture:	$R_i + P_j \xrightarrow{k_3} P_i + P_{j-k} + R_k.$	$k_3 \sum R_i \sum P_j$
Termination of chain		
by recombination:	$R_i + R_j \xrightarrow{k_{4,a}} P_{i+j},$	$k_{4,a} \sum R_i \sum R_j$
by disproportion:	$R_i + R_j \xrightarrow{k_{4,b}} P_i + P_j,$	$k_{4,b} \sum R_i \sum R_j$
monomolecular:	$R_i \xrightarrow{k_{4,c}} P_{i-0} + R_0,$	$k_{4,c} \sum R_i$
by recombination inside "cell":	$R_i + R_j \xrightarrow{k_{4,d}} P_{i+j},$	$k_{4,d} \sum R_i$
by disproportion inside "cell":	$R_i + R_j \xrightarrow{k_{4,\text{é}}} P_i + P_j.$	$k_{4,e} \sum R_i$

where P, R and M represent the inactive polymer molecule, the active macromolecule (radical) and the monomer, respectively; i, j and k denote the number of monomer sections in the molecule.

The chain transfer can be intra- and intermolecular. The concentration of radicals at that time is proportional to the volume of the sample. The termination of the chain by recombination and by disproportion inside the "cell" is kinetically monomolecular, since it is limited by the diffusion of an active centre towards the "cell". The reverse polymerisation reaction is considered to be an equilibrium process. An increase in temperature reduces the influence of the reaction. Above $T_{crit.} = \Delta H_p / \Delta S_p$ the equilibrium should shift completely in favour of the degradation process.

Differential equations have been set up for the time rates of the change of inactive molecules, active centres and concentrations of degradation products. A solution has not been derived to the set of differential equations that give a general description of the thermal destruction of different polymers (from the group under consideration). Usually simplified schemes with preferred mechanisms of initiation and chain termination are used. Theoretically derived ratios are then compared with those established experimentally. To clarify the mechanism of the destruction process it is particularly important for the MWD changes during the process to be analysed. Theoretical calculations of degradation rate variation and the average degree of polymerisation of the polymer residue during the course of the process demonstrate that for polymers with different structures, complex relationships are observed between structure parameters and the degree of transformation. These relationships are determined by the degradation mechanism [35]. Berlin and Enikolopian [36] analysed both different types of initial MWD (monodispersed, most probable ($\bar{P}_w / \bar{P}_n = 2$) and wide distribution) and mechanisms of initiation by random scission and from one end of the chain. All the termination mechanisms are considered by the authors as a reaction of the active centre with an inhibitor. The rate of the termination reaction in this case is equal to $k_{inh.}$ [inh.] [R_j]. This approach allowed the authors to describe different types of destruction processes, including a multistage polymerisation, which is characterised by the breaking away of only one monomer fragment from a macroradical during each disruption of the macromolecule.

Boucher [37] made a similar analysis of the kinetic models of polymeric degradation. He described the degradation process on the basis of an analysis of the relationship between the rate constants for bond breakage (end, middle, in a dimer) in polymers with various MWDs.

Kinetic data on polymeric degradation and the results of research on the change in molecular weight of a polymer during the process could be used successfully for a solution of the reverse problem, i.e. establishing the initial MWD of polymers [36].

The contribution of elementary reactions in a polymeric degradation process can vary not only as a result of the chemical nature of a polymer but also as a result of the conditions of process completion [38]. The results of studies on how molecular weight and temperature affect the formal-kinetic depolymerisation mechanisms of polymethylmethacrylate can be shown as an example. In the low-temperature region (338-366°C) the rate of depolymerisation decreases with an increase in the molecular weight of the polymer. The initiation of a chain under these conditions occurs preferentially from the end of the chain and the termination of the chain is bimolecular. In the high-temperature zone (≥ 463°C) the rate of depolymerisation increases with an increase in molecular weight. It was concluded that the break of macromolecules according to the laws of probability (random scission) is predominant and that the termination is monomolecular as a result of fast depolymerisation up to the end of the chain and volatilisation of a low-molecular radical.

The length of a kinetic chain ν during the thermal degradation of polymers can either be equal to or substantially less than a physical chain as a result of the loss of an active centre. It is known that ν is proportional to the ratio of the sum of the probabilities of chain initiation, chain transfer and chain termination reactions to the total sum of the probabilities of chain termination and chain transfer reactions (corresponding to the ratio of rates for the above mentioned reactions). The increase in temperature must lead to a decrease in ν which results from the fact that the activation energy of termination reactions and chain transfer reactions is larger than the activation energy of chain initiation reactions, and the role of the first two reactions increases with a rise in temperature.

It is obvious from the theory that the rate of decrease of the molecular weight of a polymer does not depend on ν with large kinetic chain length values and that it is proportional to ν for small kinetic chain length values. Table 3.4 shows the influence of the polymeric degradation mechanism on the liberation rate of volatiles [35, 39]. The most essential factor is that, for long kinetic chain lengths, the rate of liberation of a volatile material diminishes linearly as a result of the degree of transformation without any dependence on the type of initiation mechanism. When initiation occurs as a result of the probability of $\nu < 1$, the depolymerisation rate curve passes through a maximum. Chain transfer increases the probability of random disruption of a macromolecule and liberation of volatile low-molecular-weight fragments, larger in size than a monomer.

The larger the size of a volatile monomer fragment L, the larger the probability of the presence of a maximum on the destruction rate curve. In addition, depending on the depth of transformation, lower values are required for the chain transfer constants k_3 [35].

The location of the maximum during degradation according to the law of probability corresponds to $\alpha_{max} \sim 0.26$ and a maximum degradation rate $W_{max} \sim 0.4\ kL$, where k is the constant of the net rate of bond destruction within a macromolecule. At the same time, for polystyrene, in which depolymerisation is followed by chain transfer, $\alpha_{max} = 0.4$. The number of products that form only one product (monomer) during pyrolysis is small. In the majority of cases the pyrolysis of first-group polymers is followed by the liberation of a range of products.

In spite of the complex composition (Table 3.4) the formation of volatile products can be described from the viewpoint that the bonds in the principal chain are disrupted and the newly formed fragments participate in chain transfer. An example is cited of this kind of analysis of the degradation of volatile polypropylene products [40]. Schematically, the process could be represented in the following way:

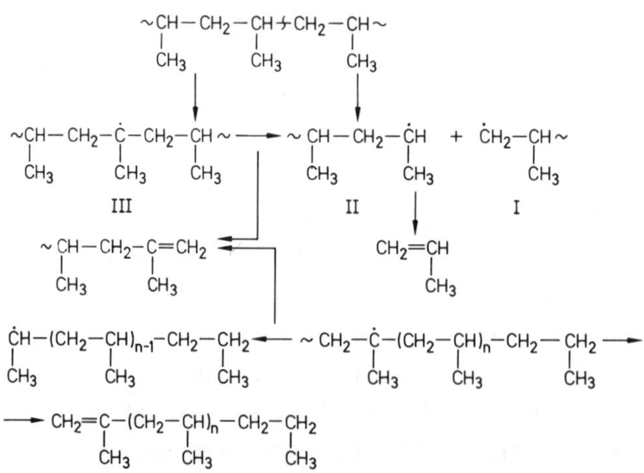

Table 3.4 Relation between the rate of volatile product liberation, depolymerisation and initial molecular weight distribution

Mechanism of chain initiation	Length of kinetic chain	Mechanism of chain termination	(MWP $\bar{P}_w/\bar{P}_n$)	Type of $W_0 = f(\bar{P}_n^0, \bar{P}_w^0, \nu)$ function	Diagram of relation between W and degree of transformation
On the end groups	$\nu \leqslant \bar{P}_n^0$		1; 2	$W_0 \neq f(\bar{P}_n^0, \bar{P}_w^0, \nu)$	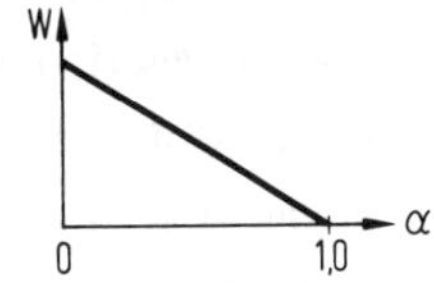
	$1 \ll \nu \ll \bar{P}_n^0$	First order of magnitude	1; 2 1; 2	$W_0 \sim \frac{1}{\bar{P}_n^0}\nu$ $W_0 \sim \frac{1}{\sqrt{\bar{P}_n^0}}\nu$	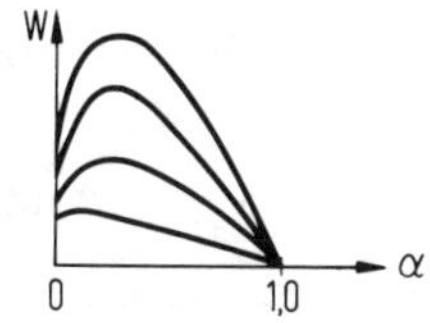
		Disproportionation recombination	2	$W_0 \neq f(\bar{P}_n^0, \bar{P}_w^0, \nu)$	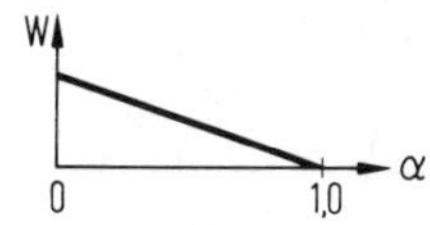
	$\nu \leqslant 1$		1	$W_0 \sim \frac{1}{\bar{P}_n^0}$	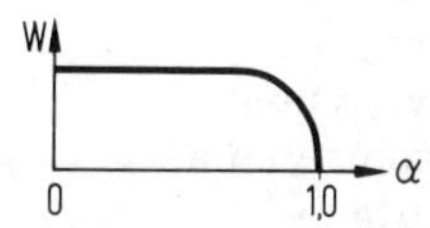
According to law of probability (random scission)	$\nu \leqslant \bar{P}_n^0$		1 2	$W_0 \sim \bar{P}_n^0$ $W_0 \neq f(\nu)$	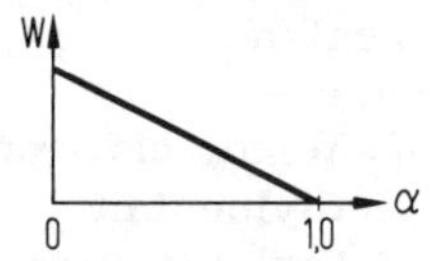
	$1 \ll \nu \ll \bar{P}_n^0$	First order of magnitude	1; 2	$W_0 \neq f(\bar{P}_n^0, \bar{P}_w^0)$ $W_0 \sim f(\nu)$	
		Disproportionation Recombination	1; 2	$W_0 \neq f(\bar{P}_n^0, \bar{P}_w^0)$ $W_0 \sim f(\nu)$	
	$\nu \leqslant 1$		2	$W_0 \sim \frac{1}{\bar{P}_n^0}\nu$	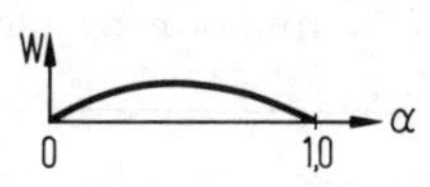

When the principal chain bonds rupture, two types of macroradicals, I and II, are formed which participate in the intramolecular chain transfer to the nth carbon in the macroradical. Table 3.5 shows the composition of volatile products from the degradation of polypropylene at 400°C and the type of radical and carbon atom that is attached during chain transfer. This data explains how the product was formed.

The information on the kinetic parameters of the thermal degradation of polymers, which includes data on elementary reactions, was obtained for only a small number of carbon chain polymers [35, 41, 42] (Table 3.6). The most complete data was obtained for polymethylmethacrylate and polytetrafluoroethylene. Activation energies of elementary initiation reactions, chain formation and chain termination reactions were estimated for

Table 3.5 Products of polypropylene decomposition at 400°C

Compound	Percentage mol.%	Type of radical	n^*
Methane	3.9		
Ethane	4.8	II	3
Propane	1.9	I	4
Propylene	21.4	II	
Isobutylene	3.0		
Butane	0.07		
2-Methylbutene	0.15		
1-Pentene	0.12		
Pentane	24.3	II	5
2-Pentene (cis,trans)	0.23		
2-Methylpentane	0.93	I	6
2-Methyl-1-pentene	15.4	II	3
2,4 Dimethylpentane	0.34		
2,4 Dimethyl-1-pentene	0.98	I	4
4-Methylheptane	2.3	II	7
2,4-Dimethylheptane	<0.1	I	8
2,4-Dimethyl-1-heptane	18.9	II	5
4,6-Dimethylnonane (threo)	0.29	II	9
4,6-Dimethylnonane (erythro)	0.36	II	9
2,4,6-Trimethyl-1-nonene (threo)	0.36	II	7
2,4,6-Trimethyl-1-nonene (erythro)	0.44	II	7

$^*n$ = number of carbon atoms in compound

Table 3.6 Kinetic parameters of pyrolysis of polymer materials

Polymer material	k_0, sec^{-1}	E kcal/mole	k_i, sec^{-1}	Comment
Polymethylmethacrylate	$\sim 10^9$	32	$k_{1,a} \sim \exp(-48\ 000/RT)$	From end of chain, large ν
			$k_{1,\sigma} \sim \exp(-99\ 000/RT)$	
	$\sim 10^{13}$	43-48	$k_2 \sim \exp(-18\ 500/RT)$	From end of chain, small ν
	$\sim 10^{16}$	58	$k_4 = 3 \cdot 10^{14} \exp(-20\ 000/RT)$	
			$k_4 \sim 10^{18} \exp(-29\ 000/RT)$	According to law of probability
Polytetrafluoroethylene	$\sim 10^{19}$	80	$k_1 = 2.1 \cdot 10^{28} \exp(-118\ 600/RT)$	
			$k_2 = 7 \cdot 10^{11} \exp(-36\ 200/RT)$	
			$k_4 = 3 \cdot 10^{13} \exp(-30\ 000/RT)$	
Poly-α-methylstyrene	$\sim 10^{22}$	65	$k_1 = 4 \cdot 10^{18} \exp(-65\ 000/RT)$	
			$k_2 \sim \exp(-42\ 500/RT)$	
Polystyrene	$\sim 10^{15}$	55	$k_{1,a} \sim \exp(-88\ 000/RT)$	
		45	$k_2 \sim \exp(-24\ 000/RT)$	
		65.3	$k_4 \sim 10^{20} \exp[-(24\ 000 \pm 2000)/RT]$	According to law of probability
Polyethylene	$\sim 10^{18}$	72	$k_1 = 4 \cdot 10^{18} \exp(-72\ 000/RT)$	
			$k_3 \sim \exp(-20\ 000/RT)$	
			$k_4 \sim \exp(-32\ 000/RT)$	High pressure, temperature 40-100°C
Polypropylene		55-61	$k_4 \sim \exp(-23\ 000/RT)$	Temperature 20-80°C

these substances. The influence of initiation mechanisms on the effective activation energy of the total pyrolysis process was demonstrated for polymethylmethacrylate. The values of activation energies for the initiation of polymer degradation as a result of random scission are close to the average value for the dissociation energy of carbon-carbon bonds. The activation energy of chain formation as a result of the destruction of radicals followed by formation of monomers corresponds to the sum total of activation energy of chain formation and the polymerisation enthalpy: $E_2 = E_p + \Delta H_p$.

The high values for the activation energy of macroradical degradation are interesting. Using EPR methods a direct investigation was conducted of the kinetics of macroradical degradation in rigid polymers at elevated pressures [43, 44]. The macroradicals were generated in polymer matrixes in different ways. The results of these investigations show the interrelation between the kinetics of elementary reactions and molecular movements in the polymer matrix. The increase in pressure further retards the molecular motion and stabilisation of the radicals. At the same time, the recombination constants and activation energies of this reaction change substantially. When the pressure is increased from 1 to 10 000 atm, for instance, the E_4 value for polymethylmethacrylate decreases from 30.0 kcal/mole (60-100°C) to 22.0 kcal/mole (105-155°C), and for polypropylene from 17.5 to 13.2 kcal/mole (80-110°C) [43].

According to the authors, the recombination of macroradicals under these conditions occurs through a sequential mechanism. It is known that the temperature ranges for the unfreezing of molecular motion in a rigid polymer correspond to regions of fast macroradical degradation. Destruction of polypropylene alkyl macroradicals, for instance, occurs in a low-temperature region (100-250°K), and the activation energy is equal to 11 kcal/mole, which is close to the activation energy of molecular motion (13 kcal/mole). In a higher-temperature region (>250°K), the activation energy increases accordingly to 48 and 58 kcal/mole [45]. High values for the activation energy of radical recombination are linked to the diffusion mechanism of radical degradation and the cooperative nature of radical diffusion. In elastomeric, high-molecular-weight compounds the destruction of macroradicals occurs with an activation energy close to the values of that observed in fluid viscous media (for polybutadiene E_4 = 4 kcal/mole).

The conditions of pyrolysis affect the character of reactions. Thus when very low-weight portions (micrograms) are used with external heating, polystyrene depolymerises with approximately 100% yield of monomer ($E_{ef.}$ = 65.3 kcal/mole). When pyrolysis of a polymer occurs in a stream of nitrogen at 600-640°K, then the pyrolysis reaction is followed by a chain-transfer reaction and the formation not only of a monomer but also of other macromolecule fragments. It should be mentioned that the effective activation energy of monomer, dimer and trimer styrene formation is equal to 55-66 kcal/mole and exceeds the activation energy values for benzene and benzene derivative formation (20-30 kcal/mole) [46].

A rise in temperature always complicates the composition of generated products, which indicates a complication in the degradation mechanism of polymeric substances and the existence of side reactions.

A change in the functional group in macromolecules leads to a redistribution of electron density and a change in bond strength, and inevitably determines the direction of the chemical reaction during heating. When the length and the branching of alkyl substituents in polyalkylmethacrylates increase, for instance, the probability of bond disruption in the side substituent also increases. During the thermal degradation of poly-tert-butylmethacrylate, there is a competing detachment of isobutylene in addition to the disruption of the main chain. At the same time, the electron-donor substituents containing nitrogen atoms strengthen the C-N-C bonds in poly-N-alkyl-maleate and prevent the destruction of the imide ring. Disruption of the principal chain bonds is observed simultaneously [47].

Polycondensation polymers, which belong to the first group, are basically compounds that contain oxygen and nitrogen heteroatoms in the principal macromolecular chains. Simple and complex polyethers based on alkyleneglycols, aliphatic dicarboxylic acids and aliphatic polyamides can be included in this group.

There is a suggestion that heterochain polymers, formed by polycondensation and accompanied by the formation of low-molecular-weight products cannot decompose in the same way as the addition type of polymers. Nevertheless, a substantial number of heterochain polymers, which can be obtained both by polymerisation and by polycondensation, are now known. In contrast, there are examples of polycondensation processes not accompanied by the formation of any low-molecular-weight products. The division of polymers into polymerisation and polycondensation classes is based on a kinetic criterion. During polycondensation the growth of every physical chain is accompanied by the disappearance of an active functional end group. This is why initially dimers are formed, followed by trimers etc., and only at the end of the reaction are high-molecular-weight products formed.

It has already been mentioned that some of the heterochain polymers can degrade according to the polymerisation mechanism. The presence of ionic admixtures accelerates this destruction [21-34]. When polyoxymethylene is carefully purified and hydroxyl end groups are replaced by methoxyl groups then the effective activation energy of polymer thermal degradation increases from 26 to 41 kcal/mole [34]. Acylation of the end-amino groups in polycaproamide decreases the rate of depolymerisation. If the amount of moisture and admixtures in a polymer is reduced, the $E_{ef.}$ of polycaproamide thermal degradation increases from 15 to 43 kcal/mole. It is claimed [48-50] that thermal depolymerisation of poly-ε-caproamide occurs under the influence of a hydrolysis, acidolysis and aminolysis reaction. When homolytic disruption of bonds occurs, the rate constant for the degradation of polyamides increases to a value of $\sim 10^{13} \exp(-60\ 000/RT)$.

The nature of active centres during the thermal destruction of heterochain polymers has not been determined. There is both heterolytic and homolytic disruption of C-O and C-N bonds. According to Ritchie [51] the decomposition of complex ethers occurs through a molecular mechanism of simultaneous β detachment (intramolecular transfer of the β-hydrogen atom to the oxygen of a carbonyl group and disruption of the C_α-O bond). The molecular mechanism of β detachment occurs through an intermediate state

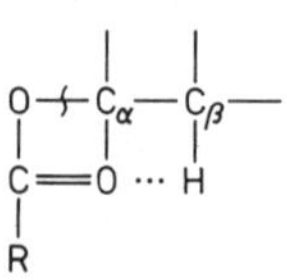

where

R acid radical

As a result, the formation of fragments with end carboxyl and unsaturated groups occurs. In fact, Camerbick and co-authors [p. 357 of ref. 51] propose an analogous mechanism for the pyrolysis of polyhexamethyleneadipylamide. The degradation of the polyamide begins with the disruption not of the $\sim$CO-$\wr$-NH$\sim$ bond, that is considered to be the least stable [50], but of the $\sim$CO-N(H)-$\wr$-CH$_2\sim$ bond in a β position relative to the carbonyl group, that is C_α-N. Molecules with end amide groups and unsaturated vinyl groups are the products of degradation. When there is thermal dehydration of end amide groups, nitrile groups can form.

The hydrolysis of polyamide under the influence of dehydrational moisture initiates the formation of molecules with end carboxyl and amino groups, which in turn are subjected to decarboxylic and deamination reactions (liberation of large amounts of ammonia and carbon dioxide).

At a high temperature, the degradation of aliphatic polyamides is accompanied by crosslinking, cyclisation and condensation reactions. However, the percentage of nonvolatile residues is comparatively small. At high pyrolysis temperatures the role of the radical reactions apparently increases. The introduction of aromatic fragments or monomer units into a polymer promotes crosslinking reactions during pyrolysis. Aromatic polyethers, polyamides and polyurethanes, which can be classified in the second polymer group, thus undergo carbonisation during pyrolysis.

Pyrolysis of Second-Group Polymers (Intramolecular Detachment of Atoms or Groups / Non-Volatile Products)

Despite the large number of polymers that have a tendency to react through intermolecular processes involving the elimination of atoms or functional groups (cyclisation, crosslinking reaction etc.) no pyrolysis theory has been developed for these polymers. At present there is an accumulation of information on the general picture of the chemical transformations that polymers in this group undergo and of the chemical mechanism of the carbonisation process.

It should be mentioned here that the pyrolysis of second-group polymers is frequently accompanied by the disruption of the bonds in the principal macromolecule chain. However, crosslinking reactions and reactions involving the formation of more thermodynamically stable systems predominate in the degradation reactions.

Pyrolysis processes involving molecules in this group proceed in the direction of the decrease of the thermodynamic potential of the system

(Gibb's energy) - $\Delta G = \Delta H - T\Delta S$, or of the decrease in the free energy of the system - $\Delta F = \Delta U - T\Delta S$. In the overwhelming majority of cases, the pyrolysis of polymers in this group is accompanied by heat liberation, that is, pyrolysis occurs with a decrease in the system's enthalpy (or internal energy). At the same time the entropy of the formed system also decreases, since, during the crosslinking reaction, the origination of fragments with conjugate bonds increases the rigidity of macromolecules and the intermolecular interaction becomes stronger, limiting molecular freedom of movement.

Among the polymers of this group is a subgroup of saturated polymers in which there is either a detachment of side substituents and formation of polyene fragments or intramolecular cyclisation and aromatisation reactions during the first stage of pyrolysis. Another subgroup includes polymers with aromatic carbocyclic and heterocyclic groups (monomer units). Bond disruption in the macromolecule chains in these types of molecules is followed by dehydration, crosslinking reaction and regrouping and formation of condensed aromatic structures.

On account of the reactions in which substituents participate and crosslinking reactions that are followed by bond disruption in the macromolecular chain and by volatilisation of low-molecular-weight degradation products, the yield of a nonvolatile carbon residue does not usually reach the theoretically computed values if it is computed from the percentage of carbon in the monomer unit. It should, however, be noted that second-group polymers are generally characterised by higher yields of nonvolatile residues (high-value char numbers) than the first group of polymers.

Saturated polymers of the vinyl series

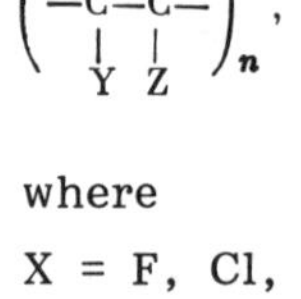

where

X = F, Cl, Br, I, OH, CH_3COO and Y, Z = H or halogen eliminate water, hydrogen halide or acetic acid at a relatively low temperature.

Elimination rates for hydrogen halides decrease in the order: iodo-, bromo-, chloro, fluoro-containing polymer. The elimination of HI from polyvinyl iodide occurs when the latter is maintained at room temperature. Polyvinyl bromide also shows considerable instability. As the number substituted by fluorine and hydrogen atoms increases in fluorocontaining polymers, the competition between the carbon-carbon scission reaction and the dehydrofluorination reaction also increases.

The kinetics and mechanism of the initial stages of pyrolysis, which take in the detachment of side substituents, have been studied in great detail for polyvinyl chloride. According to a large number of studies, the detachment of hydrogen chloride and the formation of polyene sequences in polyvinyl chloride occurs according to a molecular mechanism. The initiation of the process occurs as a result of the presence of weak bonds in molecules. Among the members of the group, conjugated carbonylallylic groups, $\sim\underset{\underset{O}{\|}}{C}-CH{=}CH-CHCl-CH_2\sim$, are found to cause instability of polyvinyl chloride. These are easily formed during the oxidation of methylene groups, which are conjugated with double bonds. The effective constant of the elimination rate for HCl, k_{prob}, from the probability theory (random scission) is several orders lower than the rate constant of HCl elimination from monomer units adjacent to the double bond or the polyene sequence ($k_{pol.}$) [52]:

$$k_{prob} = 2.5 \times 10^4 \exp(-21\,500/RT) = 8 \times 10^{-8} sec^{-1} \quad (175°C)$$
$$k_{pol.} = 8 \times 10^{15} \exp(-35\,000/RT) = 7.5 \times 10^{-3} sec^{-1} \quad (175°C)$$

An analogous situation was discovered earlier [21] in the case of intramolecular elimination of acetic acid from polyvinyl acetate:

$$k_{prob} = 1.4 \times 10^{10} \exp[(-33\,200 \pm 2\,000)/RT] = 4.6 \times 10^{-6} sec^{-1} \quad (224°C)$$
$$k_{pol.} = 3 \times 10^{9} \exp[(-20\,400 \pm 2\,000)/RT] = 3.76 \times 10^{-1} sec^{-1} \quad (224°C)$$

If hydrogen chloride does not have enough time to disperse from the reaction zone, then an autocatalytic dehydrochlorination of the polymer occurs [53]. The effective activation energy of dehydrochlorination of chlorocontaining polymers is equal to 28-35 kcal/mole. Dehydrochlorination of polyvinyl chloride is accompanied by the formation of small quantities of benzene ($E_{ef.}$ = 32 kcal/mole), which occurs according to a radical mechanism.

A study of the pyrolysis of deuterated and regular polyvinyl chloride mixtures indicated that the formation of benzene and other aromatic hydrocarbons occurs chiefly as a result of intramolecular cyclisation of polyene fragments with disruption of the carbon-carbon bonds of the chain [54]. Radical reactions involving carbon-carbon bond destruction are accelerated when the temperature is increased to 350-400°C. The intramolecular and intermolecular reactions of polyene fragments and aromatic products were also observed in this temperature range [55]. The non-volatile by-products of polyvinyl chloride pyrolysis melt at 400°C and dissolve in organic solvents. Different alkylaromatic compounds, unsaturated hydrocarbons and condensed aromatic substances were discovered in the volatile products [56].

When compared to polyvinyl chloride, the pyrolysis of polyvinylidene chloride is characterised by a substantially higher enthalpy change value for the second stage of transformation of the polyene structures (18 and ~122 cal/g respectively [22]). This is due to the fact that in the latter case structural reactions and new bond formation reactions prevail to a greater extent over carbon bond scission reactions.

The formation of a carbon skeleton structure occurs at this stage of chlorocontaining polymer pyrolysis. During pyrolysis polyvinylidene-chloride forms an infusible and insoluble by-product. When carbonised products of polyvinyl chloride and polyvinylidenechloride undergo high-temperature treatment, carbon materials with graphite and amorphous carbon structures are formed. The ability of carbonised products to undergo graphitisation when heated up to high temperatures (2000-3000°C) at atmospheric pressure is linked to the tendency of carbon-containing organic substances to display so-called mesophase formation during carbonisation in the 400-600°C range. The mesophase state of a substance (likewise a liquid-crystal state) correlates with the ordering and inter-orientation of molecules in the polycyclic aromatic structure [57].

Very high temperatures can develop on the surface of carbonising polymers during combustion. According to the data [58], the temperature on the surface of a burning carbonised phenolic polymer sample, for instance, is 1500°C, which means that processes of carbon structure formation are sufficiently complete. It is known that the latter has a very considerable influence on the reactivity and rate of reaction between the carbonised product and the gaseous oxidiser [59].

Polyacrylonitrile can serve as an example of how the development of a conjugated bond system occurs, not as a result of side substituent detachment, but as a result of side substituent interaction and intramolecular cyclisation. Cyclisation of polyacrylonitrile is initiated by the formation of methyleneimino groups. This formation is the result of an intermolecular transfer of tertiary hydrogen atoms to the nitrogen of a nitrile group. During the accumulation of sections with conjugated C=N bonds, the effective activation energy of the polyacrylonitrile cyclisation process decreases from 61 to 5 kcal/mole [60].

Cyclisation can be initiated by electrophilic centres in the polymer. Acetonitrile, acrylonitrile, propylene, ethylene, hydrogen cyanide, ammonia, and other compounds have been discovered among the volatile products of polyacrylonitrile degradation. However, as the degree of C=N coupling increases, the composition of the degradation products changes. Hydrogen cyanide is a principal degradation product of polyacrylonitrile that has undergone preliminary thermal treatment at 200°C in an inert atmosphere. The formation of hydrogen cyanide is caused by the elimination of nitrile groups that did not react during the cyclisation:

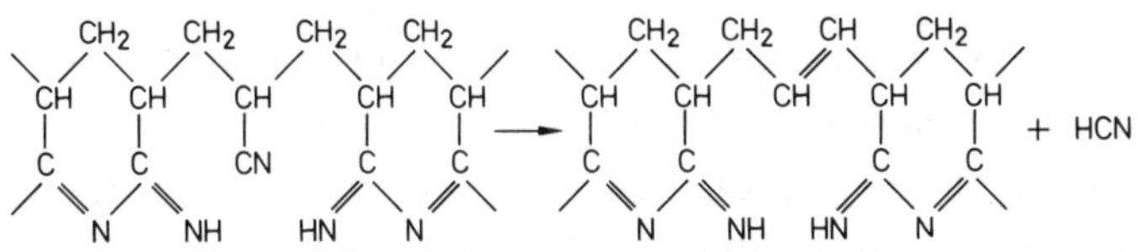

According to Hay [61], ammonia is liberated either as a result of an aromatisation reaction of end groups in "ladder" polymers or as a result of more thermodynamically favoured interactions between neighbouring amino groups:

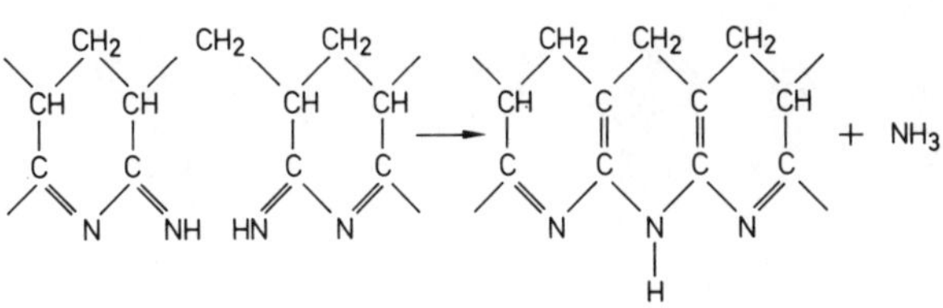

After a polymer has been heated to 500°C, dehydration of the polymer and the formation of conjugated nitrogen-containing polycyclic fragments occurs. Liberation of ammonia is observed at temperatures higher than 600°C.

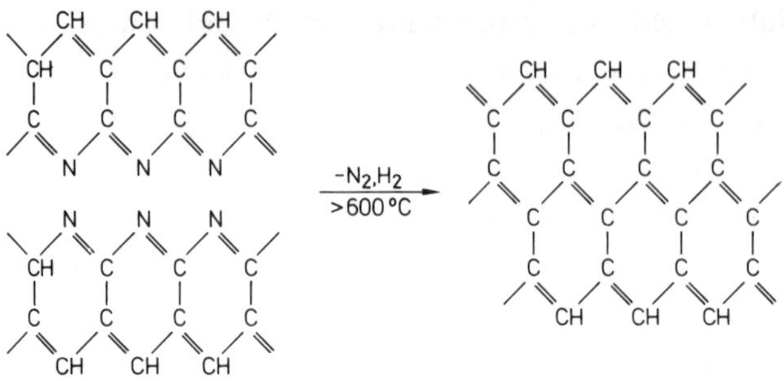

Preliminary crosslinking of polyacrylonitrile molecules favours the intramolecular cyclisation reaction and the probability of carbon-carbon bond scission in the principal macromolecule chain, and hence a decrease in the yield of combustible volatile degradation products.

In practice, this method is used for obtaining high-molecular-weight and highly stable carbon polyacrylonitrile fibres through weak preliminary oxidation of the polymer. The influence of oxidation on the nature of chemical transformations in the 300-1000°C range has been studied in detail by Watt and Green [62].

Intermolecular cyclisation as a result of the reaction of neighbouring groups has been studied on the basis of statistical theory [63]. The theory predicts that up to 18% of the functional groups do not participate in the cyclisation reaction. Experimental proof of this computation has been obtained not only for polyacrylonitrile but also for polyvinyl methyl ketone. During the pyrolysis of the latter, water is eliminated and ladder-structured fragments with conjugated C=C bonds are formed.

Cellulose and its derivatives constitute a subject of great interest among the heterochain polymers of the first group. Cellulose is a polysaccharide made up of β-glucoside sections:

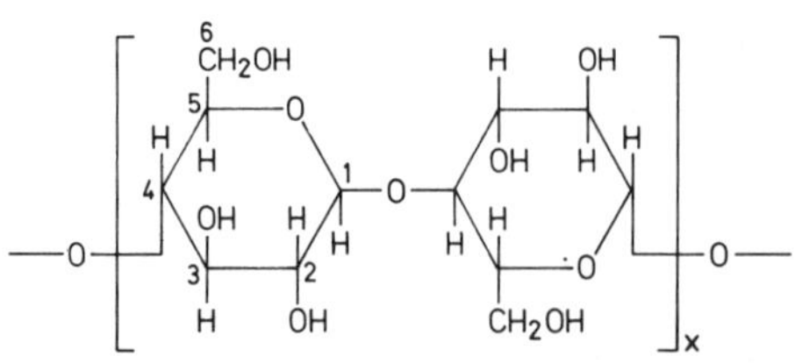

The presence of hydroxyl groups causes a strong intermolecular interaction that is due to hydrogen bonds. Hydroxyl groups in cellulose are very reactive and their reaction with different chemical reagents is widely used in practice for the purpose of obtaining various modified materials, including materials with reduced flammability.

One of the main decomposition products of cellulose thermal degradation is levoglucosan; the gaseous products developed include H_2O, CO, CO_2, H_2 and saturated and unsaturated hydrocarbons, and in the tarry residues are found alcohols, ketones and aldehydes [17, 64]. All these products readily ignite and burn, causing increased combustibility of cellulose materials. The problem of reducing the flammability of cellulose materials is linked to the type of chemical process that occurs during the pyrolysis of polymers. In the case of pyrolysis of cellulose producing intramolecular degradation, the reactions of the disrupted bonds of the principal chain are suppressed. The ring degradation reactions with liberation of combustible chain fragments are also quenched and a sharp reduction is observed in the flammability of cellulose materials.

Primary alcohol groups have a pronounced effect on the stability of the cellulose ring. Oxidation of the ring leads to rapid destruction of the polymer. Transformation of this group into an ether group or blockage of the group in any other way causes a decrease in levoglucosan yield and furthers the intramolecular degradation of the glucoside segments [65]. Thermostable high-molecular-weight compounds with sections of two or three conjugated bonds were obtained through a slow dehydration of cellulose, for example [66].

Tang and Bacon [67] assume that a carbon microstructure is generated during pyrolysis of cellulose in the 300-500°C range. During that period there is dehydration and homolytic disruption of the least stable C-O-C and C-C bonds inside the ring. Subsequent recombination of the short-lived free radicals, which represent four carbon atom fragments, is regarded as a "polymerisation" reaction of these construction "bricks". The yield of carbonised residue from the four carbon atoms inside the glucose section must amount to 29.6% of the molecular mass of the section. In practice, this magnitude varies in a 28-38% range.

A direct correlation exists between cellulose macromolecule orientation and the preferred orientation of the base carbon layers. Two schemes of "polymerisation" of C_4 remainders in glucoside cycles are shown in Fig. 3.4.

During a reaction in the longitudinal direction, the C_4 residues line up in parallel with the macromolecular axis, recombine with the formation of

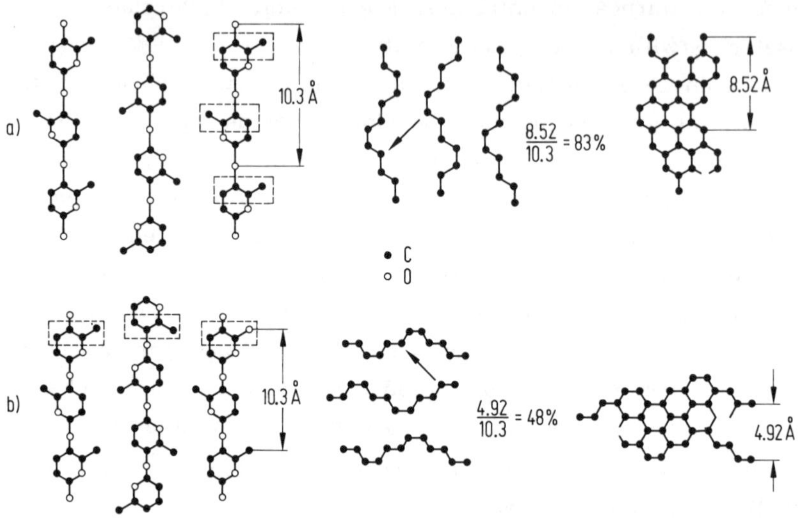

Fig. 3.4 Diagram of graphite-like layer formation during (a) longitudinal and (b) cross-sectional polymerisation of the C_4 remainders of the glucoside cycle.

polymer chains and, by reacting with similar neighbouring chains, form hexagonal layers of char. Longitudinal shrinkage during high-temperature treatment must be approximately 17% in this instance. During the "polymerisation" in a cross-sectional direction, shrinkage must be more than ~52%. Thus experimental results indicate that a longitudinal "polymerisation" process is preferable.

With the help of X-ray diffraction analysis of carbonised cellulose products it was discovered that the polymeric char crystallites are identical in size to the crystallites of the original cellulose.

It should be mentioned that the morphological structure of the carbonised residue during pyrolysis of polymers is generally a function not only of the original substrate morphology but also of the conditions of the process itself (mass and heat transfer during pyrolysis).

When cellulose-based char materials are obtained, the char yield can be increased by treating the original substrates with compounds that prevent oxidation of the cellulose and play the role of catalytic dehydration agents (a mixture of borax and boric acid, diammonium phosphate, different salts etc.). As a rule, such treatment also reduces the flammability of cellulose materials.

The second group of polymers that are readily carbonised during pyrolysis is characterised by the presence of aromatic carbocyclic and/or heterocyclic segments in the principal macromolecular chain. Polymers of this group are usually represented by rigid-chain, relatively highly thermostable, heat-resistant substances. The thermal stability and the nature of the chemical reactions that occur during pyrolysis of these polymers are a function of the nature of the substituents and the cross-linkage groups situated between the carbocycles and heterocycles. Carbo-chain polymers that can be included in this group are: polyxylene, polybenzyl and polycarboarylenes (polyphenylene, polynaphthalene etc.) and crosslinked phenol-formaldehyde polymers. Heterochain polymers in this group are aromatic simple and complex polyethers, polyamides, polyurethanes, polyazophenylenes, polyimides and other polyheteroarylenes.

Despite the wide variety in this group of polymers, they display major common mechanisms of pyrolysis, and it is possible to deduce the interdependence of thermal properties and chemical composition:

1. The thermal stability of polymers and the yield of carbonised residue during pyrolysis increases as the relative percentage of aromatic groups per ring inside the principal macromolecule chain rises.
2. When the degree of heterocycle aromatisation increases, the thermal stability of heterocycle polymers also increases.
3. The least stable bonds of the crosslinkage groups which connect aromatic nuclei or heterocycles are the first to break during pyrolysis.
4. Substituents of aromatic nuclei and heterocycles that participate in intermolecular crosslinking reactions or in intramolecular cyclisation reduce the rate of liberation of volatiles and increase the yield of char residue. Otherwise, the contribution of substituent groups to char formation is small. Detachment of substituents and their removal in the form of volatile products is observed. The thermal stability of polymers decreases at the same time.
5. The thermal stability of the aromatic heterochain and heterocycle polymers depends to a large extent on the stability of the hetero-atomic bonds in the hydrolytic ionic reactions.
6. When the number of mobile hydrogen atoms capable of participating in chain transfer and disproportionation reactions in a macromolecular section decreases, the thermal stability of the polymer increases.

7. Isomerism of aromatic rings influences the thermal stability and the tendency of polymers to form char residue during pyrolysis. Meta-isomers are less stable and characterised by smaller char numbers than para-isomers. Isomerism of heterocycle polymers affects their hydrolytic stability in the same manner.
8. Polymers with ladder and three-dimensional network structures are more thermally stable and are characterised by larger char numbers than their linear analogues.
9. Heteroatoms with unpaired *p* electrons (N, O, S) are partially removed from the carbonised products of polymer pyrolysis at temperatures above 600°C. This removal of heteroatoms from a carbon skeleton can occur during heating in the 1300-1700°C range. In general, the yield of char residue during high-temperature treatment is greater for carbocyclic aromatic polymers than for heterocyclic ones.

The transformation of polymers into carbonised products is a multi-stage process. Pyrolysis of polymers at temperatures of 600-700°C occurs as a result of the disruption of the least stable bonds in the principal chain (or bonds in the heterocycles). A larger quantity of volatile products is liberated at that time. A carbon-rich, non-volatile residue can either be fusible (meltable) and dissolve in organic solvents or infusible and insoluble. In the latter case this means that crosslinking processes of fragments are energetically favoured.

Subsequent heating of the nonvolatile residue of pyrolysis formed during the first stage leads to a further aromatisation as a result of recombination and dehydrocondensation. Low-molecular gaseous compounds are generally part of the volatile products of high-temperature pyrolysis stages. Normally an analogous picture is observed during the pyrolysis of first group polymers. The difference is that second-group polymers already contain completed aromatic units for the formation of condensed hexagonal carbon layers. In first-group polymers, aromatic segments are formed as a result of different reactions of the side substituents.

We shall now discuss the pyrolysis of a number of polymers containing aromatic and heterocyclic groups in the macromolecular chain. The order and extent of examination is arbitrary and not related to the practical significance of the polymers or to any of their specific thermal properties or combustion characteristics. The examples given illustrate the general principles mentioned above for the pyrolysis of second-group polymers.

Phenol-Formaldehyde Polymers. A large amount of work has been dedicated to the study of the structure of phenol-formaldehyde oligomers, their setting processes and their pyrolysis. By using the same initial reagents in different ratios and different catalysts, it is possible to obtain products that differ in their percentage of functional groups, their isomeric composition, their degree of branching and their reactivity [68]. The direction of the pyrolysis reactions and the properties of the char products formed depend to a certain degree on the initial polymeric structure [69].

Intensive research has been conducted in the past into the kinetics and mechanism of phenol-formaldehyde polymer pyrolysis in connection with the development of heat-resistant and char materials with a variety of valuable properties. Different methods have been used to evaluate the kinetic parameters of the destructive processes for phenol-formaldehyde polymers. Values were obtained for the effective activation energy from the results of measurements of the liberation rate of volatile products under isothermal and dynamic conditions during polymer pyrolysis. The values fluctuate in the 60-70 kcal/mole range [17, 70-73]. These differences in the experimentally obtained $E_{ef.}$ are due to discrepancies in averaging the parameters of complex processes for different temperature ranges.

When polymers were heated to 1000°C in an inert atmosphere with a constant heating rate of 12 degrees/min, one author of this book discovered that the whole pyrolysis process of phenol-formaldehyde polymers can be accurately subdivided into four stages in the following temperature ranges: lower than 180°C, 180-360°C, 360-740°C and above 740°C. In the low-temperature zones of polymer heating (lower than 360°C) the formation of adsorped moisture, carbon monoxide, carbon dioxide and also water, due to the partial dehydration of the polymer, is observed.

The formation of CO and CO_2 is linked to the destruction of the oxygen-containing groupings, which emerge as a result of the oxidation of resits (C-stage resins) in the process of their production and through their exposure to air. The percentage of CO and CO_2 increases as the thermo-oxidative stability of the polymers falls. The results obtained tally with the radical mechanism of phenol-formaldehyde polymer oxidative destruction, which was proposed by Conley and coauthors [74, 75].

The most complicated process in terms of the nature of reactions taking place is pyrolysis in the 360-740°C range. At this stage an intensive bond disruption occurs in the principal polymer chain, which leads to the formation of a large number of low-molecular-weight products. These products condense at room temperature. Also, gaseous substances are formed at this stage. Among these products, phenol, toluene, cresol, xylenols, saturated and unsaturated aliphatic hydrocarbons, carbon monoxide and carbon dioxide and hydrogen have been discovered. This process is accompanied by water formation. Gaseous products are partially formed as a result of aromatic nuclei destruction. The effective activation energy for a gross process of volatile product liberation in the 360-460°C range is 39.0 kcal/mole and in the 460-740°C range, 45.0 kcal/mole (by the TGA method).

The effective activation energy for the gross process of liberation of gaseous volatile polymer decomposition products in the 460-740°C range is 26 kcal/mole (manometric analysis method). According to mass-spectroscopic data, liberation of methane and carbon monoxide in the same temperature interval occurs with an $E_{ef.}$ of 36.5 and 31.5 kcal/mole respectively.

Similar characteristics of the formation sequence of phenol-formaldehyde polymer pyrolysis products (Fig. 3.5) have been determined by Ouchi and Honda [76]. Shulman and Lochte [73] determined the rate and activation energy for the formation of different polymer degradation products using

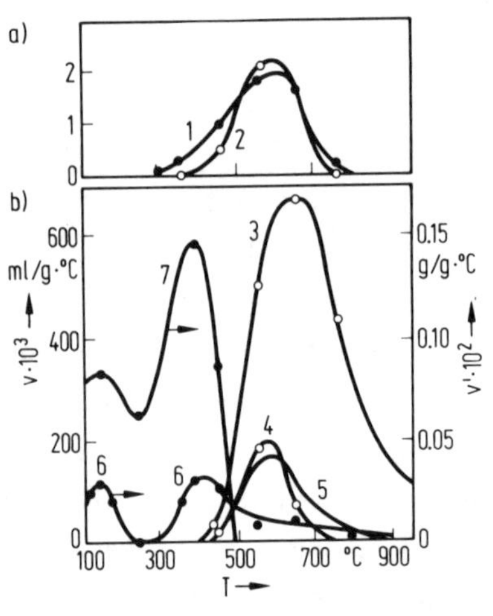

Fig. 3.5 Relationship between the rate of formation of phenol-formaldehyde polymer pyrolysis products and temperature (a) = (1) CO_2; (2) C_2H_6; (b) = (3) H_3; (4) CH_4; (5) CO; (6) H_2O, and low molecular weight products (7). Ratio Ph/form = 1, catalyst = NH_4OH.

mass spectroscopy. It is interesting that the formation of phenol products occurs with high activation energies (75 ± 25 kcal/mole) and that the liberation of carbon monoxide in the 400-600°C and 600-750°C ranges occurs with $E_{ef.}$ = 28 and 56 kcal/mole respectively. This latter number, however, corresponds to a different mechanism of carbon monoxide formation from polymeric fragments.

The average magnitudes of gross process $E_{ef.}$ were computed as a function of pyrolysis temperature with consideration to the rate of degradation product formation at a given temperature, the molar concentration of the volatile products and the formation activation energy. When the temperature rises from 300°C to 400°C the activation energy increases from 16 to 26 kcal/mole and, in the 450-725°C interval, $E_{ef.}$ fluctuates in the 34-48 kcal/mole range.

An analysis of pyrolysis products leads to the conclusion that, along with the homolytic disruption of the carbon-carbon bonds in the macromolecule and the formation of phenol and its derivatives and low-molecular-weight fragments, there are also intramolecular and intermolecular dehydration reactions taking place. Three types of dehydration reaction are possible, resulting in the formation of a nonvolatile residue of condensed hexagonal units inside the structure:

1. Phenolic hydroxyl groups interact and form xanthene chains. This tendency is favourable when *o,o'*-isomers are present in a polymer. Detachment of a hydrogen atom from a CH_2 group in a xanthene section and recombination of xanthyl radicals with subsequent detachment of oxygen and dehydration leads to the formation of aromatic, rubicene-type fragments.

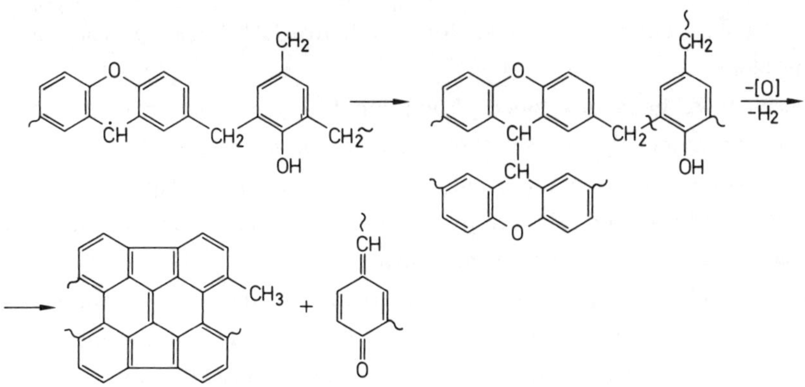

2. Intramolecular reaction between hydroxyl groups and hydrogen atoms from neighbouring aromatic nuclei. This reaction leads to the formation of fluorenic rings. This tendency must predominate during the transformation of fragments with *o*- and *n*'-isomeric units. Dehydrocondensation of the fluorenic rings also causes initiation and development of hexagonal carbon layers.
3. Intermolecular reactions occur between hydroxyl and methylene groups. In this case, triphenylmethane-type segments are formed. Subsequent dehydration and dehydrocondensation provide conditions for the origination and development of hexagonal carbon layers:

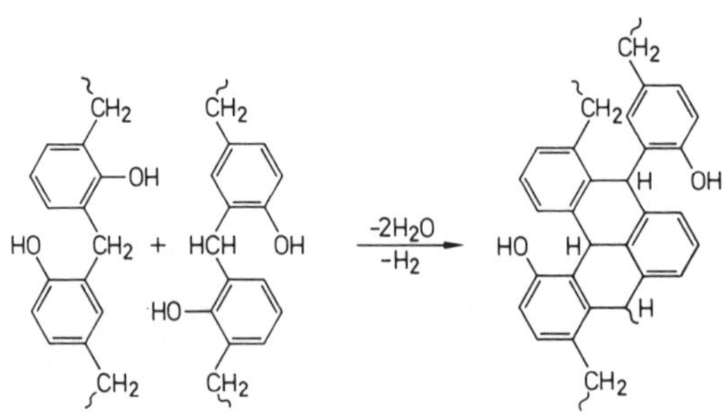

The formation of carbon is observed at temperatures above 400°C and especially between 700°C and 800°C. The aromatisation of nonvolatile residues becomes more pronounced in this temperature range.

The carbon structure of phenol-formaldehyde polymer chars formed at this stage of pyrolysis is generally disordered. There are only small regions of orderliness inside the hexagonal aromatic carbon layers. The sizes of the crystallites inside a pyrolysis product of phenol-formaldehyde polymer at 900°C, for instance, are: L_a = 31-70 Å, L_c = 17-29 Å, d_{002} = 3.605-3.704 Å. The char number for a bakelite "A"-based polymer at 900°C is 55%.

Schemes have been developed to explain the participation of a large number of methylene groups in the synthesis of a carbon skeleton [77]. Hydroxyl groups increase the yield of char by participating in the cross-linking of polymers during their pyrolysis. Arenoformaldehyde analogues have smaller magnitudes of char numbers.

The thermochemical yield of the char residue during the pyrolysis of polymers, Y_c, is shown in reference [78] as a function of the chemical structure. Y_c increases with a rising number of aromatic equivalents (hydrogen atoms) in the polymer's structural segment. In turn, the relationship between char residue and the amount of polymer degradation products affects the ignition and combustibility of the latter. Winkler and Parker [79] derived a formula for the determination of the thermochemical yield of the char residue:

$$Y_c = 72{,}06\left[\frac{N}{M} - \left(\frac{N}{M} - \frac{N - P_0}{M'}\right)\left(1 - \frac{H}{S_0}\right)^2\right], \qquad (3.3)$$

where

N total number of phenolic nuclei in the original novolac-type oligomers

M molecular mass of a middle structural segment

$M' = 94N + 12(N - 1)$ average molecular mass of the oligomer

P_0 number of nuclei in branched groups per oligomer molecule (generally corresponds to number of branches)

H functionality of crosslinking agent (for a hexamethylenetetramine, $H = 12$)

S_0 total functionality (number of active centres) in the original oligomer

Oligomers containing N number of trifunctional phenol nuclei and $N - 1$ number of conjugative methylene groups have the same number of active centres $S_0 = 3N - 2(N - 1)$; the term $[1-(H/S_0)]^2$ is equal to the quantity of sol-fraction in the hardened (set) polymer. Therefore, knowing the structural parameters of the phenol-formaldehyde polymer (from experiments) it is possible to predict the thermochemical yield of the char residue during pyrolysis.

A statistical approach to the synthesis and hardening (setting) of a novolac oligomer using hexamethylenetetramine allowed the authors to evaluate the essential structural parameters of the crosslinked polymer and to calculate the char yield during pyrolysis (Fig. 3.6). Figure 3.6 shows how the degree of novolac oligomer hardening (average numbers $N = 7.58$ and $P_0 = 2.74$) affects the char yield.

Equation 3.3 is based on the assumption that during pyrolysis of a polymer the branching groups do not participate in the formation of the carbon skeleton. A statistical method has been used for calculation of the average number of phenolic nuclei in novolac oligomers [80]. These calculations were based only on synthesis data (correlated for phenol and formaldehyde and the amount of reacted formaldehyde). Theory predicts that the maximum yield of char residue, or the char number (CN), must

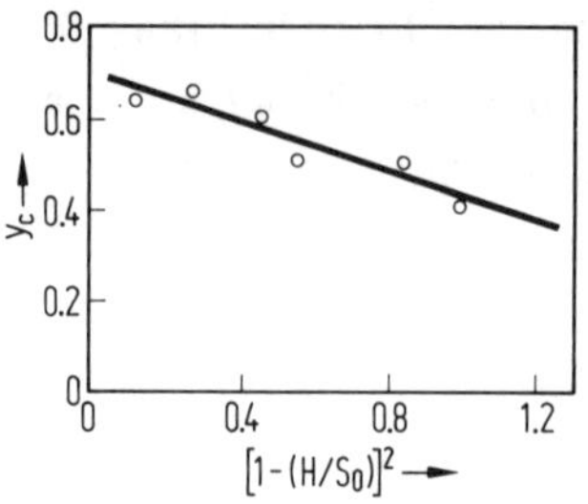

Fig. 3.6 Effect of degree of reinforcement of phenol-formaldehyde oligomer on char yield (Y_C) at 800°C (in the coordinates of equation 3.3).

be consistent with an equimolecular phenol and formaldehyde ratio [81]. However, in practice, a maximum CH (~67.9%) is attained at a ratio of 1:(1.1-1.3).

Polyethers. Before discussing the pyrolysis of aromatic complex polyethers we shall discuss the polyalkyleneterephthalates, which occupy an intermediate position between aliphatic and aromatic polymers.

The most detailed study of this class of polymer has been conducted for polyethyleneterephthalate (PET). Pyrolysis of PET, like that of aliphatic analogues, starts from a disruption of the weakest carbon-oxygen bonds inside the ether groups. It is assumed that this disruption occurs according to the law of probability (random scission) and with a molecular mechanism through a transitional state involving a six-membered ring [82, 83]:

$$\sim C_6H_4C\begin{matrix}\nearrow O\cdots H\searrow \\ \searrow O-CH_2\nearrow\end{matrix}CH-OCOC_6H_4\sim \rightarrow \sim C_6H_4COOH + CH_2{=}CHOCOC_6H_4\sim .$$

Subsequent transformation reactions of the formed intermediates lead to the emergence of a large number of degradation products. Among them were found terephthalic and benzoic acids and their anhydrides, benzene, acetaldehyde, small quantities of ethylene, methane, carbon monoxide and carbon dioxide, and water. The effective activation energy of PET pyrolysis, calculated according to the formation of individual products and the disappearance of functional groups in the condensed phase, fluctuates in a 16-58.7 kcal/mole range. The upper limit for $E_{ef.}$ draws near to the value of the dissociation energy of the C-O bond in the complex ether group (60 kcal/mole). According to Ritchie [51], the formation of acetyl-dehyde also occurs according to a molecular mechanism. It is associated with the so-called disproportionation reaction between vinylbenzoate and acidic fragments:

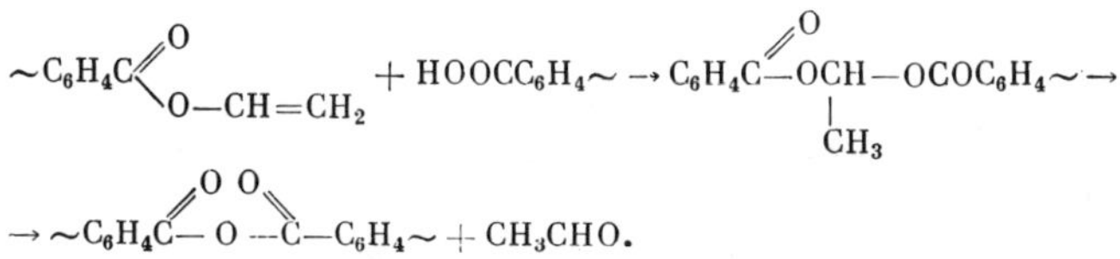

The course of this reaction with a quantitative yield of benzoic acid anhydride [51] was clearly shown by using the reaction between vinylbenzoate and benzoic acid as a model.

In the same way, the disruption [84] of bonds of complex-ether groups is induced. This is associated with the formation of secondary radicals when a hydrogen atom breaks away from the methylene groups of a glycol constituent. Secondary radicals disintegrate according to the scheme:

$$\sim C_6H_4COO\dot{C}H-CH_2COOC_6H_4\sim \rightarrow \sim C_6H_4COOCH{=}CH_2 + \\ + \dot{O}COC_6H_4\sim .$$

The vinylbenzoate fragment also disintegrates according to the radical mechanism:

$$\sim C_6H_4COOCH{=}CH_2 \rightarrow \sim C_6H_4\dot{C}{=}O + O{=}CH-CH_2.$$

Radicals disappear during a chain transfer on to methylene groups. At this time carboxyls, benzaldehyde groups and acetaldehydes are formed. Another scheme for the radical process of PET destruction was analysed by Kovarskaya [85]. This scheme proposed homolytic disruption of C-O and C-C bonds in the glycol component. In fact, Delbourgo and coauthors [86] hold to this scheme, describing the formation of many volatile products and enrichment of the residue structure by aromatic nuclei:

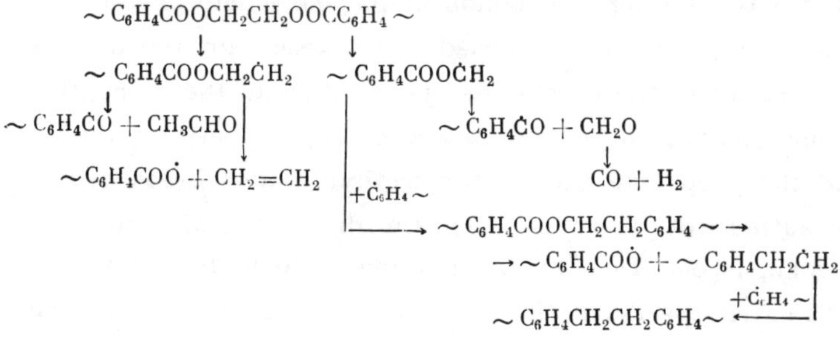

It appears that the accumulation of aromatic nuclei in the residue can also occur as a result of a direct attack of the phenyl radicals on the terephthalate groups of a polymer:

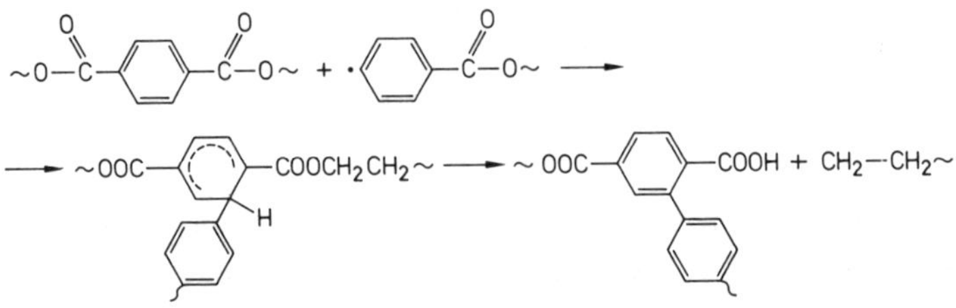

During the high temperatures of pyrolysis in combustion conditions (when PET burns in air the surface temperature reaches 430-480°C) the decomposition of PET is carried out according to a complex mechanism in which molecular reactions compete with radical ones.

During pyrolysis of aromatic complex polyethers (polycarbonates, polyacrylates) at 250-500°C, the disruption of the principal chain bonds is accompanied by intensive crosslinking. For instance, when a polycarbonate is heated at a temperature close to 300°C this leads to the formation of up to 90% crosslinked insoluble product with a mass loss of 10-20% [41]. The higher the degree of polymer crosslinking, the less the losses of volatile substances observed during any subsequent heating. The residue is increasingly enriched by aromatic nuclei, and the carbon percentage also increases sharply.

A survey has been conducted of the early works on the thermal decomposition of different aromatic complex polyethers [87]. The majority of works show that disruption, primarily of C-O bonds, occurs in complex-ether groups during the pyrolysis of such polymers. Different admixtures or additions of an ionic nature speed up the destruction process.

Lee [88] analysed the strength of bonds in polycarbonate on the basis of 4,4'-dioxydiphenylpropane. He proposed two schemes for the destruction process. According to these schemes, in addition to the disruption of C-O bonds in ether groups, carbon bonds are disrupted between the aromatic ring and the propylene group, and methyl groups break off. According to the authors of reference [89], the disruption of carbon bonds in the principal chain is catalysed by protonating compounds. It was discovered [90, 91] that the disruption of carbon bonds in the principal chain can be induced to a small degree by the addition of radicals to a phenylene ring:

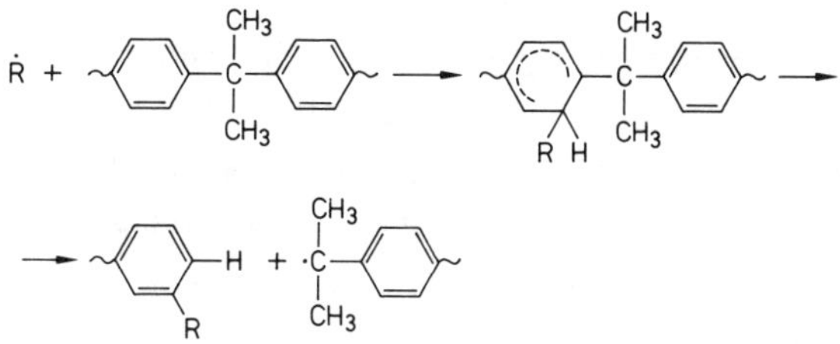

or to the carbon of a carbonate group:

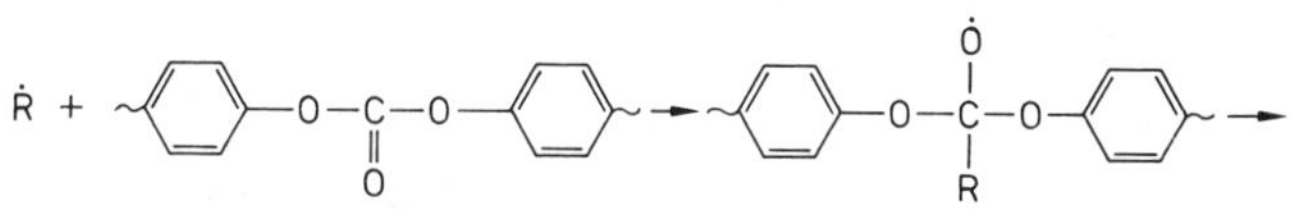

with subsequent breaking up of the formed macroradical

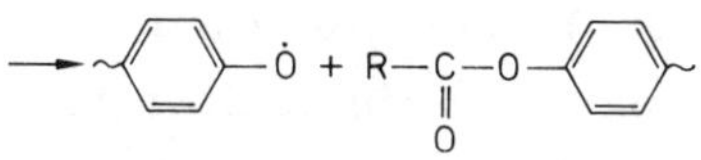

Isopropenylphenol, isopropylphenol, cresol and large amounts of methane and carbon dioxide have been found in the products of the thermal decomposition of polycarbonates based on 4,4'-dioxdiphenylpropane at 400°C.

Aromatic Polyamides. When compared with aliphatic polymers the aromatic polyamides have higher thermal stability and lower combustibility. High-modulus thermostability fibres obtained from poly-*m*-phenyleneisophthalamide (called Nomex in the US and Phenylon in the USSR), for instance, have an OI (Oxygen Index) of above 27%. They withstand the effect of a high-temperature flame for short periods and become carbonised.

When the pyrolysis of aromatic polyamides was examined, it was discovered that the thermal stability of polymers is essentially affected not only by the isomerism of the substituents' position inside the original diamines and dicarboxylic acids but also by the nature of the crosslinkage groups between the aromatic nuclei in the amino and acidic constituents [92].

The thermal stability of polymers changes systematically with an enhancement of the π-conjugation between phenylene rings. Polymers with para-isomer groups are more thermostable than ones with meta-isomer fragments. However, the latter have higher thermal oxidative stability. It is interesting that isomerism of substituents in the amino components shows up to a greater extent on the hydrolytic stability of aromatic polyamides than the isomerism in the acid component. The rate of hydrolytic destruction for aromatic polyamides diminishes when a total positive charge of nitrogen and hydrogen atoms in the amide group decreases, which is why the substituents that increase the electron density of a carbon atom (decrease $\delta+$) promote the increase of polymer stability.

Intensive degradation of aromatic polyamides usually occurs at heating points situated above 330-370°C. At a lower temperature a change in molecular weight has been observed (an increase, then a decrease in the intrinsic viscosity of a polymer) and water and carbon dioxide are formed. This indicates a competing condensation reaction of functional groups and disruption of the principal chain bonds and also shows that there is a decarboxylation of the end acid groups.

It is assumed that at high temperatures (>330°C) there are reactions combined with homolytic disruption of the principal chain, in addition to hydrolytic reactions with amino group disruption. Benzene, benzonitrile, benzoic acid, ammonia, hydrogen cyanide, CO, H_2 and others have been discovered among the volatile products of aromatic polyamide thermal destruction [93-94]. Conclusions based on the results of quantum-chemical calculations of bond stability in aromatic polyamides indicate that a homolytic disruption of the Ph-N bond occurs during pyrolysis (the energy of the Ph-N bond is ~77 kcal/mole; the energy of N-C and C-Ph bonds is ~83 and 97 kcal/mole respectively [95]. This explains the higher thermal stability and resistance to flame ignition of the aromatic polyamide, which has the structure

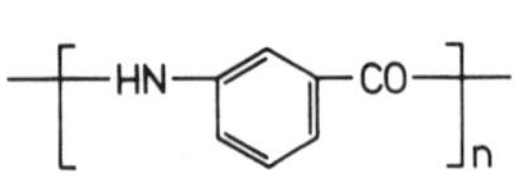

in contrast to poly-*m*-phenyleneisophthalamide:

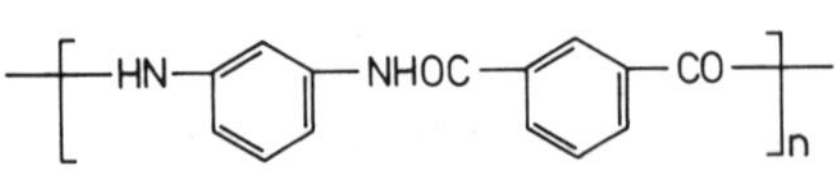

With the increase in pyrolysis temperature the contribution of homolytic disruption becomes more essential. Fragments with carbamide groups are formed as a result of Ph-N bond disruption, and the dehydration of these leads to the formation of nitrile groups.

Homolytic disruption of bonds in the principal macromolecular chain can be induced, apparently as a result of the detachment of relatively loosely held hydrogen atoms of the amido group:

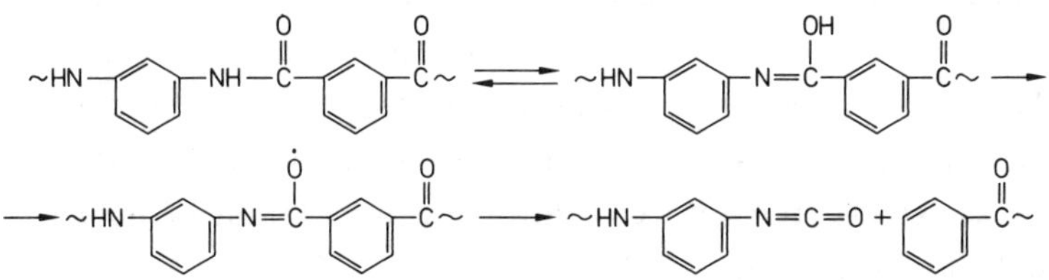

Hydration and carboxylation of isocyanate fragments cause formation of amino groups; the recombination of phenolic radicals causes the accumulation of aromatic nuclei in nonvolatile residues.

Under normal conditions, amido groups in aromatic polyamides are not significantly enolised but, at increased temperatures, the probability of such tautomeric regrouping increases [96]. Tautomeric rearrangement of amido groups can also be a cause for the crosslinking of polymers as a result of dehydration of hydroxyl groups. It is interesting that in the presence of nitrogen oxide at 350°C, the crosslinkage of polymers and the formation of water and carbon dioxide increase [97].

The effective activation energy of thermal destruction of aromatic polyamides varies on account of their structure in a 35-55 kcal/mole range (the $E_{ef.}$ value of the polymer phenyleneisophthalamide at 400-450°C, for instance, is 54 kcal/mole) [97]. The ageing and partial destruction of polyamides further decrease their thermal stability and increase their flammability. Thus it was discovered that the oxygen index for fabric on a base of aromatic polyamide (Nomex) decreased after photochemical ageing from 27.1 to 25% when the samples were placed horizontally and from 23.2 to 20.5% when they were placed at an angle of 45° [98]. In this way, the flammability of a polymer material increases after prolonged ageing. It can burn if exposed to air.

Polyheteroarylenes. The comparative characteristics of the thermal stabilities of different polyheteroarylenes are given in a monograph by

V.V. Korshak [99]. Thermostable polymer materials on a base of polyheteroarylenes (polyamides, polybenzimidazoles, polyoxydiazoles, polyquinoxalines etc.) are now more widely used in industry. Many of them have a unique stability record as far as ignition is concerned and their physical and chemical properties are retained for a long time when the polymers are exposed to chemically active media and high temperatures.

Noticeable decomposition of polyheteroarylenes has been observed at temperatures of above 400-500°C, whereas the role of acyclic fragments of the macromolecular chain is essential in the initial states. Thus it was discovered during a study of the thermal degradation of polyimides that, in the presence of moisture, the rate of hydrolytic destruction of polyamidoacid fragments increases [94, 100].

Generally, the polyamidoacids are present in the solid phase in a dissociative state, which is why they can undergo intramolecular degradation, causing a decrease in the molecular weight of the polymer [101]:

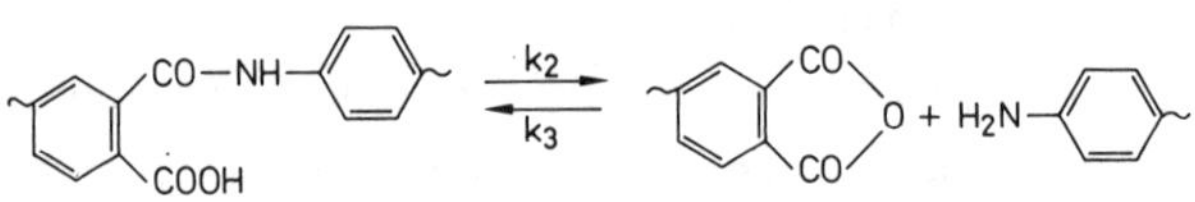

The effective activation energy values for the thermal imidisation of polyamidoacid on the base of aniline fluorene and 3,3'-4,4'-tetracarboxydiphenyl-2,2'-propane (E_1), for intramolecular degradation (E_2) and for the reverse reaction of amidoacid synthesis (E_3) are: $E_1 = 17 \pm 2$ kcal/mole, $E_2 = 22 \pm 2$ kcal/mole, $E_3 = 6 \pm 2$ kcal/mole. As a result, it is expected that the contribution of the intramolecular reaction of acyclic fragment dissociation will increase at high temperatures.

An analysis of the pyrolysis products of polyimides shows the homolytic disruption of bonds inside an imido cycle, accompanied by an intensive crosslinking process [94, 102]. Disintegration of the imido cycle leads to the formation of carbon monoxide and carbon dioxide. It is presumed that tautomeric regrouping from imido bonds into isoimido bonds plays an important role in the formation of CO_2 [103]; the effective activation energy for the liberation of carbon dioxide reaches 30 kcal/mole and that of carbon monoxide approaches 57-65 kcal/mole. The $E_{ef.}$ values change very little during heating in the presence of oxygen. During pyrolysis of polyimides the destruction also affects the aromatic cycles, a condition that is particularly obvious when isotope analysis is applied [100]. However, the process of crosslinking and enrichment of nonvolatile residue by aromatic nuclei prevails, which is why polyheteroarylenes as a rule have high char numbers. The thermal stability of aromatic polyheteroarylenes and their flame resistance essentially depend on the presence of labile hydrogen atoms inside the structural monomer unit of the macromolecules.

It is known that polyimides based on pyrazinetetracarboxylic acid dianhydride and diaminothiodiazole

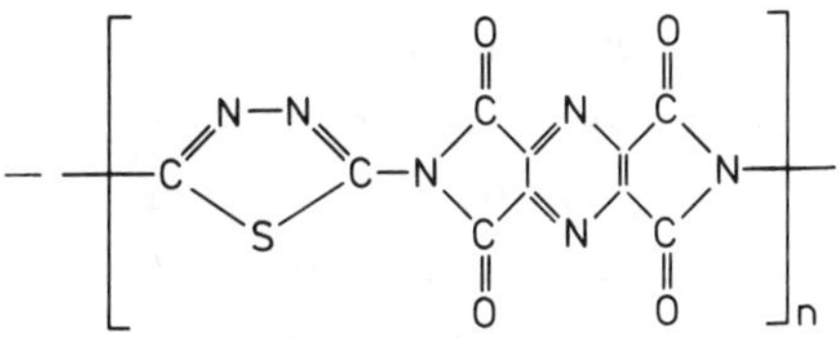

do not burn or carbonise when subjected to a flame and keep their useful properties for a long time when heated in an air atmosphere at a temperature of 600°C. At the same time, their analogues based on pyromellitic acid dianhydride

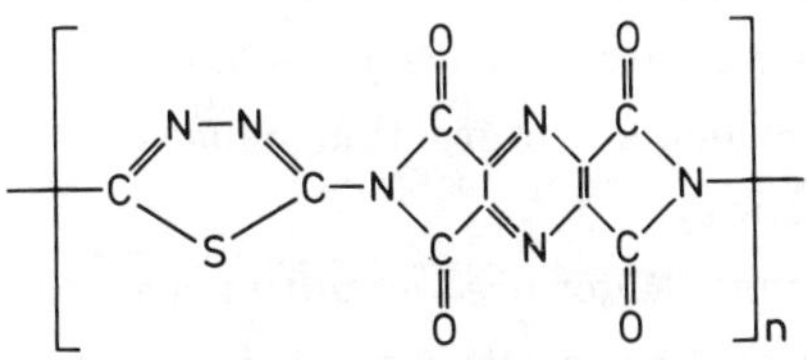

containing 0.67% hydrogen are intensively carbonised when heated above 320°C and completely disintegrate in an air atmosphere at 400°C over a period of 25 hours [104]. These results are in agreement with the results of a study on the thermal oxidative degradation of polypyromellitimide in the presence of oxygen isotopes [105]. It was discovered that oxidation of aromatic nuclei and their destruction at 450°C occurs at an even higher rate than the destruction of imido cycles.

An "aromaticity index" was derived for heterocycles on the basis of a semi-empirical calculation using the method of linear combination of atomic orbitals. A correlation was found between the latter and the thermal stability of polymers containing these cycles [106-107]: with increasing π-conjugation between cycles, the thermal stability of the polymers increases.

Moos and Skinner [108] assume that the atomic ratio H:C in a polymer and the total percentage of hydrogen in the macromolecular structure play a determining role both for the thermal stability and for the combustibility of polymers. According to them, the combustibility of polymers depends to a greater degree on the percentage of hydrogen in their composition than on the percentage of flame-quenching component (P, Cl). Apparently, the correlation discovered in these studies is a reflection of the general principle: the presence of weak bonds, the possibility of radical formation and of chain transfer and the possibility of disproportionation as a result of hydrogen cause a decrease in the thermal stability of polymers and an enrichment of the volatile decomposition products by a fuel component (with an increased hydrogen percentage). Disruption of π-conjugation between aromatic rings and heterocycles and the reduction in intermolecular interaction always has a negative effect on the thermal stability of polymers and the yield of nonvolatile char residue and, as a result, the rates of polymeric combustibility.

Polycarbarylenes that do not contain hinged crosslinks between the rings should, in general, be characterised by larger char number values than polyheteroarylenes. Nevertheless, no high-molecular-weight compounds of this type, suitable for reprocessing into articles have been obtained. Chain rigidity and strong intermolecular interaction causes insolubility and infusibility of these compounds.

During a study of the pyrolysis of benzene oligomers, naphthalene oligomers, anthracene oligomers and their copolymers, it was discovered that a disruption of the chains occurs followed by dehydrocondensation of the rings and the formation of aromatic condensed fragments. In the 500-660°C range, the decomposition has an effective activation energy of 60-80 kcal/mole. The volatile products contain hydrogen, methane and other hydrocarbon gases (C_2, C_3). This shows that aromatic rings are destroyed. The yield of a carbonised residue for insoluble anthracene-based oligomer at 900°C is ~82-85%. With an increase of molecular weight and a decrease in oligomer branching, the yield of nonvolatile residue increases steadily [109]. The nature of the molecular weight distribution of oligoarylenes reflects their microstructure and their physical and chemical properties [110].

The ignitability and combustibility characteristics of polymer materials are closely connected with their thermal stability and with the correlation between the amount of volatile and the amount of char residues during pyrolysis. The structure of the volatile products (the correlation between combustible and noncombustible components) and the physical and chemical properties of the char layer are also very significant.

Fish and Parker [111] studied ignitability, combustibility and the ability to form smoke for a large number of polymer foam materials, with char numbers from 20% to 76% (polyurethanes, polyisocyanurates, polyimides and others). According to these authors, when the thermochemical yield of the char residue increases during pyrolysis of polymers, not only do all of the studied characteristics decrease but the relative toxicity of the gaseous products also decreases.

VanKrevelen [112] made a comparative analysis of the ignitability of different polymers from the first and second groups using carbonised residue yield and the composition of volatile products. This work represents an attempt to answer an important question: it is possible to predict the degree of combustibility of a polymer material on the basis of its chemical structure?

If one considers only pyrolysis products that are not inhibitors of flame reactions, then it appears that a correlation exists between the yield of char residue during pyrolysis and the oxygen index of the polymers (Fig. 3.7). This correlation can be expressed by a simple empirical relation:

$$\mathrm{OI} \cdot 100 = 17{,}5 + 0{,}4 \cdot \mathrm{CN}, \qquad (3.4)$$

where

OI oxygen index of the polymer

CN char number at 850°C

The analysis of a vast number of polymers allowed the important conclusion to be drawn that the inclination towards carbonisation (formation of char during pyrolysis) is an additive property of the polymer. Quantitatively it can be evaluated from the contributions of separate functional groups that build up the polymer section. Each group has its own characteristic contribution to the formation of char residue during the pyrolysis of polymers (Table 3.7).

A group's contribution is expressed by carbon equivalents that each individual group contributes to the formation of char residue per polymer section. In this case, the larger the number of C equivalents, the greater

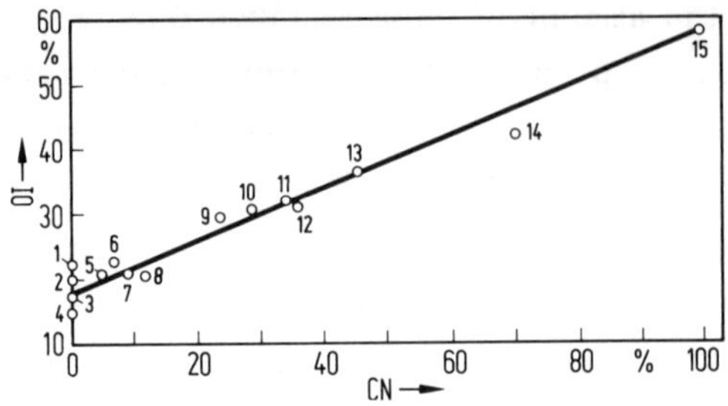

Fig. 3.7 Correlation between the oxygen index (OI) of polymers and yield of char residue during pyrolysis (CN). (1) polyformaldehydes; (2) polyethylene, polypropylene; (3) polystyrene, polyisoprene; (4) polyimide; (5) cellulose; (6) polyvinyl alcohol; (7) PTFE; (8) PAN; (9) polyphenyleneoxide (fire-resistant); (10) polycarbonate; (11) Nomex, fire-resistant; (12) polysulphone; (13) polyphenolformaldehyde; (14) polyimide; (15) carbon.

the inclination towards formation of char of that particular functional group. Because the ability to form a carbonised residue during pyrolysis is an additive property, a polymer is characterised by high-value char numbers. The latter can be computed according to the following formula:

$$CN = \frac{12 \cdot 100 \left[\sum_i (IFC)_i\right]}{M} \% , \qquad (3.5)$$

where

$(IFC)_i$ a group's inclination to form char in carbon equivalents per section of polymer

M molecular weight of a principal polymer section

Using this formula, VanKrevelen [112] calculated the values of the char numbers and the oxygen indexes of different polymers and compared them with experimentally derived ones (Table 3.8). Thus the combustibility of polymers without groups that can act as inhibitors of the gas-phase reactions may be predicted on the basis of their chemical structure.

One of the promising directions in decreasing the combustibility of fuel materials is the control over the pyrolysis process and the shifting of its general course through an increase in the yield of carbonised residue. Later we shall discuss some of the chemical aspects of this problem.

Table 3.7. Group contribution in the formation of polymer char-residue. (IFC = inclination to form char)

Group	IFC in C-equi valents	Group	IFC in C-equi valents
Aliphatic groups		Heterocyclic group	
-CHOH	0.33		
other	0	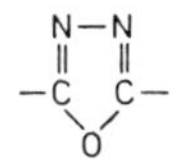	1
Aromatic groups		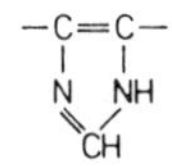	3.5
	1		
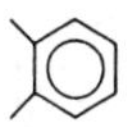	2	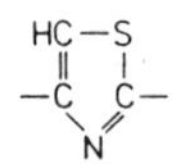	3.5
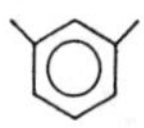	3	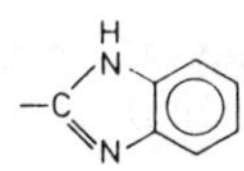	3.5
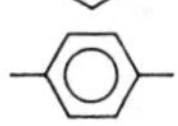	4		
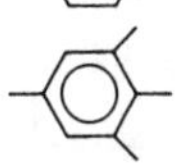	6	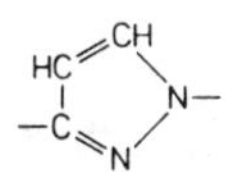	7
	6	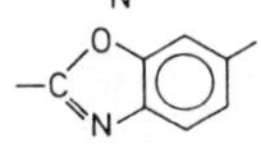	7
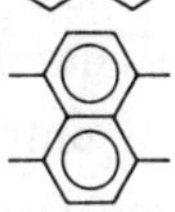	10	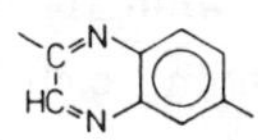	9
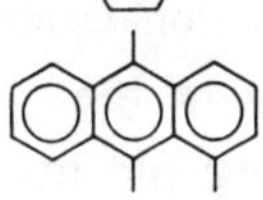	14	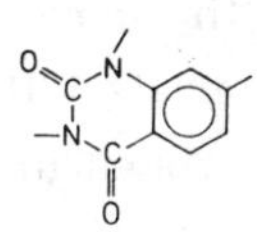	11
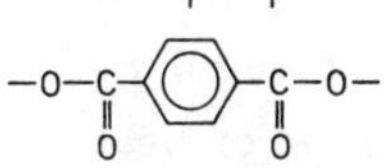	1.25		
		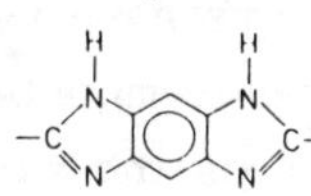	10
Groups directly bonded with aromatic nuclei (disproportionation correction)		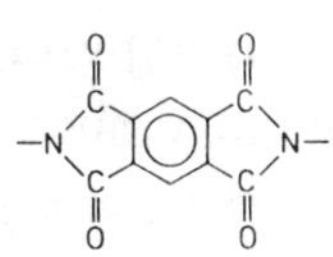	12
$>CH_2$ and $>CHCH_3$	-1	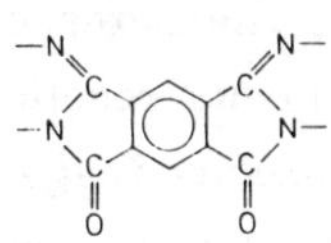	10
$-CH_3$	-1.5		
$>C(CH_3)_2$	-3		
$-CH(CH_3)_2$	-4	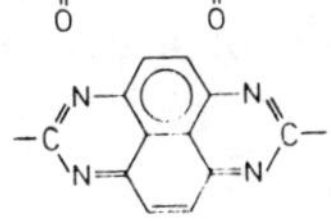	15

Table 3.8 Experimental and calculated values for polymer char numbers (CN) and oxygen indexes (OI)

Polymer	CN	%	OI	%
	Calcul- ation	Exper- iment	Calcul- ation	Exper- iment
Polyoxymethylene	0	0	17.5	15.3
Polymethylmethacrylate	0	0	17.5	17.3
Polyethylene	0	0	17.5	17.4
Polystyrene	0	0	17.5	18.3
Cellulose	5	5	19.5	19.9
Polyethyleneterephthalate	8	8	20.7	20.6
Polycarbonate	24	24	27.5	29.4
Polyamide Nomex (with flame retardant)	30	35	29.5	29.8
Polyphenylene oxide (with flame retardant)	30	28	29.5	30.5
Polyphenol-formaldeyhde resin (with flame retardant)	45	45	35.5	35.5
Polybenzimidazole	67	70	43.5	41.5
Carbon	100	100	57.5	56-64

At this point we wish to emphasise the characteristics of the char layer and its participation in combustion reactions. It is important not only to know at what rate the carbonised layer is formed during diffusion combustion of polymer materials but also its rate of combustion and how the correlation between zones in a condensed phase changes during the process. These problems have not actually been worked out for regular conditions of polymer combustion. In the latter case, the flow of heat from the flame towards the surface is insufficient for gasification of the formed carbonised layer by direct carbon evaporation (sublimation of carbon is observed at temperatures higher than 4000°C). The controlling stage of a carbonised layer's gasification (for the conditions under consideration) should be the heterogeneous oxidation reaction.

For a heterogeneous reaction, the interaction of the surface of the solid stance with the gaseous reagent plays an important role. This is why the rate of combustion of a carbonised layer will be determined by how an oxidation reaction occurs: only on the surface of a carbonised layer or through its full thickness as a result of the presence of capillary channels or pores. In other words, consideration must be given to the effect of the morphological structure of a polymer's carbonised layer. The morphology of the latter also determines the thermal and mass transfer during the combustion of polymers.

When carbonised polymers burn, the structure of the combustion wave is more complex. In the condensed phase, several zones can be distinguished: the original product zone, the polymer heating and pyrolysis zone, the carbonised layer zone and a superficial carbonised layer zone, where an intense exothermal reaction of heterogeneous combustion occurs.

The structure of a combustion wave in the gaseous phase is determined by the nature of the volatile pyrolysis products, by the rate at which they diffuse or penetrate the carbonised layer, and by the nature of the products of the heterogeneous reaction that enter the gaseous phase.

The most interesting works here are those that present a theoretical analysis of the ablation of a carbonising polymer when subjected only to a thermal flow, i.e. the effect of mass flow due to the mechanical erosion of the polymer by a high-speed gaseous flow is eliminated. Matsumoto and coauthors [113], for instance, analysed unidimensional models of nonstationary and stationary ablation processes for carbonising polymers. It is assumed in the model of nonstationary ablation that the zone of pyrolysis and the carbonised layer zone have capillary channels. The porosity of the carbonised layer is constant. The original polymer is not porous but in a polymer pyrolysis zone porosity changes from zero up to the value of the char porosity.

A solution to the set of equations that are a function of the conservation of energy and the conservation of mass laws for each reacting zone at appropriate limiting conditions will allow an evaluation to be made of the temperature distribution within the condensed phase zones, the linear rate at which the carbonisation zone and the pyrolysis front advance and the change in the width of the carbonisation zone during a nonstationary period.

This approach to the analysis of gasification of carbonising polymers is also valid at lower rates of heat flow, caused by diffusion flames striking the sample surfaces, than the heat flow rates investigated by Matsumoto and others.

Heterogeneous Oxidation of Carbon

Carbon reacts with gaseous compounds. It is known that the oxidation and gasification of carbon occurs when carbon reacts with water vapour, oxygen and carbon dioxide. The rate of such reactions becomes noticeable at increased temperatures. Strong oxidisers such as fluorine react energetically with carbon at room temperature and cause self-ignition of the carbon.

A large amount of research is being dedicated to the study of the mechanism and kinetics of the interactions between carbon-containing materials and oxygen, water vapour and carbon dioxide on account of their great practical importance. In addition to the allotropic forms (graphite, diamond), there are a number of polymorphous modifications of carbon which differ in crystallographic and morphologic structure, and physical and chemical properties, such as reactivity and catalytic activity. The polymorphous structures generally comprise a single group and are called, though not at all accurately, amorphous carbon.

In the presence of excess air, the carbon materials with predominantly amorphous, disordered structures begin to oxidise at a noticeable rate at a temperature nearing 350°C. Graphite and so-called glass carbon, a material with a very low porosity towards gases, begin to oxidise at 400°C and 450°C respectively.

The reaction with water vapour and carbon dioxide occurs at even higher temperatures. The heterogeneous reaction between solid carbon and gas occurs under the condition of continuous molecular or convective diffusion of the reagents towards the reaction surface and the continuous reverse diffusion of the reaction products. Once delivered to the surface, the reagent is adsorbed on the surface or, generally chemisorbed on active surface centres. The formation of the surface complex may be accompanied by liberation of gaseous reaction products. Vaporisation of carbon occurs either as a result of the destruction of the surface complex due to the effect of heat, or when the complex collides with a new molecule of gaseous reagent. After the desorption and withdrawal of the reaction product, new centres develop on the surface. These centres have a propensity towards the latter specific reaction with a gaseous reagent.

Depending on temperature conditions, the heterogeneous reaction between carbon and gases may occur in the kinetic or the diffusion region.

In the latter case, the role of a limiting stage could be assumed either by the transfer of gases and reaction products inside the pores of the carbon material (internal diffusion region) or by volume diffusion of gaseous substances through a gaseous boundary layer towards the external surface of a solid substance (external diffusion region). At low temperatures, the diffusion of gases and reaction products does not limit the oxidation of carbon.

It is natural that when scientists develop kinetic parameters for the heterogeneous oxidation of carbon they consider the process under conditions that either exclude, take in or control the diffusion factors.

A major contribution to the development of contemporary concepts of the heterogeneous oxidation of carbon and its combustion has been made by Zeldovich, Predvodihtelev, Frank-Kamenetskii, Hitryn, Kantorovich and many others. Reviews of theoretical and experimental literature in the area are reflected in many monographs and symposia, in particular in [114, 115].

It is now generally accepted that the most important part in carbon vaporisation is played by superficial complexes formed during reactions of carbon with gases. With the aid of the isotopic method (using ^{18}O and ^{14}C) it was discovered that the dissociation of complexes and the exchange reactions between the complex and gases actually determine the carbon vaporisation process [115, 116].

Surface carbon-oxygen complexes were first discovered by Shyhlov and coworkers [117] in the late 1920s when they studied the absorption of oxygen by coal. A structure for the oxide and anhydride forms was assigned to the surface groups. Later, ketones, lactones, ketenes, chromenes, quinoids and other groups capable of different chemical reactions were discovered during the study of surface oxygen complexes within different oxidised carbon materials [118].

It is claimed that the nature of the carbon-containing gases formed during decomposition of the complex is intimately connected to the structure of the complex. In fact, the name "complex" in this particular case is a conventional term, because in the indicated surface oxygen groups there are normal covalent bonds between the oxygen and carbon atoms but none of the weak interaction forces characteristic of complexes. However, the chemisorption of oxygen-containing gases must undoubtedly include an intermediate stage of formation of true complexes with active adsorption centres on the carbon surface. Surface complexes of gas and carbon can be multicentred, i.e. molecules of gas react simultaneously with a number of carbon atoms. Active centres can be isolated or adjacent to each other.

It must be mentioned that during the study of the heterogeneous oxidation and gasification of carbon materials most of the samples that were used contained more than 98% carbon (graphites, diamonds, black soots, carbon threads, fibres, coals). It is assumed that carbon atoms from the amorphous, disordered part of the carbon structure, peripheral carbon atoms from the base layers in crystalline grains and defects in the carbon crystalline lattice are the most susceptible to an attack of oxygen. With optical and electronic microscopy it was established that the oxidation of carbon atoms occurs with relative ease within the base planes of the graphite lattice and on the base dislocations of the crystal structure.

In the model of carbon structure proposed by Coulson, the edge carbon atoms in the base plane $(10\bar{1}l)$ may assume the bivalent state s^2p^2. This kind of carbon hybridisation has a higher reactivity by contrast to atoms with sp^3-, sp^2 and sp-electron hybridisation. At temperatures below 1000°C, oxidation of carbon atoms of graphite actually occurs faster in the plane $(10\bar{1}0)$ than in the plane $(11\bar{2}0)$. It is interesting that in the first case, carbon atoms are aligned in a "zigzag" configuration during the oxygen attack:

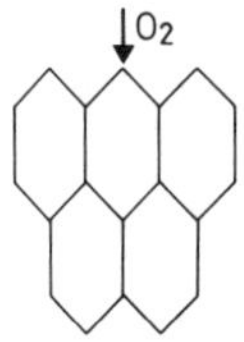

and that in the second case the carbon atoms are aligned on the end of the plane in a "chair" shape:

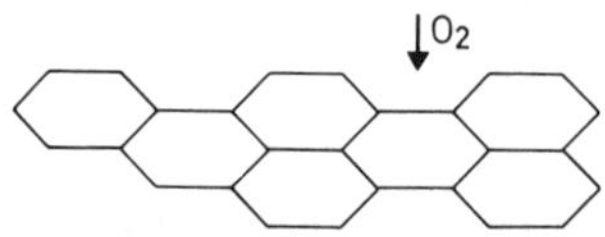

$$k_{(10\bar{1}0)} = 10^{9.77}\exp(-66\ 000/RT)$$

The oxidation rate constants in both planes are equal to:

$$k_{(10\bar{2}0)} = 10^{8.44}\exp(-62\ 000/RT)$$

With a temperature increase, the rates of oxidation in both directions approach each other [p. 153 of ref. 115]. The distance between edge carbon atoms in the zigzag and "chair" configurations is different, which is why it appears that the nature of carbon-oxygen complexes may be different. It is generally assumed that with a temperature increase, the less active centres of carbon begin to participate in the oxidation process.

Carbonising polymers form a nonvolatile residue during pyrolysis in the temperature range 400-700°C. The latter residue contains less than 98% carbon. For instance, polyvinylidene chloride at 430°C forms char with 72.1% C and a C:H ratio of 80. Char of polyvinylchloride at 560°C contains 95% C, C:H = 35; cellulose char at 565°C = 92.9%, C:H = 44, char of polyvinyl benzene at 640°C = 97.5%, C:H = 46 [119].

An appreciable number of paramagnetic centres were detected in the char residues. At low pyrolysis temperatures (high percentage of hydrogen atoms) these centres have broken σ-bonds and can react with free radicals.

It was determined by the EPR-spectroscopy method that at increased temperatures (~300°C), physical adsorption of oxygen on the char residues of polymers is virtually unobservable, although chemisorption of oxygen does occur [120]. The areas of the carbon skeleton in polymer chars, enriched by hydrogen-containing groups, are the most sensitive to oxidation reactions.

According to the data [58], under the conditions for diffusion combustion of carbonising polymers, thorough combustion of the char layer occurs at temperatures below 2000°C. Apparently, under such conditions the various reactions between the char and the gases should be taken into consideration. In particular, the total combustion of the carbonised layer must be considered not only as a result of the reaction with oxygen, CO_2 and H_2O but also as a result of the direct attack by atoms and radicals, which diffuse from the gas phase towards the surface. In addition, carbon residues can catalyse the reactions of thermal degradation of carbon-containing gases, which diffuse through pores or channels inside the char layer of the gas phase. The type of reaction that actually occurs is indicated in a number of cases by the formation of soot deposited on the char residue.

It has been demonstrated conclusively that when carbon is oxidised by molecular oxygen, both carbon monoxide and carbon dioxide are formed.

Ratios between the monoxide and dioxide in the reaction products depend on the temperature of the process. At temperatures below 1200°C, the ratio of $CO:CO_2$ approaches 1. The rate of oxidation is proportional to the concentration of oxygen. At temperatures above 1800-2000°C the ratio approaches 2 and the rate of oxidation is of zero order in the oxygen concentration. It is interesting that the ratio of $CO:CO_2$ does not depend on the pressure of the surrounding atmosphere or on the type of carbon material.

A number of schemes have been studied to describe the formation of products from the oxidation of carbon. The schemes differ in (a) type of surface carbon-oxygen complexes, (b) number of active centres and molecules of oxygen that participate in chemisorption and reaction with these centres and (c) mechanism of dissociation of the complexes.

If the details of the specific stages of the process are neglected, the general reactions of carbon oxidation can be expressed by the following equations:

1. When one oxygen molecule reacts with two active carbon centres:
   $2C + O_2 = 2CO$
2. When two oxygen molecules react with three carbon centres:
   $3C + 2O_2 = 2CO + CO_2$
3. When three oxygen molecules and four carbon centres participate in the reaction:
   $4C + 3O_2 = 2CO + 2CO_2$

The first reaction was called a carbon combustion reaction by Z.F. Chuhanov. This reaction, which leads only to the formation of carbon monoxide, was observed when the oxidation process was conducted in a dry atmosphere. It is assumed that this reaction prevails at very high temperatures. The source of carbon monoxide may be the carbonyl groups of quinoid rings. Just like other carbon oxidations, the indicated reaction may occur at a slow rate and under combustion conditions, i.e. under conditions of a progressive thermal self-acceleration of the oxidation reaction. For heterogeneous combustion not to be complicated by the gas phase (flame) combustion of carbon monoxide, the carbon monoxide must be rapidly removed by a gas flow.

The third reaction mainly occurs at temperatures below 1200°C, leading to an equimolecular ratio of carbon oxides formed.

Initially, oxygen molecules undergo chemisorption:

$$xC + 2O_2 \rightarrow C_x(2O_2)(chem.)$$

then, when the surface complex collides with the third oxygen molecule, decomposition of the complex occurs:

$$C_x(2O_2)(chem.) + O_2 \rightarrow (x-4)C + 2CO + 2CO_2$$

When the temperature increases, the probability of the second reaction occurring increases. The second reaction differs from the third reaction in that the destruction of the surface complex occurs directly through heating:

$$C_x(2O_2)(chem.) \rightarrow (x-3)C + 2CO + CO_2$$

Different researchers discovered that carbon oxidation at high temperatures occurs with an effective activation energy equal to 55-62 kcal/mole. This value is associated with the activation energy of decomposition of the surface oxygen-carbon complex compounds. If rapid removal of reaction products from the surface is not provided then a gasification (vaporisation) reaction between carbon and carbon dioxide can occur.

It should be mentioned that, according to the data [87], during the pyrolysis of phenolphthalein-based polyarylates, the liberated carbon dioxide is partially reduced on the char residue. The percentage of carbon in the latter approached 92% at a pyrolysis temperature of 500°C.

On a purified surface of carbon, the reduction of carbon dioxide to carbon monoxide was observed at 450°C when the surface oxygen-carbon complex is formed. The destruction of the complex leads to the formation of products and the vaporisation of carbon. In fact, this process might be considered as secondary to the combustion of polymers.

The reaction between carbon and carbon dioxide is expressed by the net equation $C + CO_2 \rightarrow 2CO$. According to different sources, the effective activation energy fluctuates in the range 59-97 kcal/mole. The reaction between carbon dioxide and carbon is, in fact, very complicated. It includes several stages of exchange reaction between oxygen and the surface, and the sorption and desorption of gaseous reagents and reaction products.

When CO_2 with a tracer carbon atom was used, it was discovered that radioactive carbon monoxide is formed according to the reaction

$$^{14}CO_2 + C_x \xrightarrow{k_1} {}^{14}CO + C_{x-1}(CO) \qquad (A)$$

Subsequent reduction of the surface by nonradioactive carbon monoxide also leads to the formation of nonradioactive carbon dioxide:

$$C_{x-1}(CO) + CO \xrightarrow{k_1'} CO_2 + C_x \qquad (B)$$

The transfer of carbon atoms from the solid phase into gaseous phases occurs as a result of surface layer destruction

$$C_{x-1}(CO) \rightarrow CO + C_{x-1} \qquad (C)$$

The exchange reaction between oxygen and the surface (A) occurs at temperatures that are approximately 200°C below those of the gasification reaction (B). At temperatures below 900°C, $k_1 \sim \exp(-41\ 000/RT)$ and $k_1' \sim \exp(-26\ 000/RT)$. The equilibrium constants at 850°C for forward (A) and reverse (B) reactions are expressed as

$K_1 = k_1/k_1' = 0.17$ [p. 221 of ref. 115]

Frank-Kamenetskii cites values of activation energies equal to 63 kcal/mole for the desorption stage of carbon monoxide (see reaction B). In the presence of steam, the gasification of carbon becomes especially noticeable at temperatures above 1000°C. It should be mentioned that a moist atmosphere accelerates the gasification of carbon by carbon dioxide. Generally, the reaction between carbon and water vapours is represented by a combination of reactions leading to the formation of CO, H_2 and CO_2 in gas form. The ratio between these gases depends on temperature conditions and on the ratios between the rate constants.

$$C_x + H_2O \underset{k_2'}{\overset{k_2}{\rightleftarrows}} H_2 + C_{x-1}(CO),$$

$$C_{x-1}(CO) \xrightarrow{k_3} CO + C_{x-1},$$

$$CO + C_{x-1}(CO) \underset{k_1}{\overset{k_1'}{\rightleftarrows}} CO_2 + C_x.$$

Under high vacuum, the products of the reaction between carbon and steam are only CO and H_2 in an equimolecular ratio. The reaction is, in general, endothermic; it has an activation energy (according to different sources) in the range of 55-90 kcal/mole.

According to N.V. Lavrov, the free radicals $\dot{H}$, $\dot{O}$, and $\dot{O}H$ participate in the heterogeneous combustion of carbon. They can be formed as a result of the dissociation of H_2O on carbon and the subsequent reaction between hydrogen atoms and CO_2 and O_2 molecules, which is why an increase in humidity accelerates the oxidation and carbon gasification process.

Indeed, atomic oxygen reacts more energetically with the carbon surface than does molecular oxygen and its rate of reaction with carbon is higher (at 370°C, by one order of magnitude) than that between $\dot{O}H$ radicals and carbon. The rate of carbon oxidation by atomic oxygen at low pressures depends to a small degree on the nature of the carbon material (pyrographite, coal, black soot) according to the data [121]. However, the chars of brown coal obtained at different temperatures differ in their ability to react with atomic oxygen [114].

In the case of diffusion combustion of carbonising polymers, the heterogeneous combustion of the nonvolatile char layer may be observed visually on account of the bright glow of incandescent carbon particles. After flaming combustion ceases on this surface layer, the heterogeneous process can continue for a fairly long time. This process occurs according to the reaction of carbon oxidation discussed above. Usually, the process is called smouldering (incomplete combustion) and it presents a serious fire hazard. Consequently, the problem of quenching reactions that involve the heterogeneous oxidation of carbon is studied seriously.

Chemical Processes in the Gas Phase

Pyrolysis of polymers during combustion leads to the formation of various products, a large number of which are in the gas phase.

The composition and quantity of gaseous fuel components determine the initiation and spread of the flaming combustion of the original polymer substance and also determine the type of mechanism for the combustion process. During pyrolysis, polymers form compounds that belong to different classes (depending on the polymer's structure): hydrocarbons, alcohols, acids, aldehydes, hydrocarbon halides and others. The burning of each individual compound has specific characteristics. The situation is more complicated when mixtures of these compounds are burned. At present, the mechanism and kinetics of gas phase reactions during combustion are not known for the majority of these compounds, despite the enormous amount of literature in this field.

In terms of the combustion of polymers, there is very little information on the type of stable products that are present in the gas phase of the flame as a result of polymer pyrolysis. There is also little information on how the concentration of these products changes from zone to zone of the flame and in what way the course of the reactions of gas phase transformations can be accelerated or decelerated.

Complex chemical processes occur in conjunction with the transformation of gaseous compounds within the flame. Such processes include cracking, oxidation and the formation of condensed products etc.

It is appropriate for the data available on stable products that have been discovered within diffusion flames of polymers to be discussed at this stage and for the most important processes that reduce the flammability of the polymer to be briefly described.

Diffusion Flames of Polymers

Polymethylmethacrylate. The composition of the gas within a diffusion flame of polymethylmethacrylate during its candlelike burning was studied for the first time by Fenimore and Jones [122]. They studied the composition of the gases using the mass spectrometer. The gas samples were taken periodically along the height of the central flame axis during combustion of the polymer in an atmosphere of 20% O_2 + 80% Ar at reduced pressure 50 mmHg). Polymeric round bar samples were used with a cross-section of 0.36 cm^2. A melt was formed on the surface of the burning polymer, and up to 40% of it flowed down the sample without vaporising. The surface of the sample had a semispherical form. A luminous flame developed 1.05 cm away from the surface of the sample.

Figure 3.8 shows the concentration profiles of the combustion products and the temperature distribution in the gaseous phase. The monomer is the major product near the surface: up to 70% of the carbon in the gas phase 0.1 cm from the surface belongs to the monomer. Hydrocarbons

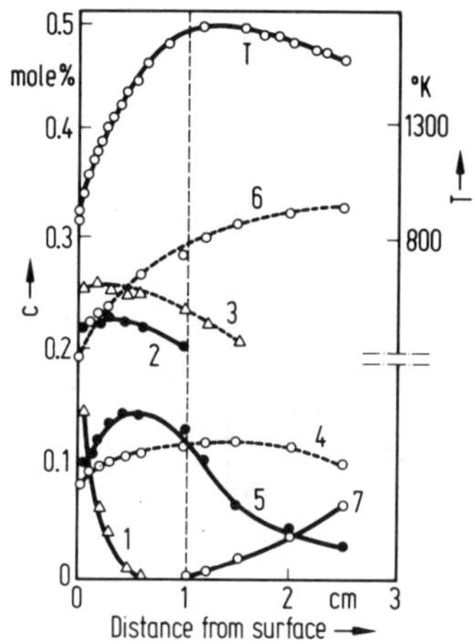

Fig. 3.8 Temperature (*T*) and concentration (1-7) profiles along the axis of a PMMA laminar diffusion flame (50 mmHg, mixture: 20% O_2 + 80% Ar). (1) MMA, (2) C_2 (+0.2 mol.%), (3) H_2 (+0.2 mol.%), (4) CO_2, (5) CO, (6) Ar(0.5), (7) O_2.

with two carbon atoms in the molecule were discovered along with monomer (marked as C_2 hydrocarbons). Methane and C_3 hydrocarbons are present in smaller quantities. It is interesting that concentrations of the monomer and C_2-C_3 hydrocarbons decrease in the preflaming zone - an indication that there is an additional decomposition of gaseous products from the destruction of polymer materials in this zone.

Fenimore and Jones did not find a significant amount of oxygen or intermediate products of hydrocarbon oxidation in the preflaming zone. They therefore concluded that oxygen does not participate actively in the decomposition of the condensed phase of the polymer or in the cracking of decomposition products in the preflame zone. Later, Burge and Tipper [18] analysed the composition of gaseous products from an analogous diffusion flame of polymethylmethacrylate burning in air at atmospheric pressure; the diameter of these samples was 25.00 mm.

Analysis of gases was carried out using gas chromatography. A wider range of products was discovered both along the central axis of the flame and in the cross-section (Fig. 3.9). Near the melt surface (distance ~ 0.1 cm), the organic compound detected in the largest quantity was methylmethacrylate; methane, ethylene, propylene, acetylene and formaldehyde were found in lesser quantities and only traces of propane, C_4

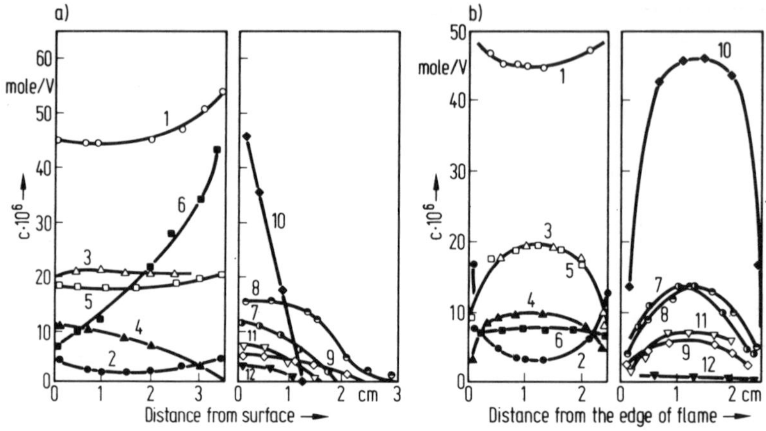

Fig. 3.9 Concentration profiles of gases (a) along the axis and (b) in a cross-sectional direction from the edge of a PMMA diffusion flame at 0-1 cm from the surface (combustion in air). (1) N_2 (0.5), (2) O_2, (3) consumed O_2, (4) CO, (5) CO_2 (6) H_2O, (7) CH_4, (8) C_2H_2, (9) C_2H_4, (10) MMA, (11) C_3H_6, (12) CH_2O. V = const.

hydrocarbons and peroxides were established. Oxygen was discovered along the total length of the flame axis and near the surface.

It is interesting that the concentration of methylmethacrylate and hydrocarbon gases increases as the probe moves towards the centre of the sample. There was no indication of the presence of such products of intermediate oxidation as aldehydes, ketones or alcohols near the surface. The change in concentration of carbon monoxide and carbon dioxide is very small in the preflaming zone. Burge and Tipper [18] estimated the time that pyrolysis products remain in this zone and the temperature conditions of the zone. Gases remain in the zone for a few milliseconds. According to these authors, when the temperature is between 500°C and 700°C and the retention times in the preflaming zone in the millisecond range, the oxidative pyrolysis of gases cannot occur to any noticeable degree. Almost the entire amount of products of the complete oxidation (H_2O and CO_2) and oxidising cracking of carbon-containing gases (which were discovered in the preflaming zone) is formed as a result of the reaction between oxygen and the polymer material surface. That is why concentrations of CO and CO_2 are virtually constant in this zone.

The principal oxidation reactions of pyrolysis products occur in the front of the flame. In the flame reaction zone there is an intensive reduction in the CO concentration and an increase in CO_2 and H_2O concentration. It should be mentioned that the maximum temperature in the flame of a polymer material detected by Burge and Tipper [18] was too low because the authors used a thermocouple probe with a large cross-sectional junction area. Oxygen was discovered in all the zones of the flame of burning polymethylmethacrylate in an oxygen-nitrogen mixture [123]. The percentage of oxygen in the flame did not change appreciably when the concentration of oxygen in the surrounding atmosphere increased from 18.5% to 48%. According to these authors, when there is such an increase in oxygen concentration in the surrounding atmosphere, an increase occurs in the oxygen consumption inside the reaction zone from 70% to 90%. It is still not clear how oxygen participates in the reaction with the surface of the polymer.

According to Madorsky ([16] p. 198) the high-temperature pyrolysis of polymethylmethacrylate in a vacuum is accompanied by secondary reactions involving bond rupture in the monomer molecule. To summarise, the percentage of low molecular gases formed increases from 3.6% to 18.2% and to 60.3% when the temperature increases from 525°C to 725°C and to 825°C respectively. The yield of monomer decreases from 96.2% to 80.6% and 37.4%.

All the products, except peroxides and formaldehydes, that were detected in the composition of the gases within the polymethylmethacrylate flame were also detected in the composition of pyrolysis gases. Apparently oxygen, which diffuses to the polymer surface, contributes to the initiation of the degradation reaction for polymethylmethacrylate through a mechanism similar to the one at lower temperatures. However, the yield of products in the oxidation process itself is low.

Polyethylene and polypropylene. Fenimore and Jones [122] discovered CO, CO_2 and unsaturated hydrocarbons in the composition of gases in the candlelike flame of polyethylene samples burning in an atmosphere of 80% Ar + 20% O_2 at a pressure of 50 mmHg (distance from the surface of melted polymer: 0.3 cm). The percentage of C_2, C_3 and C_4 hydrocarbons was 0.24, 0.06 and 0.04 mol.% respectively. The temperature on the surface of the burning melt reached under these conditions was ~900°K. A more detailed analysis of polyethylene diffusion flames was made by Burge and Tipper [18]. They studied samples of low-density and high-density polyethylene. In the case of high-density polyethylene, there was practically no branching in the macromolecule's chain. For low density there were 27 methyl groups per 1000 carbon atoms.

When low-density polymers in the shape of round bars are burning in an air atmosphere at normal atmospheric pressure, a layer of melt is formed on their surface. Above the melt at a distance of up to approximately 1 mm from the surface a nonluminous cylindrically-shaped zone of blue flame was observed; this had a width of some 3-4 mm, and above it was a cone of luminous yellow flame with a height of up to 3 cm. The maximum temperature was detected at a distance of 1.5-2.5 cm from the surface of the melt along the central axis of the flame. The surface temperature was approximately 400°C.

Among the components of gases near the melt surface (0.1 cm), a large amount of nitrogen (~70%), carbon monoxide (~5%), carbon dioxide (~10%) and water (~5%) were discovered. Low-molecular hydrocarbons amount to approximately 10% of the total quantity of gas. Among these low-molecular hydrocarbons the percentage of ethylene was the highest. Also, traces of ethane, propane and C_4 hydrocarbons were discovered, and near the leading edge of the flame there were traces of peroxides and formaldehydes. The percentage of oxygen did not exceed 2% near the surface.

Fig. 3.10 shows concentration profiles of gaseous products discovered along the length of the central axis in a diffusion flame of low-density polyethylene and also in the cross-section, 0.1 cm from the polymer melt. At ascending positions within the flame, a decrease of hydrocarbon and carbon monoxide concentrations was observed. The percentage of N_2, CO_2 and H_2O increases.

High-density polyethylene samples differed from a low-density polymer in that during the combustion of the former, the resulting melt was more viscous and its spreading was not as pronounced. Within the flame, above the melted surface, more hydrocarbons and less water were detected. The concentration of oxygen inside the flame is minimal 1.5 cm away from the melting surface.

The authors observed that when the concentration of oxygen in the surrounding atmosphere increases to above 40%, the formation of soot becomes more intense inside the flame. As a result, the luminosity of the yellow flame becomes more intense. The concentration of oxygen near the flame base in this case diminishes in comparison with burning in an atmosphere of air. Apparently, the process is closely connected to the

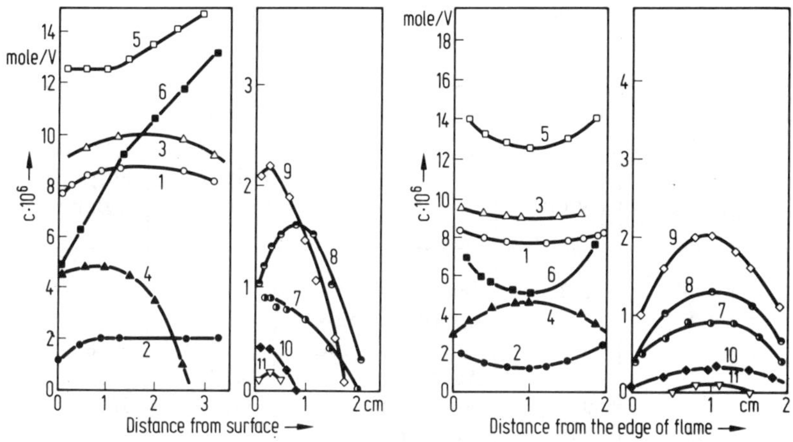

Fig. 3.10 Concentration profiles of gases (a) along the axis and (b) in a cross-sectional direction from the edge of a polyethylene diffusion flame at 0-1 cm from the surface (combustion in air). (1) N_2 (0.1), (2) O_2, (3) O_2 (0.5) consumed, (4) CO, (5) CO_2 (6) H_2O, (7) CH_4, (8) C_2H_2, (9) C_2H_4, (10) C_2H_6, (11) C_4H_8. V = const.

increase in consumption of oxygen in the reaction flame zone. The temperature of the polymer material surface also increases.

The composition of gases was examined by mass spectrometry [19] using a quartz probe at a distance of less than 0.1 mm from the surface in the centre of a polyethylene diffusion flame in an air atmosphere. The rate of flow of an oxidising gas directed towards the surface was 60 cm/min and the diameter of the samples studied was about 5 mm. The derived data for the composition of gases is given in Table 3.9. Along with saturated and unsaturated carbon compounds, a considerable amount of CO, CO_2 and H_2O was detected. Oxygen was also found near the surface. The percentage of CO, CO_2 and H_2O in terms of the amount of oxidised methylene groups (25.5 mol.%) corresponds to the consumption of 1.28 moles of oxygen per mole of CH_2 groups. In other words, one quarter of the polymer molecules from the surface must undergo oxidation under these conditions of polyethylene combustion.

Results derived in reference [19] are in general similar to those of Burge and Tipper. The conclusion about oxygen participating in the decomposition of the condensed phase during combustion of polyolefins is also similar. The same conclusion comes from an examination of more

Table 3.9 Composition of gases in polyolefin diffusion flames close to the surface of melt

Compound	Mol.%		Compound	Mol.%	
	Polyethylene	Polypropylene		Polyethylene	Polypropylene
Methane	0.65	1.1	Octenes	-	1.9
Ethylene	5.8	2.7	Carbon monoxide	3.4	3.2
Ethane	-	0.5	Carbon dioxide	11.0	8.8
Propylene	1.3	3.4	Water	14.6	12.1
Allenes	1.2	0.1	Nitrogen	55.5	52.9
Butenes	1.7	2.0	Oxygen	1.6	1.2
Pentanes	0.7	7.4	Argon	0.6	0.6
Hexanes	1.3	-	Aromatic hydrocarbons		
Heptanes	0.4	1.1		0.2	1.0

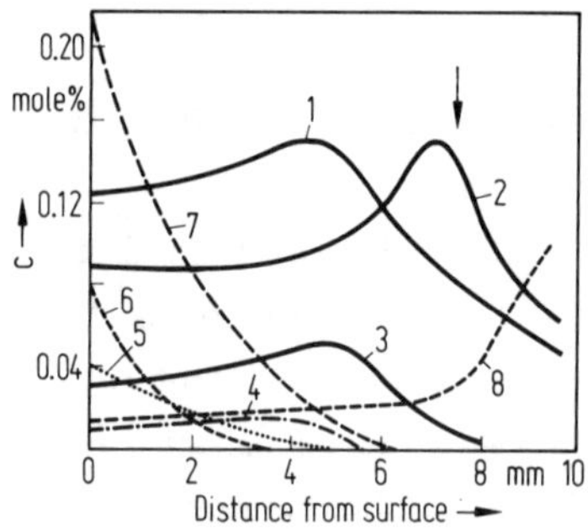

Fig. 3.11 Concentration profiles of gases along the axis of a polypropylene diffusion flame (combustion in an air atmosphere) (1) H_2O, (2) CO_2, (3) CO, (4) CH_4, (5) C_3 hydrocarbon, (6) C_4 hydrocarbon, (7) ΣHC, (8) O_2. The arrow indicates the point that corresponds to T_f, max.

detailed data on polypropylene combustion products (Table 3.9 and Fig. 3.11) and from an analysis of a polymer's surface layers, which have been mentioned before.

When a sample of polypropylene was burned in an air atmosphere, the temperature of the surface reached 375°C (under the above-mentioned conditions), and the maximum temperature of the flame approached 1300°C.

In the gas phase, the increase in the percentage of H_2O and CO was accompanied by a rapid decrease in the concentration of hydrocarbon components. The increase in the concentration of CO_2 in the zone of maximum flame temperature corresponds to the depletion of carbon monoxide. Unlike Burge and Tipper, Stuetz and coworkers [19] think that fuel gases undergo oxidative degradation inside their preliminary heating zone in the diffusion flame and that this happens before they reach the front of the flame, that is, before they reach the reaction zone with a maximum temperature gradient.

If one compares data from Tables 3.5, 3.9 and Fig. 3.11, and also takes into account the dilution of flaming gases by nitrogen, then it is obvious that the qualitative composition of the products near the surface of the burning polypropylene is similar to the composition of the products of the polymer's thermal destruction at 400°C. This composition is enriched by the presence of oxygen-containing compounds. There are no products with a number of carbon atoms in the molecule greater than eight. Mostly one finds ethylene. Yet it is known that the rate of decomposition of aliphatic hydrocarbons increases with an increase in chain length in proportion to the magnitude $\sim 2(n - 2)$, where n is the number of carbon atoms in the molecule. The degradation of hydrocarbons is accompanied by the formation of ethylene [124]. It is obvious from Fig. 3.11 that the initial rate of disappearance of C_4 hydrocarbons in the polypropylene diffusion flame is larger than the rate of disappearance for C_3 hydrocarbons.

Diffusion flames of polyolefins in a counterflow of oxidising gas (air) are discussed in reference [125]. Samples with diameters of 21 mm were tested. Blue horizontal (flat) flames with widths of 0.5-0.8 mm emerged at a distance of 2.0 mm from the sample surface in the case of polyethylene, and at 2.5 mm for polypropylene. The maximum temperature inside the polyolefin flame reached 1500°C, and near the surface, 480°C. Close to the surface, nitrogen (~72%) and oxygen (~1.5%) were discovered. Despite the presence of a strong counterflow of oxidiser, the percentage of oxygen in the zone from the surface of the melt up to the lower part of the flat luminous flame is low. In general, there are the same tendencies towards disappearance of pyrolysis products and formation of combustion products as have been observed in other studies.

There is thus no doubt that the degradation of organic compounds in the zone before the front of the flame occurs quite energetically.

Polyoxymethylene. Formaldehyde, carbon monoxide and carbon dioxide were discovered in equal quantities by Fenimore and Jones [122] inside a polyoxymethylene diffusion flame near the melt surface at a temperature of ~750°C. Hydrogen was found in a quantity four times lower than CO. A more detailed study of polyoxymethylene (polyformaldehyde) combustion was carried out by Brown and Tipper [126]. They studied the composition of gases inside the laminar candlelike diffusion flame of polyformaldehyde samples burning in an atmosphere of 80% N_2 + 20% O_2. The diameter of the samples was 12.7 mm. The temperature near the surface reached ~400°C, and the maximum temperature of the flame was 1325°C.

Figure 3.12 shows the concentration profile of the gaseous products along the central axis of the sample and along the height of the flame at a radial distance of 1.14 cm from the centre of the sample. The surface of the sample is consumed faster near the edge of the flame than in its centre; hence in Fig. 3.12 the combustion products in (b) are indicated by points with a negative coordinate in relation to the zero position of the surface along the central axis.

Polyformaldehyde depolymerises easily, forming formaldehyde. At temperatures above 600°C, the latter decomposes forming carbon monoxide and hydrogen. The concentration of formaldehyde inside the flame near the surface of the melt did not exceed 15% because the flow of combustible pyrolysis products was diluted by nitrogen and combustion products, which were diffusing towards the surface. Quantitative ratios of gases, discovered 0.1 cm from the surface of the melt in the centre of the sample

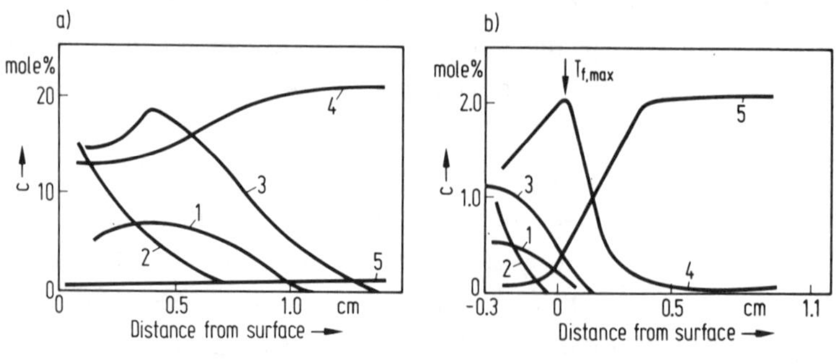

Fig. 3.12 Concentration profile of gases inside the diffusion flame of polyoxymethylene along the central axis, (a), and at 1.14 cm from it, (b). (1) H_2, (2) CH_2O, (3) CO, (4) CO_2, (5) O_2.

were (in mol.%) $CH_2O:CO:CO_2:H_2$ = 12.8:13.8:12.9:4.9. These ratios are similar to the one discovered earlier [122].

Brown and Tipper analysed isoconcentration contours of pyrolysis products for polymers and oxygen inside the flame and concluded that formaldehyde undergoes partial oxidation near the sample's surface. Near the surface ~0.3 mol.% O_2 has been detected. The maximum percentage of H_2 and CO is attained 0.5 cm from the surface along the central axis of the diffusion flame. The ratio of $CO:H_2$ is equal to ~2.7. The temperature of this region reaches 1100°C and if one takes into consideration the fact that rate constants for reactions between OH radicals, H_2O and CO differ (at 720°C, say, by five times) and also that the diffusion coefficient for H_2 is larger than that of CO, then the reason for the excessive percentage of CO in this flame zone becomes clear.

The maximum percentage of carbon dioxide was detected in the front of the flame where the maximum temperature was registered. However, a complete combustion, which corresponds to the stoichiometric reaction between formaldehyde and oxygen (mixed with nitrogen) was not attained inside the flame front under the conditions of combustion studied.

Polyethyleneterephthalate. The laminar diffusion combustion of polyethyleneterephthalate is discussed in [127]. The authors used samples with diameters of 21 mm. The combustion of samples took place in an air atmosphere. Flat diffusion flames were obtained as a result of a counterflow of oxidiser. The distance between the oxidising gas jet and the sample's surface was 12.8 mm.

During stationary combustion the melt is formed on the surface of the sample, there are yellow flames in close proximity to the surface and a blue zone above it. The width of the glowing zone inside the flat flame varied within the 1.0-1.8 mm range, depending on the flow rate of the oxidiser. The width of the blue flame zone was ~0.2 mm. When the oxidising flow was substantially increased, formation of a carbonised layer on the surface was observed and the flat structure of the flame was disrupted.

The authors discovered that ~0.6 mm from the surface of the melt, along the central axis of flame, the composition of gas is identical to the composition of pyrolysis products of polyethyleneterephthalate at 480°C (the surface temperature during combustion). Acetaldehyde is an exception; it is present in a small quantity.

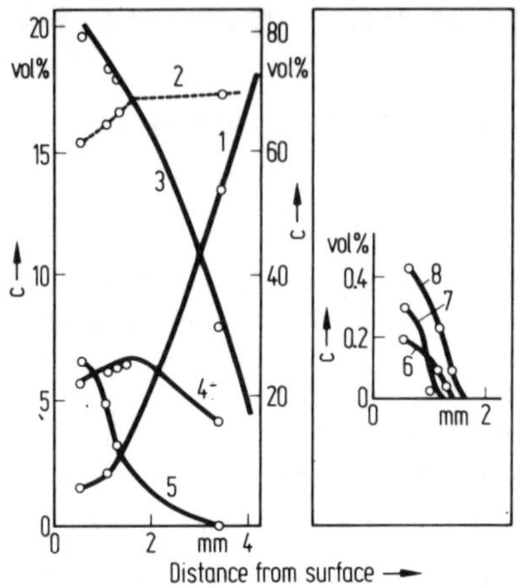

Fig. 3.13 Concentration profiles of gases along the axis of a plane (horizontal) diffusion flame of polyethyleneterephthalate. (1) O_2, (2) N_2, (3) CO_2, (4) H_2O, (5) CO, (6) CH_4 (7) CH_3CHO, (8) C_2H_4. The rate of air flow is 5.8 l/min; the rate of thorough combustion of the sample is 2.6 mm/min.

Figure 3.13 depicts the concentration profiles of the combustion products along the height of the central axis inside the diffusion flame of polyethyleneterephthalate. Nitrogen and oxygen diffuse through the front of the flame towards the surface of the polymer. The percentage of nitrogen 0.6 mm from the surface of the melt is 63%; oxygen, 1.5%. Inside the blue flame zone an increase in H_2O concentration and a decrease in CO concentration were observed. Outside the blue flame zone, at a distance of 2 mm, there was practically no carbon monoxide. Hydrocarbons and acetaldehyde were almost completely consumed in the blue flame zone. It should be mentioned that the percentage of carbon dioxide in the composition of the flaming gases of polyethyleneterephthalate constantly diminished along the height of the flame (in contrast to other polymers). Assuming that the percentage of nitrogen is 60% near the surface of the melt and taking into consideration the dilution of the gas mixture by nitrogen, the authors calculated the composition of gases near the surface: CH_3CHO = 4.0%, C_2H_4 = 1.6%, CO = 11.6%, CH_4 = 0.2%, CO_2 = 18.8%, H_2= 3.4%. There are processes taking place in the gas phase that are linked to the cracking of organic compounds - the products of polymer pyrolysis at a very small distance from the surface of the melt.

Delbourgo and coworkers [86] studied the composition of polyethylene-terephthalate pyrolysis products under conditions that simulated temperature gradients over the height of the flame. It was shown that under these conditions there is very intense decomposition of the gaseous products of polymer pyrolysis. In fact, CO and CO_2 are formed as a result of acetaldehyde decomposition and decarboxylation of aromatic acids.

Polytetrafluoroethylene (PTFE). PTFE does not burn under normal conditions. For flaming combustion to take place, PTFE requires high concentrations of oxygen in the surrounding atmosphere. Fenimore and Jones [122] studied the composition of gases in a PTFE flame burning in an atmosphere of 95% O_2 + 5% N_2. The characteristic feature of the combustion process for this polymer is a complete absence of oxygen (Fig. 3.14). Among the gaseous components, CF_4, COF_2, carbon monoxide and carbon dioxide were detected. These authors believe that the products detected are formed as a result of heterogeneous reactions between oxygen and the surface of the polymer. Tetrafluoroethylene under similar conditions forms a diffusion flame with the same gaseous composition but a smaller percentage of COF_2.

The ratio of (CO + CO_2):CF_4:COF_2 is 1:0.9:0.4. This ratio did not change when the flow of oxidiser fluctuated [122]. Since oxygen was not discovered near the surface, it was subsequently assumed that an important role is played by atomic fluorine diffusion out of the flame in the

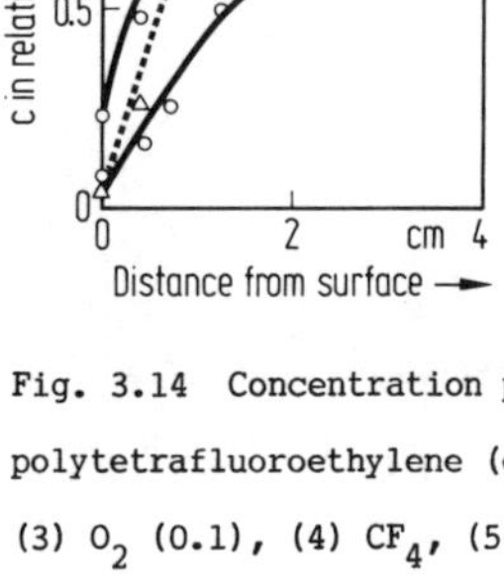

Fig. 3.14 Concentration pofiles of gases along the axis of a diffusion flame of polytetrafluoroethylene (combustion in an atmosphere of 95% O_2 + 5% Ar). (1) Ar, (2) CO_2, (3) O_2 (0.1), (4) CF_4, (5) COF_2.

degradation of the C phase during PTFE combustion. A calculation of the equilibrium composition for combustion products of PTFE in an oxygen atmosphere and under adiabatic conditions indicates the presence of 6-10% atomic fluorine. Experimental confirmation was obtained of its presence inside the flame [128].

Polymer based on Epoxy Resin, strengthened by *m*-Phenylenediamine. The diffusion flames of carbonising polymers are very difficult to study because the stable structure of the flame breaks up. The situation becomes even more complicated by soot-formation processes inside the flame. If probes are used for sampling the products their openings are rapidly clogged by solid particles.

With the help of mass spectroscopy, a study was made of the composition of gases during candlelike combustion of polymer samples based on epoxy resin ED-20 and also of the gaseous composition of the flame spreading over the surface of a flat polymer sample [129]. An analysis was conducted of a candlelike diffusion flame along the height of the flame and in a radial direction over its cross-section 1-2 mm from the surface of the sample.

Figure 3.15 shows a change in the concentration of stable combustion products along the central axis in the diffusion flame of a polymer. During combustion in an atmosphere of 28% O_2 + 72% N_2, the maximum temperature inside the flame (~100°C) was attained approximately 5 mm from the sample surface. The temperature inside the condensed phase at a

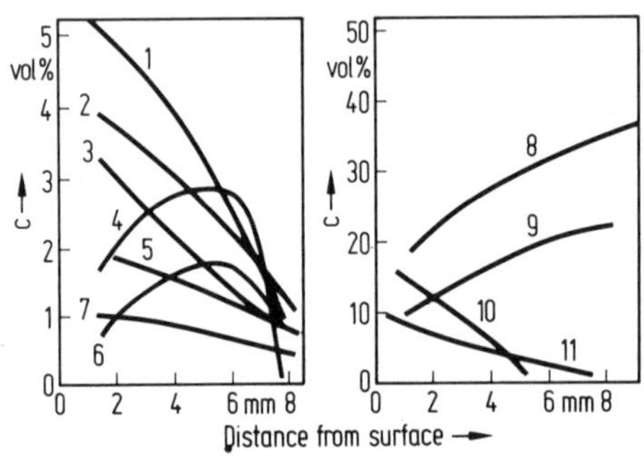

Fig. 3.15 Concentration profile of products along the axis of the diffusion flame of an epoxy resin. (1) C_2H_4, (2) C_2H_2, (3) C_6H_6, (4) CH_3CHO, (5) HCHO, (6) C_3H_6, (7) $CH_3C_6H_5$, (8) CO_2, (9) CO, (10) O_2 (11) CH_4.

depth of 1.5 mm from the surface was 600°C. The sampling of gases was done during a short period of stable combustion. It was difficult to perform the analysis since soot particles settled over the probe. Samples 20 mm in diameter were used for studying the compositions of flaming gases.

Inside the flame of an epoxy polymer strengthened by *m*-phenylenediamine, carbon monoxide and carbon dioxide, benzene, toluene, propylene, acetylene, methane, formaldehyde and water were discovered. It is presumed that styrene and ethyl- and isopropylbenzene are present in very small concentrations (less than 0.02% by volume). Oxygen is located in a zone near the surface of the sample (10-12%). Its concentration decreases when the maximum temperature zone is attained.

An analysis of the products along the surface of the samples at a height of 1-2 mm indicates that near the combustion surface the flame has a turbulent nature. According to the authors of the above-mentioned work, oxygen is drained into the jet of the flame near the surface and consumed during the oxidative pyrolysis of the polymer's degradation products. On the basis of results obtained by these authors, it is difficult to depict the structure of an epoxy polymer diffusion flame. However, as in the case of other polymers, it is obvious that there are pyrolytic reactions of organic compounds inside the preflame zone. The presence of oxygen and inert gas near the surface of the burning polymer indicates that the preflame zone is a region in which there is a partial preliminary mixing of the fuel components and an oxidant for the destruction of a polymer. Thus the laminar diffusion flame of a polymer may be considered as a premixed flame in a very fuel-rich system.

The sampling and analysis of gases from polymer diffusion flames presents certain difficulties. The methods used for these purposes have their limitations, and this affects the results obtained. In fact, in each of the examples a large number of initial and intermediate compounds are to be expected both near the surface and in the low-temperature flame zone.

Different methods of spectral analysis, that allow data to be obtained on the composition of intermediate particles (atoms, radicals, ions) inside the flame, have not been used in the study of polymer diffusion flames.

Despite the limited information about fuel gas composition inside the flame of polymers, a few groups can be clearly distinguished in the compounds detected. They are saturated and unsaturated aliphatic and oxygen-containing compounds, such as aldehydes. Acetylene is present in the flames of the majority of the above-mentioned polymers. In the case of polymers that also contain other elements besides C, H, and O, the presence of nitrogen-containing carbon compounds and other compounds may be expected.

Before the fuel components burn to final products they participate in complex pyrolysis reactions that are complicated by the presence of large amounts of oxygen. Pyrolysis of all carbon compounds occurs in the direction of the formation of more thermodynamically stable substances.

The thermodynamic stability of carbon compounds in a gaseous state may be characterised by the value of the free energy of formation (ΔG_f^0) and by its temperature dependence at 1 atm. According to the accepted ideas, a negative value is assigned to the energy of formation ΔG_f^0 if heat is liberated from the system into the surrounding atmosphere during the formation of a compound. Where heat is absorbed during the formation of compounds, a positive value is assigned to the energy of formation. The smaller the magnitude of the formation energy for carbon compounds at a given temperature, the more stable the compound will be under these conditions. When the value of the free energy of formation of a compound increases, its thermodynamic stability diminishes.

For solid carbon $\Delta G_f^0 = 0$. The most stable carbon compounds in a wide temperature range (1400-1500°K) are CO_2 and CF_4, which have ΔG_f^0 values at 1000°K of -95 and -187 kcal/mole respectively. One very stable compound is carbonyl fluoride (COF_2) with $\Delta G_f^0 = -140.0$ kcal/mole [128]. A comparison of the thermodynamic stabilities of different carbon compounds at different temperatures is used for the study of the probable sequence of reactions and of their thermal decomposition.

The change in Gibb's free energy is a measure of the driving force of a pyrolytic process. A pyrolytic reaction with the formation of one or other intermediate and final products is possible if the change in Gibb's free energy $\Delta G_r < 0$ varies as a result of the reaction. Thus organic acids undergo decarboxylation, alcohols undergo dehydration, and aldehydes form carbon dioxide during their decomposition, for instance $CH_2O \xrightarrow{(1)} H_2 + CO$ or $CH_3CHO \xrightarrow{(2)} CH_4 + CO$. The relatively fast disappearance of acetaldehyde (formed during the combustion of polyethyleneterephthalate) was discovered by Delbourgo and co-workers [86]. It is obvious from the data cited above that the major products in the polytetrafluoroethylene flames are COF_2, CF_4 and CO_2. Methane is the most thermodynamically stable of the saturated aliphatic hydrocarbons. The higher members of the homologous series have a positive ΔG_f^0, even at temperatures greater than 400°K. It should, however, be kept in mind that nonequilibrium systems (from the thermodynamics point of view) often appear to be fairly stable in practice.

The real rate of a substance's transformation depends on the condition of surmounting the energy barrier between two states and also on the mechanism of the process. Nevertheless, it was clearly shown by experiments that the real thermal stability of hydrocarbons at high temperatures increases in a sequence concurrent with thermodynamic stability: *n*-paraffins < branched paraffins < cycloparaffins < aromatic hydrocarbons < polycyclic aromatic hydrocarbons. The intermediate state between the paraffins and aromatic hydrocarbons at temperatures above 600-700°C is occupied by olefins.

A thermodynamic approach to the study of the pyrolysis of carbon compounds by calculating the energy of the bonds in the molecule explains the tendencies observed towards dehydrogenation and the formation of olefins during pyrolysis, cyclisation of hydrocarbons with the formation of aromatic compounds and tendencies towards dehydrocondensation reactions of aromatic compounds, which lead to origination of polycyclic aromatic compounds and soot formation.

Overall, it may be said that the general direction of a substance's pyrolytic reactions is similar for both the condensed and the gaseous states. The differences are due to kinetic mechanisms which are connected with the mechanism of the entire pyrolysis process and its intermediate stages and to temperature and time conditions; differences become apparent in the morphological features of the formed carbon.

Vast numbers of works cited in different monographs and symposia [1, 124, 130, 131] are devoted to the study of the kinetics and mechanism of gas phase reactions for the degradation of different carbon compounds. Radical reactions, which can take place according to both chain and non-chain mechanisms, play a large role in the cracking of organic compounds. We do not think it necessary to discuss in detail the degradation mechanism of carbon compounds; those with interest are referred to the cited literature.

At this point it would seem appropriate to mention some examples of the processes occurring in the gaseous phase which are important from the viewpoint of the flammability of polymer materials. First, we are interested in what kind of reactions occur during polymer combustion; second, in the formation of smoke and, finally, in the possibilities that exist for regulation and control of these reactions in order to quench the smoke formation process and slow down the combustion of the material.

Smoke Formation during the Combustion of Polymers

Smoke liberated during the combustion of polymer materials is a suspension of condensed particles in a mixture of gaseous combustion products and surrounding atmospheric medium. Depending on the chemical structure of the polymer materials and the combustion conditions, the suspension can consist either of condensed liquid droplets of the combustion products (mist) or of solid particles, possibly with additional condensation of products from the gas phase flame reaction on the surface of these solid particles.

Both chemical and pyrophysical processes are the basis of smoke formation during the combustion of polymer materials. Solid or liquid condensed combustion products are often formed as a result of various homogeneous reactions of pyrolysis and oxidation of substances. First, the gaseous products are formed, and later the products undergo condensation according to the usual physical mechanism. Condensation of the combustion products can take place both in flame zones and outside the flame.

The formation of condensed products by a chemical mechanism during the combustion of polymers is a more complex process. In this case the reactions leading to the formation of condensed solid particles occur most intensively within the high-temperature flame zone or near the flame. A typical example of the formation of condensed solid products by a similar mechanism for the polymer flame is the formation of soot. At the same time, the formation of soot indicates incomplete combustion of the organic substance.

As a rule, polymer materials are multi-component systems, which is why the formation of smoke during combustion could be caused not only by carbon liberation (soot) inside the flame but also by the simultaneous formation of solid particles of oxides of metals and metalloid elements, through chemical and physical condensation of various pyrolysis and combustion products.

To control the process of smoke formation in polymers it is necessary to have an understanding of the mechanisms of the process in a standard system under standard conditions. The information available on the character of smoke formation during the combustion of polymer materials is insufficient, and a systematic, in-depth scientific investigation is necessary.

An analysis of published data on the formation of smoke during the combustion of polymer materials shows that the quantity and nature of smoke depends on the chemical structure of the polymers, the composition of the materials, the percentage of oxygen in the surrounding atmosphere and the nature of the combustion mechanism. Some types of material release more smoke when they undergo incomplete combustion (cellulose materials, fire-proofed polyurethanes); other types exhibit increased smoke formation during flaming combustion (polyvinyl chloride, polyethers, polystyrene) [132].

During incomplete combustion of polyvinyl chloride a thick fog generally forms which contains small droplets of hydrochloric acid. During flaming combustion, a smoke that contains mainly solid soot particles and, to a lesser degree, condensed drops of hydrogen chloride, is formed. A similar change in the concentration of gaseous hydrogen chloride and in the density of smoke in the cell was detected during the smouldering and flaming combustion of polyvinyl chloride [54]. The presence of moisture in the surrounding atmosphere is conducive to the condensation of hydrogen chloride.

The processes of formation of condensed carbon particles (soot) inside a flame are the most intrinsic to smoke formation during the combustion of polymers. The presence of a luminous yellow flame near the surface of burning polymers indicates that the formation of carbon is initiated in a low-temperature zone of flame. Flame luminosity is caused by dispersed radiation of solid carbon particles. Table 2.2 shows that diffusion flames of aliphatic polyamide and polyoxymethylene are the most transparent by comparison with those of polyolefins; that is, they display the least inclination towards formation of carbon (soot).

If particles of soot do not have time to burn within the high-temperature flaming reaction zone, then smoke is liberated from the tip of the flame. During diffusion combustion of polymers almost identical mechanisms were observed in terms of the formation of soot and the liberation of smoke in the combustion of low-molecular organic compounds. This is to be expected because low-molecular pyrolysis products participate in the formation of carbon inside the polymer's flame.

It is known that the capacity to form soot during the combustion of organic carbon compounds depends on the oxidiser-to-fuel ratio. In the case of premixed flames, the formation of soot is observed only when the oxygen-to-fuel ratio is less than a critical value, determined by the nature of the fuel. Oxygen-containing compounds are characterised by lower critical values of the oxidiser-to-fuel ratio than saturated and unsaturated hydrocarbons [133]. In the case of diffusion flames, the situation favours the formation of condensed carbon particles more than in the case of premixed flames.

Low-molecular hydrocarbons form a series in respect of soot formation during diffusion combustion (as already mentioned above): *n*-paraffins < branched paraffins < cycloparaffins $\lesssim$ olefins ≤ cyclo-olefins < diolefins < acetylene hydrocarbons < alkylbenzenes < naphthalene derivatives < higher polycyclic aromatic hydrocarbons. There is no formation of soot inside the diffusion flames of methanol, urotropin and hexatransdecalin [131].

The nature of the smoke liberated during diffusion combustion of polymers depends to a large extent, as already stated, on the polymer structure and on the conditions of pyrolysis and the oxidation processes. It is difficult to judge the capacity of polymers to form soot from the magnitude and rate of change of only the optical density of combustion products but general tendencies are evident.

Polymers that do not form hydrocarbon compounds inside a flame do not exhibit an inclination towards formation of soot (polyformaldehyde, polytetrafluoroethylene). If, during the pyrolysis of polymers, the formation of aliphatic hydrocarbon compounds (polyethylene) predominates, then these polymers are less inclined to form soot than polymers that form aromatic hydrocarbons (polystyrene, ABS). During the combustion of polymethylsiloxane elastomers there is much less liberation of sooty smoke than in the case of siloxane elastomers with phenylic substituents, and especially by comparison with hydrocarbon rubbers [132, 134].

A similar picture is observed when aliphatic and aromatic polyamides are compared. It is interesting that, in air, phenolic polymers and cellulose materials have less of a tendency to soot formation during diffusion combustion than do polyolefins (see Table 2.2). This may be linked to the fact that during diffusion combustion of phenolic polymers and cellulose materials a large amount of oxygen-containing products enter the gaseous phase and these products undergo oxidative degradation more rapidly. Epoxy polymers show a very strong capacity for soot formation and smoke liberation. Not only unsaturated aliphatic but also aromatic hydrocarbons have been disovered in their flames.

The mechanism of soot formation in diffusion flames of hydrocarbon gases has not been definitely determined, yet it is generally accepted that the soot formation process has a radical-chain nature. Important roles in the formation of carbon are played by acetylene and ethynyl radicals, C_2H, formed from it. Besides ethynyl radicals, other carbon radicals ($\cdot\dot{C}H$, $\dot{C}H_2$, $\dot{C}H_3$, $\dot{C}(O)H$, $\dot{C}_2H_3$) and also carbon atoms (C, C_2, C_3) were found inside the flames of hydrocarbon compounds [2; 135]. Moreover, the presence of active positive and negative ions has also been discovered.

The existing theories of soot formation inside the flames of organic substances are based on the study of how the various active particles (radicals, atoms, ions) participate in the processes of nuclei formation and nucleation of condensed particles, in the processes of their growth and in their total combustion. Numerous studies of flames of hydrocarbon compounds show that formation of carbon is observed in a zone at a certain distance from the front of the flame or in a zone adjacent to it depending on the nature of the fuel and the temperature conditions inside the flame. It is assumed that the mechanism of soot formation can change considerably at that time. In particular, the most probable reactions are polymerisation reactions at a relatively low flame temperature ($< 1000°C$) of unsaturated and aromatic hydrocarbons. In this case, the formation of polycyclic aromatic hydrocarbons should precede the formation of condensed carbon particles.

Aromatic hydrocarbons are not stable at high temperatures. A partial decomposition of benzene rings takes place together with the formation of acetylene, ethylene and hydrogen. The formation of condensed carbon particles in this case is a result of the direct participation of acetylene, ethynyl and C_2 radicals in the formation of a coplanar carbon skeleton (Table 3.10).

It should be mentioned that, in the low-temperature flame zone, where carbon particles are found, various resinous products are also generally discovered. The relationship between the rate of soot particle formation and the rate of resinous product formation along the height of a methane diffusion flame indicates the important role of the resinous products in the formation of condensed particles [137].

Monoaromatic and polyaromatic hydrocarbons, their alkyl and alkenyl derivatives and partially hydrogenated cyclic polyacetylene compounds were discovered inside the luminous flame zones of various hydrocarbons [138]. As is known, all these compounds can react readily with radicals. Similar reactions lead to an increase in molecular weight and to the formation of compounds with low partial vapour pressures.

Numerous factors point to the participation of free radicals in soot formation in the gaseous phase during the combustion of various substances. Additional agents such as as chlorine, methyl bromide, carbon tetrachloride and sulphur trioxide promote the formation of soot in a flame; additives such as nitrogen oxide, carbon monoxide, sulphur dioxide, amylnitrate, methyl alcohol and tetraethyl lead slow down or suppress the formation of condensed carbon particles [133]. The effect of hydrogen and small concentrations of aromatic and unsaturated organic compounds may be explained in terms of a free radical mechanism.

Table 3.10. Schemes of the reactions leading to formation of soot in flame

Reaction	Comment
Pyrolysis of Hydrocarbons	(124)
$C_nH_{2n+2} \rightarrow \dot{C}_nH_{2n-1} \rightarrow C_mH_{2m} + \dot{C}_{(n-m)}H_{2(n-m)-1}$	
$\dot{C}_{n-m}H_{2(n-m)-1} \rightarrow \dot{R} + CH_2{=}CH_2$ ($\dot{R}$ mostly $\dot{C}H_3$)	
$CH_4 \rightarrow C_2H_6 \xrightarrow{-H_2} C_2H_4 \xrightarrow{-H_2} C_2H_2 \xrightarrow{-H_2} C$	
Formation of intermediate product C_2H_6 from CH_4	(151, 155)
$2\dot{C}H_3 \rightleftarrows C_2H_6$	
$\ddot{C}H_2 + CH_4 \rightarrow \dot{C}H_3 + \dot{C}H_3 \rightarrow C_2H_6$	

Table 3.10. (continued)

Reaction	Comment
Formation of radicals from CH_4	
During pyrolysis	
$CH_4 \rightarrow \dot{C}H_3 + \dot{H}$	
$CH_4 \rightarrow \ddot{C}H_2 + H_2$	
Inside flame	
$CH_4 + \dot{H} \rightleftarrows \dot{C}H_3 + H_2$	
$CH_4 + \ddot{O} \rightleftarrows \dot{C}H_3 + \dot{O}H$	
$CH_4 + \dot{O}H \rightleftarrows \dot{C}H_3 + H_2O$	
Reactions of acetylene	
$\begin{cases} C_2H_2 \rightleftarrows \dot{C}_2H + \dot{H} \\ C_2H_2 + \dot{H} \rightleftarrows \dot{C}_2H + H_2 \end{cases}$	
$\dot{C}_2H + C_2H_2 \rightarrow C_4H_2 + \dot{H}$	
$\dot{C}_2H + C_4H_2 \rightarrow C_6H_2 + \dot{H}$	
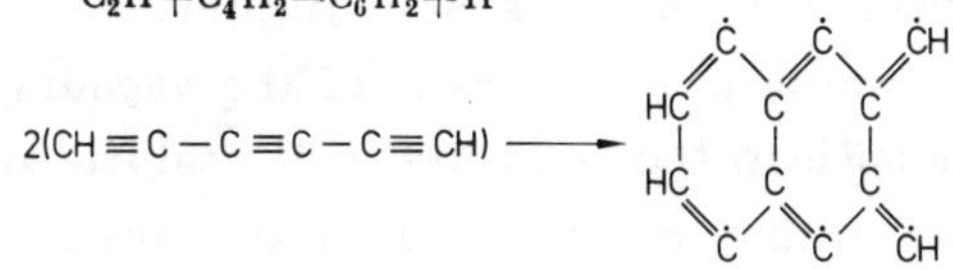	
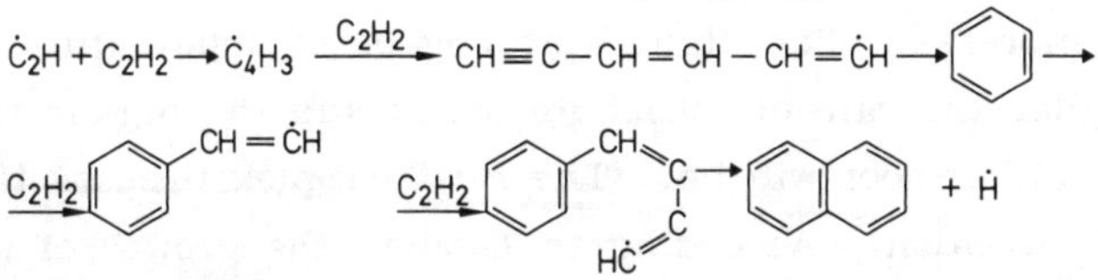	
Dehydrpolycondensation of aromatic hydrocarbons	(142, 156)
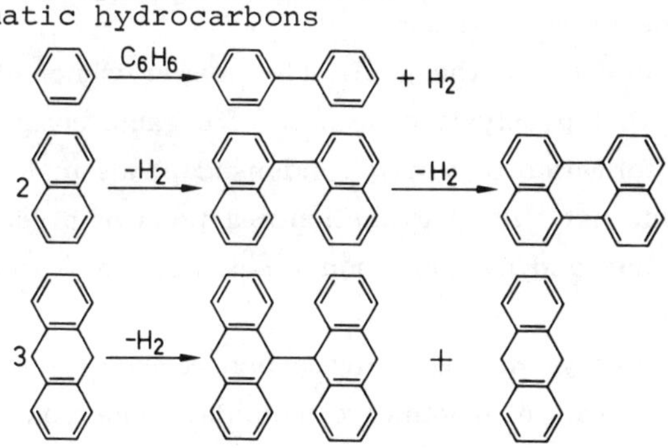	
Formation of soot particles' nuclei	M and M^+ - hydrocarbon and its ion; s, S^{++} -soot particles (neutral and charged)
$M + M \rightarrow S$ $S \rightleftarrows S^+ + \bar{e}$	
$M + M^+ \rightarrow S^+$ $M \rightleftarrows M^+ + \bar{e}$	
$M^+ + M^+ \rightarrow S^{++}$	

The effect of the introduction of organic compounds on the acceleration of soot formation inside a flame of acetylene increases in the following order [139]: benzene < cyclooctatetraene ≦ styrene < naphthalene < toluene < 2-methylnaphthalene < 1,3-butadiene ≤ phenanthrene < anthracene < 2-methylanthracene. Pyrene is less effective than anthracene. When a hydrogen atom is substituted by a methyl group the effect is greater than for an increase in the number of condensed rings in the molecule.

The accelerating effect of the compounds mentioned on the formation of soot inside the acetylene flame increases in the same order as the increase in affinity of the compounds towards the methyl radical (ability of the radical to remove the hydrogen atom from the carbon). It is assumed that the formation of solid carbon particles inside the flame is determined by the rates of such processes as the formation of nuclei of a new phase in the gas and the growth of particles and their gasification.

Tesner [140] shows that a specific degree of supersaturation is required in a system for nuclei to form (i.e. the ratio of the vapour pressure of the condensing substance to the pressure of its saturated vapour). This degree of supersaturation can be reached at a specific rate of temperature increase. The higher the degree of supersaturation in a system, the higher the rate of nuclei formation and the higher the dispersion of the resulting soot will be. The most complex process is the mechanism of nuclei formation. According to Tesner, the growth of nuclei and soot particles also occurs as a result of the direct destruction of hydrocarbon molecules on the formed surface.

How does the new phase develop in the gas? The answer is not clear. One thing that is certain is that pyrolysis processes of organic compounds inside flames lead to to the formation of a new condensed phase in the gas. These processes include not only degradation reactions of molecules but also molecule reconstruction and consolidation reactions, i.e. reactions of synthesis.

According to the "drop" theory, higher hydrocarbon compounds are formed as a result of the pyrolysis of organic compounds. The molecular weight and concentration inside local gas phase zones grow until the so-called dew point is reached (the partial pressure of the substance in the gas reaches the value of the saturated vapour pressure). This condition leads to condensation and to the formation of drops. Subsequent chemical reactions of dehydropolycondensation of the compounds in the liquid phase lead to the formation of solid particles with a high percentage of carbon [141].

The authors of reference [142] actually observed in an acetylene flame the formation of a fog made up of light-yellow droplets which darkened when the temperature was increased to 800°C and then turned into black solid particles. The percentage of carbon in the latter reached ~96% and the percentage of hydrogen ~4%. Parker and Wolfhard [143] and later Frazee and Anderson [139] assumed that intermediate unstable components are formed as a result of pyrolysis of hydrocarbons. "Drops" which lead to the formation of soot particles consist of molecules of higher hydrocarbons that have not undergone condensation under the influence of Van der Waals forces. They represent a complex or cluster of highly active intermediate components. According to Frazee and Anderson, the cluster consists of highly unsaturated particles or molecules of a free-radical type.

Recently a theory was developed stating that positively charged hydrocarbon fragments are the nuclei of soot particles inside the gaseous phase. Processes of nucleation and condensation of compounds occur more rapidly in the presence of positively charged hydrocarbon ions. It was discovered that the larger positively charged ions inside the soot-forming flames of hydrocarbons belong to the higher hydrocarbons. With an increase in size of the ionic particles along the flame's height, there is a proportional increase in the soot particle size distribution.

According to the calculations of Japanese scientists [145], the charged particles favour the decrease of vapour pressure over the surface of the substance approximately in proportion to the square of the electric charge and in inverse proportion to one quarter of the drop radius. A benzene molecule in the presence of charged particles, for instance, can form drops under conditions that are far from a state of saturation. A polycyclic aromatic hydrocarbon with 14 aromatic rings ($C_{42}H_{16}$) is capable of forming drops inside the flame zone at ~750°C in the presence of a triply charged ion. It should be mentioned that the formation of carbon particle nuclei is prominent in the zone of flame reactions, where the concentration of active particles (radicals, ions) is the greatest. Inside the front of the flame, the concentrations of active particles (radicals, atoms, ions, electrons) differ from the concentrations at thermodynamic equilibrium, which indicates an absence of equilibrium in the zone of flame reactions.

The formation of charged particles in a flame can take place as a result of collisions between atoms and molecules with a high kinetic energy, which is why the higher the temperature the greater the probability of ionisation. The thermal emission of electrons and their "sticking" to neutral atoms or molecules leads to the formation of positively and negatively charged particles. The lower the ionisation potential of an atom or a molecule, the easier the detachment of an electron.

However, a nonequilibrium concentration of charged particles in the front of the flame leads us to the conclusion that there are other ways for these particles to form, apart from the collision of electrically neutral particles. Chemoionisation is one of the important ionisation processes inside a flame.

Chemoionisation of hydrocarbon fragments inside a flame occurs as a result of specific chemical reactions. Reactions between methine radicals and oyxgen atoms are important for the formation of positive hydrocarbon ions inside the flames of hydrocarbons.

$$\cdot\dot{\underset{\cdot}{C}}H + \dot{\underset{\cdot}{O}} \rightarrow CHO^+ + e^-$$

This reaction is exothermal in the case of particles in their ground electronic state. A similar reaction between oxygen atoms with larger hydrocarbon radicals is probable. A correlation was determined inside a stoichiometric hydrocarbon flame between the degree of flame ionisation and the number of carbon atoms in a molecule of fuel [146].

Such positively charged ions as C_n, H_n^+ (n = 3 to 10) and H_3O^+ were discovered in the reacting zone of hydrocarbon flames [147]. The processes of charge transfer during the ionic molecular reactions could lead to the generation of new positively charged hydrocarbon ions.

Reactions between radicals and the H_3O^+ ion:

$$\dot{R} + H_3O^+ \rightarrow R^+ + H_2O + \dot{H}$$

and metal ions:

$$\dot{R} + Me^+ \rightleftarrows R^+ + Me$$

can be classified as reactions causing the ionisation of hydrocarbon radicals and are important from the viewpoint of soot formation. Formation of H_3O^+ ions can occur according to the exothermic reaction $CHO^+ + H_2O \rightarrow H_3O^+ + CO$. A survey has been carried out of the ionic processes inside flames [148].

Without considering the chemistry of the process, Howard [149] proposed the following physical model of hydrocarbon condensed particle formation inside a flame. In the homogeneous gaseous phase, positively charged hydrocarbon ions of the $C_nH_n^+$ type constitute the nuclei of a new phase. Neutral molecules gather around the ions. During the initial stage of the growth of such clusters, ions keep their positive charge. The aggregation of ions is not possible until a critical size of particle is reached on account of the electrostatic repulsive forces. This size depends on the kinetic energy of the particles and corresponds to the size of soot crystallites (20-30 Å). An agglomeration of crystallites and a growth of the surface then follows as a result of a heterogeneous reaction. At that time, spherical carbon particles are formed. A redistribution of charges occurs during their agglomeration. Particles can either become negatively charged or neutral. The surface reactions are accompanied by a recombination and emission of electrons. Dehydration increases the electrical conductance of the particles, affecting both the electrostatic forces of their interaction and the secondary aggregation processes of the particles. In this scheme the important factor is that the retention of the charge at the stage of soot crystallite growth implies the presence of ion-molecule or ion-radical reactions with the participation of both positively and negatively charged ions.

Similar-type homogeneous gas phase reactions within the flames of hydrocarbons have not been sufficiently studied, and the rate constants for the observed reactions are not known [131]. In the case of heterogeneous reactions, the situation is even more complicated.

In the light of all the concepts discussed about the formation of carbon in the flame, it would seem that action on the nucleation process and on the processes of oxidation or complete combustion of soot particles and their "precursors" is particularly effective for quenching the carbon formation process inside the flame. The heterogeneous oxidation of carbon has already been discussed in a previous chapter.

Various low-molecular organic substances enter the gaseous phase during the diffusion combustion of polymer materials as a result of pyrolysis of the material. Such substances undergo additional decomposition inside a preflaming zone and also inside high-temperature flame zones: they react with oxygen.

It should be mentioned that the origination of a new phase in a flame (formation of carbon) opens up the possibility of different heterogeneous reactions. For instance, the growth of soot particles may be possible on account of the reaction (decomposition of carbon monoxide on a surface): $2CO \rightarrow CO_2 + C$ (solid). The carbon surface can catalyse the decomposition of hydrocarbon gases. The decomposition of acetylene in the presence of "primers" (admixtures of sooty carbon) has an explosive character.

An analysis of the published data reveals the following chief chemical processes that lead to soot formation in a flame:

1. Gas phase decomposition of aliphatic hydrocarbons with the formation of acetylenes.
2. Polymerisation radical reactions with participation of acetylene, which lead to formation of low-molecular polyenes and their transformation into aromatic hydrocarbons.
3. Dehydropolycondensation of aromatic hydrocarbons with partial disintegration of aromatic nuclei (mainly those in a non-planar position) up to acetylene and its derivatives.
4. Chemoionisation of carbon compounds and their radicals (favourable for compounds with a low ionisation potential, I. With an increased number of π electrons, I drops to 4-5 electron volts for polyacetylene [150].
5. Growth of carbon hexagonal layers as a result of either ion-molecular radical or ion-radical reactions with acetylene, polyenes, $\dot{C}_2H$ and other radicals participating.
6. Oxidation of soot particle "precursors" and their complete combustion.

Almost all of these processes are accompanied by dehydrogenation. However, it was discovered that hydrogen retards the formation of soot during the pyrolysis of hydrocarbons.

The first of the processes mentioned occurs in the low-temperature flame zone. The next two also start in a low-temperature region but they are most intense near the front of the flame and inside the oxidation zone.

According to Hommann and Wagner [138], there are two types of aromatic cyclic hydrocarbons inside a hydrocarbon flame as a result of the formation of soot: reactive ones and nonreactive ones. The latter do not participate in the formation of soot. They undergo condensation on the surface of carbon particles inside a postflaming zone during cooling. Generally, they are very easy to detect with mass spectroscopy during the heating of soot in a vacuum at high temperatures.

Reactive polycyclic aromatic hydrocarbons are the precursors of carbon. They are formed as a result of process 2 in the flame of aliphatic hydrocarbons, and as a result of processes 2 and 3 in the flame of aromatic compounds. Both ionised and uncharged aromatic hydrocarbons may participate in the formation of soot nuclei [144]. Examples of reaction schemes for a number of chemical processes leading to the formation of soot are given in Table 3.10. During the combustion of polymer materials, different compounds - pyrolysis products, to be specific - vaporise. All such compounds undergo a series of complex transformations inside preflaming and other flame zones before they are transformed into final combustion products.

Despite the vast variety of organic fuel compounds i.e. products of polymer pyrolysis, many combustion mechanisms and the chemical aspects of the combustion process can be illustrated by the combustion of hydrogen, carbon monoxide and methane, according to reference [131].

Combustion of hydrogen. The mechanism of hydrogen combustion has been studied in the most detail. The combustion of hydrogen is considered to be a process that simulates the combustion of a large number of fuel gases. The mechanism of hydrogen combustion has been interpreted from the viewpoint of branched radical chain reactions, which were initially developed by N.N. Semenov.

The total process of combustion of hydrogen may be represented as a combination of elementary initiation reactions, propagation reactions and branching and termination reactions. Hydrogen atoms, oxygen atoms and $\dot{O}H$ and $H\dot{O}_2$ radicals are the active centres of the combustion process. All of them were discovered in the zone of hydrogen combustion when concentrations of $\dot{H}$-, $\dot{O}$-, and $\dot{O}H$ centres were evaluated quantitatively.

When fuel-rich mixtures burn, the concentration of hydrogen atoms exceeds that of oxygen atoms. The concentration of hydroxyl radicals is much lower than the concentration of $\dot{O}$ and $\dot{H}$. $H\dot{O}_2$ and hydroxyl radicals are less active than $\dot{O}$ and $\dot{H}$ as active centres [157].

Depending on the experimental conditions (pressure, temperature), either of the elementary reactions may predominate and determine the total rate of the process. The most important elementary processes which indicate the mechanism of hydrogen combustion are represented in Table 3.11. In the same table the kinetic parameters of the reaction are given.

A method of separating out limiting stages has been proposed [158]. This method is based on mathematical modelling of the chemical process of hydrogen combustion. The authors studied the change in the hydrogen combustion mechanism as a function of pressure variation within the 10^{-1} to 10^{-5} mmHg range and temperature variation within the 500-1200°K range. The homogeneous reactions in a given volume were taken into account in the study, e.g. reactions beyond the first hydrogen ignition limit. As presented in Table 3.11, reactions 0 and 0′ indicate the initiation of the oxidation chains; 1, 5, 16, 17 = propagation of the chains; 2-4, 6 = branching; 7-10, 15, 20-23 = different reactions for the termination of active centres. Reaction 19 and reverse reaction 18 represent a replacement of active centres by less active ones.

High temperatures (>1500°K) are usually attained inside the reaction zone of the hydrogen flame at atmospheric pressure. Under these conditions $H\dot{O}_2$ and H_2O_2 concentrations are negligibly small. The formation of the active centres $\dot{H}$, $\dot{O}$, $\dot{O}H$ is explained by reactions 0, 1-4. In addition, according to the authors of work [159] the reaction $H_2 + O_2 \rightleftarrows H_2\dot{O}$ is possible with a rate constant of formation of the radical ($\dot{O}$) $k \sim 3 \times 10^{13} \times \exp(-52\,500/RT)$.

In the absence of heterogeneous destruction of active centres under these conditions, the destruction of the centres is the result of trimolecular reactions 7, 8, 10; also the trimolecular reaction for the less active centre formation $\dot{H} + \dot{O} + M \rightarrow \dot{O}H + M$ is assumed to be the mechanism of chain termination. Rate constants for trimolecular reactions depend on the nature of the M molecule. For a hydrogen-oxygen system in the case of combustion of fuel-rich systems, the M molecule is represented by the hydrogen molecule. In Table 3.11 the parameters of reactions with $M = H_2$ are given.

Table 3.11 Elementary reactions of hydrogen combustion and their kinetic characteristics [1, 158]

No. of reaction	Reaction	Forward k_0, molecule/cm³·sec	Forward E kcal/mole	Reverse k_0, molecule/cm³·sec	Reverse E kcal/mole
0	$H_2+O_2 \rightleftarrows 2\dot{O}H-18.6$ kcal	$2.52\cdot10^{12}$	39	$8.14\cdot10^{10}$	21
0'	$H_2+O_2 \rightleftarrows H\dot{O}_2+\dot{H}-56.3$ kcal	$4.92\cdot10^{13}$	56.8	$\sim10^{13}$	0
1	$\dot{O}H+H_2 \rightleftarrows H_2O+\dot{H}+14.7$ kcal	$2.24\cdot10^{13}$	5.24	$9.90\cdot10^{13}$	20.3
2	$\dot{H}+O_2 \rightleftarrows \dot{O}H+\underset{\cdot}{O}-16.6$ kcal	$1.55\cdot10^{14}$	16.73	$1.16\cdot10^{13}$	0.705
3	$\underset{\cdot}{O}+H_2 \rightleftarrows \dot{O}H+\dot{H}-1.9$ kcal	$2.46\cdot10^{13}$	9.84	$1.07\cdot10^{13}$	7.9
4	$\underset{\cdot}{O}+H_2O \rightleftarrows \dot{O}H+\dot{O}H-16.6$ kcal	$8.70\cdot10^{13}$	18.46	$8.65\cdot10^{12}$	1.49
5	$\dot{H}+O_2+M \rightleftarrows H\dot{O}_2+M$	$3.60\cdot10^{15}$	0	$3.01\cdot10^{15}$	47.8
6	$\dot{H}+H\dot{O}_2 \rightleftarrows \dot{O}H+\dot{O}H+37.8$ kcal	$1.70\cdot10^{14}$	2.0	$2.76\cdot10^{3}$	40.8
7	$2\dot{H}+M \rightleftarrows H_2+M$	$3.60\cdot10^{15}$	0	$1.46\cdot10^{16}$	104
8	$\underset{\cdot}{O}+\underset{\cdot}{O}+M \rightleftarrows O_2+M$	$1.10\cdot10^{15}$	0	$2.67\cdot10^{16}$	118
9	$\dot{O}H+\dot{O}H+M \rightleftarrows H_2O_2+M$	$1.11\cdot10^{16}$	-1.92	$7.4\cdot10^{17}$	47
10	$\dot{H}+\dot{O}H+M \rightleftarrows H_2O+M$	$4.70\cdot10^{16}$	0	$8.59\cdot10^{17}$	119
11	$\dot{O}H+\underset{\cdot}{O}+M \rightleftarrows H\dot{O}_2+M$	$\sim10^{13}$	-	$7.00\cdot10^{14}$	60.5
12	$\dot{H}+H\dot{O}_2 \rightleftarrows H_2O+\underset{\cdot}{O}+54.5$ kcal	$\sim10^{13}$	0	$1.64\cdot10^{13}$	55.7
13	$\underset{\cdot}{O}+H\dot{O}_2 \rightleftarrows OH+\dot{O}_2+54.5$ kcal	$\sim10^{13}$	-	$\sim10^{13}$	53.5
14	$2H\dot{O}_2 \rightleftarrows H_2O_2+O_2+40.4$ kcal	$\sim10^{13}$	0	$1.30\cdot10^{14}$	40
15	$\dot{O}H+H\dot{O}_2 \rightleftarrows H_2O+O_2$	$\sim10^{13}$	0	$2.19\cdot10^{14}$	71.8
16	$H_2+H\dot{O}_2 \rightleftarrows H_2O+\dot{O}H+52.5$ kcal	$1.55\cdot10^{14}$	28.5	$1.13\cdot10^{14}$	82.3
17	$H\dot{O}_2+H_2O \rightleftarrows H_2O_2+\dot{O}H-30.7$ kcal	$1.32\cdot10^{13}$	33.4	$7.95\cdot10^{12}$	1.6
18	$H\dot{O}_2+H_2 \rightleftarrows H_2O_2+\dot{H}-16.0$ kcal	$1.55\cdot10^{14}$	28.5	$4.17\cdot10^{14}$	11.8
19	$\dot{H}+H_2O_2 \rightleftarrows H_2O+\dot{O}H+68.4$ kcal	$4.17\cdot10^{14}$	9.0	$1.16\cdot10^{14}$	79.5
20	$\dot{H} \rightarrow$) Heterogeneous				
21	$\dot{O} \rightarrow$) termination				
22	$\dot{O}H \rightarrow$) (destruction on the				
23	$H\dot{O}_2 \rightarrow$) wall)				

Combustion of carbon monoxide. Oxidation of carbon monoxide has recently been studied in depth like the oxidation of hydrogen. It is known that the ignition and combustion of carbon monoxide is affected to a large degree by water vapour, and hydrogen and some of its compounds. When there are no moisture vapours the carbon monoxide does not ignite, even at a very high temperature (~1400°C).

It was discovered that the development of a chain process of carbon monoxide oxidation in the presence of hydrogen is connected to the initial origination of active centres, according to reactions O and O′ (see Table 3.11). According to Lewis and Elbe [3], the origination of active centres in the presence of moisture could be caused by the reaction $H_2O + CO \rightleftarrows CO_2 + H_2$ (surface), which is catalysed by the surface.

The subsequent reaction between molecular hydrogen and oxygen leads to the production of an active centre and the participation of elementary reactions. Thus the mechanism of carbon monoxide oxidation inevitably includes a large number of elementary reactions from the oxidation-of-hydrogen process.

The accelerating effect of moisture vapour on the oxidation of carbon monoxide has been linked to forward reaction 4 and reverse reaction 1 (see Table 3.11). Hydroxyl radicals are formed as a result of these reactions. When these radicals attack carbon monoxide molecules, formation of the basic product of the oxidation reaction, CO_2, occurs:

$$CO + \dot{O}H \overset{K}{\rightleftarrows} CO_2 + \dot{H}$$

The rate constant for this forward bimolecular reaction (at 300-1000°K) is $k_+ = 7.1 \times 10^{12} \exp(-7700/RT)$ and the equilibrium constant is $K = k_+/k_- = 2.34 \times 10^{-3} \exp(109\,100/RT)$ ([159], p. 330). It is assumed [3] that at relatively low temperatures and pressures, the formation of carbon dioxide may take place according to the reaction:

$$H\dot{O}_2 + CO \rightleftarrows CO_2 + \dot{O}H$$

if a noticeable role of heterogeneous reactions is possible.

In the hot flame, however, the basic reaction of CO_2 formation is not the second but the first reaction given above. The direct attack of carbon monoxide by an oxygen atom:

$$CO + \ddot{O}(+M) \rightarrow CO_2(+M)$$

must be excluded from consideration, since such a reaction implies the violation of spin conservation [159].

Combustion of Methane. Methane is the most stable hydrocarbon of pyrolysis of organic compounds formed during the combustion of polymers. Thus the combustion of methane can be discussed as a process that to some extent simulates the gas phase flame reaction during the combustion of polymers. Properties similar to those in the mechanism of hydrogen combustion are usually attributed to the combustion of methane, although it is certain that the first process is more complex. The quantitative data however, are, incomplete.

During the combustion of hydrocarbons, thermal degradation processes occur, in addition to oxidation, and intermediate compounds form in different ways.

Active centres during the oxidation of hydrocarbons (similar for $\dot{H}$, $\dot{O}H$ and $H\dot{O}_2$) are $\dot{R}$, $R\dot{O}$ and $R\dot{O}_2$ (for methane, $\dot{C}H_3$, $CH_3\dot{O}$ and H_3COO accordingly). However, in the chain process of hydrocarbon substrate combustion a large role is played not only by alkyl, alkoxy- and peroxyhydrocarbon radicals but also by all active centres which are specific to the chain oxidation of hydrogen. This is connected with the fact that thermal degradation and oxidative pyrolysis of hydrogen is accompanied by dehydrogenation, the detachment of hydrogen atoms from molecules. The participation of the individual active centres in the process of methane oxidation depends on the establishment of this process (temperature, pressure). During the high-temperature combustion of methane, the reaction with participation of $H\dot{O}_2$ and $R\dot{O}_2$ radicals need not be taken into account, as is the case in hydrogen combustion when the mechanism of the process is examined. In hot flames, methane is oxidised to formaldehyde which thermally degrades to CO and H_2 if there is not enough oxygen in the front of the flame. Depending on how far the flame front has advanced, CO and H_2 are oxidised and thorough combustion occurs in the high-temperature reaction zone [114, 159].

The situation specific to the combustion of gas-rich mixtures is realised with the diffusion flames of polymers. In this case, the concentration of hydrogen atoms in a preflame, low-temperature zone increases the concentration of $\dot{O}$ atoms and $\dot{O}H$ radicals. The basic dominating reaction under these conditions, which forms a large number of methyl radicals from methane, is the reaction:

$$CH_4 + \dot{H} \rightleftarrows \dot{C}H_3 + H_2$$

The rate constant of the forward reaction is $k_+ = 5 \times 10^{14} \exp(-13\,500/RT)$ and the equilibrium constant is $K = k_+/k_- = 27.8 \exp(7850/RT)$ [159].

The initiation of the oxidation chain for methane during the combustion of fuel-rich mixtures occurs as a result of the thermal dissociation of methane with the formation of $\ddot{C}H_2$ and $\dot{C}H_3$ radicals, according to N.V. Lavrov (see Table 3.10). The reaction between methyl radicals and atomic oxygen is regarded as one of the basic reactions that leads to the formation of formaldehyde:

$$\dot{C}H_3 + \ddot{O} \rightarrow H_2CO + \dot{H}$$

and also the reaction between methyl radical and molecular oxygen:

$$\dot{C}H_3 + O_2 \rightarrow H_2CO + \dot{O}H$$

It is assumed that this is a two-stage reaction. First a peroxide radical is formed ($\dot{C}H_3 + O_2 \rightarrow CH_3O\dot{O}$) and then the radical instantaneously decomposes ($CH_3O\dot{O} \rightarrow H_2CO + \dot{O}H$). The formation of formaldehyde as an intermediate compound during the combustion of methane has been confirmed many times, in particular by using the method of tracer atoms [160]. If there is a deficiency of oxygen, the thermal degradation of formaldehyde inside the flame occurs through the following intermediate reactions:

$$\begin{array}{l} H_2CO \rightarrow H\dot{C}O + \dot{H} - 77 \text{ kcal/mole} \\ H\dot{C}O \rightarrow CO + \dot{H} - 34 \text{ kcal/mole} \\ H_2CO + \dot{H} \rightarrow H\dot{C}O + H_2 + 27.2 \text{ kcal/mole} \\ \underline{H\dot{C}O + \dot{H} \rightarrow CO + H_2 + 70.2 \text{ kcal/mole}} \\ 2H_2CO \rightarrow 2CO + 2H_2 - 13.6 \text{ kcal/mole} \end{array}$$

In the presence of oxygen, additional reactions involving participation of O_2, $\ddot{O}$ and $\dot{O}H$ radicals ought to be included in the mechanism of formaldehyde decomposition at high temperatures:

$$H_2CO + \dot{O} \rightarrow H\dot{C}O + \dot{O}H$$
$$H_2CO + \dot{O}H \rightarrow H\dot{C}O + H_2O$$
$$H\dot{C}O + \dot{O}H \rightarrow CO + H_2O$$
$$H\dot{C}O + \dot{O}H \rightarrow CO + \dot{O}H + \dot{O}$$
$$H\dot{C}O + \dot{O} \rightarrow CO + \dot{O}H$$

The summary net reaction for the oxidative degradation of formaldehyde is represented by the equation $H_2CO + 1/2\ O_2 \rightarrow CO + H_2O + 55.95$ kcal/mole.

The results of the study using water vapour containing isotropic oxygen, ^{18}O, formed a basis for the assumption that either molecular or atomic oxygen is the source of oxygen for the formation of carbon monoxide during the combustion of methane. Neither water vapour nor hydroxyl radicals from vapour serve as a source of oxygen for carbon monoxide [159].

The thorough combustion of carbon monoxide inside a methane flame at high temperatures occurs as a result of reactions between CO and hydroxyl radicals (see Chap. 2). Thus the mechanism of methane combustion under consideration includes elementary reactions of hydrogen oxidation, carbon monoxide oxidation, reactions of decomposition and oxidation of formaldehyde, and reactions with methane molecules and its primary radicals ($\dot{C}H_3$, $\ddot{C}H_2$).

A mechanism of methane oxidation has been studied in [161, 162], and in all 86 elementary reactions, with the participation of 17 different reactive species, were taken into account to describe the mechanism. The studies cited also evaluated data on the kinetic parameters of reactions with methane, hydrocarbon radicals, formaldehyde, and formaldehyde radicals participating in the reaction. Depending on the conditions of methane combustion, neither of the elementary reactions affects the characteristics of the process, according to calculations. In particular, reactions of excited oxygen molecules with the participation of such intermediate reagents as H_2O_2 and O_3 are not involved in the high-temperature oxidation of methane.

More complex transformations of methane molecules into C_2 carbon compounds and others, such as reactions leading to the formation of solid carbon inside a flame, are not generally taken into account when the mechanism of methane combustion is discussed. In a real situation, oxidation can occur at each stage of the transformation of methane molecules into condensed carbon products. The intensification of this process leads to a reduction in smoke liberation during the combustion of methane.

Chemical Aspects of the Reduction in Combustibility of Polymer Materials and Smoke Liberation during their Combustion

When chemists have developed new polymer materials with reduced combustibility they have successfully used an empirical approach to the problem, proceeding on the following grounds: In order to quench or retard the process of combustion of materials it is necessary to decrease and minimise the percentage of fuel component, increase the stability of the substance during heating and actively influence the decomposition of the substance and the flame processes.

The use of inorganic compounds for this purpose is the oldest example of the use of combustion retardant agents for materials of organic origin, dating back even to before our time.

Nowadays, different compounds are used that have the capacity to retard or terminate the combustion of polymeric materials. This is viewed as the most general approach to solving such an important problem. A change in elementary composition, i.e. in the chemical structure of the polymer, by introducing reactive compounds with flame-retardant functional groups (antipyrines), seems to be a very effective way of decreasing the combustibility of polymer materials.

A large number of different types of flame retardants are now known but chemists continue to synthesise new and more effective substances with more complex structures and compositions.

It is necessary to classify our knowledge about the action mechanism of different ingredients (including flame retardants) so that a rational selection of ingredients can be made. We should not neglect the fact that chemical factors are tightly linked to the physical properties of heat and mass transfer when retardation of combustion quenching processes is discussed. The retardation of oxidation chain reactions, for instance, also implies a deceleration in the rate of heat liberation.

If the combustion process of organic materials is assumed to be of a multistage nature then we should expect active intervention in the kinetics of chemical transformation at each stage to give apparently positive results in terms of retardation or even suspension of the combustion process. The most efficient way to actively interfere with the chemistry of the combustion process is to use different flame retardants, as stated.

Apparently all flame retardants with different structures can be subdivided into groups according to how the compounds (or the products of their transformation) participate in either of the stages or how they affect the combustion process as a whole in the zone of the combustion wave. In this case, flame retardants should be subdivided into substances that affect the chemistry of the processes in the condensed or gaseous phases or the processes in phase contact areas.

The following characteristics are used as microscopic criteria of the action mechanism of flame retardants within either the condensed or the gaseous phase of a combustion wave in polymeric materials [163]:

1. Change in the composition of volatile pyrolysis products in the presence of flame retardants
2. Change in the yield of nonvolatile residue (char)
3. Ability of flame retardants to precipitate out from a polymer substrate during the combustion process
4. Dependence of the combustion retardation effect on the nature of the oxidiser and on the structure of the polymer substrate
5. Sensitivity of the combustion retardation effect to changes in the pressure of the surrounding medium

Flame retardants with an action mechanism linked to gas phase flame reactions do not generally affect the change in composition of the volatile products of polymer pyrolysis and the yield of char. The flame quenching component enters the gas phase during the combustion process regardless of the nature of the substrate. Gas phase reactions are especially sensitive to changes in pressure and to the nature of the oxidiser. The retardation of combustion changes in the presence of flame retardants with a gas phase action mechanism when there is a change in the nature of the gaseous oxidiser. Various halogen-containing organic substances and metal salts that can enter the vapour phase at the temperatures reached on the burning surface of polymer materials can be included among such flame retardants.

It is most difficult to determine which flame retardants affect the mechanism of heterogeneous chemical reactions in terms of the macroscopic properties. In the overwhelming majority of cases, the effects of flame retardants on the combustion of polymeric materials are numerous. Flame quenching elements and functional groups that can affect the pyrolysis of polymers (reactions in the condensed phase) as well as flame reactions and heterogeneous oxidations, may be present together within the molecular structure of a flame retardant. It is thus difficult to elucidate the classification of flame retardants according to action mechanism and according to primary participation in reactions at certain stages of combustion. However, it is appropriate to characterise the general mechanism of action of different flame retardants in terms of their general behaviour during the pyrolysis and combustion of polymer materials. Such determinations must be based solely on the principal relationship between the mechanism of inhibitor action and the chemical structure of the substances and the direction of their transformations as a result of the effect of heat and oxidising agents.

Here we would like to point out the chemical aspects of combustion retardation through the use of flame retardants and discuss the most important data obtained from the study of retarded combustion with model systems.

The inhibitor mechanism of flame reactions in the presence of different additives has been the most studied. Such standard systems as laminar diffusion or premixed flames of hydrogen-oxygen or methane-oxygen (air) with specific kinetic characteristics of the combustion reactions are usually applied for this purpose.

Mechanism of Flame Retardant Action

Halogen-containing Organic Compounds. There are three types of halogen-containing compounds that are used for purposes of reducing the combustibility of polymer materials: derivatives of compounds with aliphatic, cycloaliphatic and aromatic structures. The nature and number of halogen atoms is varied in each type of structure. The compounds can represent either the reactive type of flame retardant (on account of the functional groups, they are able to become a part of the polymer structure) or the nonreactive type (these do not react chemically with a macromolecule). Low- and high-molecular compounds are used as flame retardants.

The effectiveness of flame retardation reactions using halogen-containing compounds with similar structures (differing only in the nature of the halogen) increases in the sequence [64]:

$$F < Cl < Br < I$$

As a function of their structure, the halogen compounds either undergo pyrolysis inside the condensed phase or evaporate and are destroyed in the gaseous phase.

Initial pyrolysis reactions of halogen-containing compounds generally lead to the formation of HX and RX_n but seldom to the formation of X_2, where X is a halogen atom. Detachment of HX from the aliphatic structured macromolecules is accompanied mostly by the formation of unsaturated systems. Transformation of these systems inside the condensed phase (C phase) causes the formation of a nonvolatile carbonised residue which, in the long run, affects the rate of combustion of the materials. We will not examine the effect of halogens on the processes in the C phase here, because such processes cannot be discussed without consideration of the nature of the polymer substrate.

The low-molecular compounds HX and $RX_n(X_2)$ enter the gaseous phase and participate in the chain-branching flame reactions of fuel gas oxidation.

The relationship between the efficiency of the substances' inhibitor action and the nature of the halogen and the effect of small concentrations of additives confirms that the inhibitor mechanism of halogens and their derivatives on the ignition and combustion of model systems has a chemical origin. It was discovered that halogens and their compounds do not affect the oxidation of carbon into carbon dioxide [164]. At the same time, they essentially inhibit the oxidation of CO into CO_2 [165] and the formation of formaldehyde during combustion of methane.

Different mechanisms have been suggested to describe the observed inhibitor effect of halogen derivatives on the flame reaction. All these mechanisms are based on the participation of different halogen-containing molecules and atoms or halogen ions in different stages of the radical chain combustion process.

Thus, according to Creitz [166], the inhibition of a flame is caused by reactions with the participation of oxygen atoms and the formation of intermediate compounds, oxyhalogens:

$$O + X + M \rightarrow OX + M + \Delta H$$

This reaction is exothermic; the enthalpies are ΔH = -64.29, -56.23 and -43.25 kcal/mole for X = Cl, Br and I respectively. Less exothermic is the bimolecular reaction:

$$O + X_2 \rightarrow OX + X + \Delta H$$

(ΔH = -6.13, -10.14 and -7.09 kcal/mole for X = Cl, Br and I).

Oxyhalogens, OX, react rapidly with active centres $\dot{H}$ and $\dot{O}H$, decreasing their concentration and thus inhibiting the rate of the oxidation process. In flames with insufficient amounts of fuel $[\dot{O}H] > [\dot{H}]$, the inhibition must be caused by reactions of the type:

$$OX + \dot{O}H \rightarrow HX + O_2 + \Delta H$$

(ΔH = -55.71, -48.07 and -44.82 kcal/mole for X = Cl, Br and I);

$$OX + \dot{O}H \rightarrow X + H\dot{O}_2 + \Delta H$$

(ΔH = +1, -7 and -20 kcal/mole for $\dot{X}$ = Cl, Br and I).

The formation of radicals $\dot{X}$ and $H\dot{O}_2$, less active than $\underset{\bullet}{O}$ and $\dot{O}H$, is equivalent to a termination of the chain. From a thermodynamic viewpoint the last reaction is hardly probable for oxychlorides (when X = Cl). The rate constants for these reactions are unknown. A number of studies have been concerned specifically with charged particles, X^-, when investigating the inhibition of a flame by halogen-containing compounds. Mills [167], for instance, believes that when an electron attacks a halogen-containing compound in a flame, the compound dissociates and forms a negatively charged halogen ion and a radical:

$$CF_3Br + e^- \rightarrow Br^- + \dot{C}F_3$$

Both the radical $\dot{C}F_3$ and the ion Br^- could participate in the inhibition process:

$$Br^- + \dot{H} \rightarrow HBr + e^-$$

The excess energy is carried away by the electron.

It is assumed that halogen-containing compounds are not effective as inhibitors in the oxidation reactions of a CH_4-N_2O flame on account of the electrons' stronger affinity for N_2O than for a halogen [168]. In spite of the proof that there are ionic species inside such a flame, this hypothesis has not been developed.

Nowadays it is generally accepted that hydrogen halide, HX, plays an important role in the inhibition of gas phase combustion. In cases where the hydrocarbon halide fragment, RX, is entered in the gas phase, an initial reaction of HX formation occurs inside the fuel-rich flames: $RX + \dot{H} \rightleftarrows \dot{R} + HX$; if X_2 enters the gas phase, then the reaction $X_2 + \dot{H} \rightleftarrows HX + \dot{X}$ occurs. These reactions indicate the inhibition reactions with the formation of less active radicals.

Hydrogen halide decreases the concentration of the active centres ($\dot{H}$, $\dot{\dot{O}}$, $\dot{O}H$, $H\dot{O}_2$) in the flame. The forward reactions of chain termination that are important for hot flames are:

$$\dot{H} + HX \rightleftarrows H_2 + \dot{X} \qquad (A)$$

$$\dot{\dot{O}} + HX \rightleftarrows \dot{O}H + \dot{X} \qquad (B)$$

$$\dot{O}H + HX \rightleftarrows H_2O + \dot{X} \qquad (C)$$

In fuel-rich flames, reaction (A) is the predominant one.

Various studies have confirmed that hydrogen atoms are drawn out of a flame in the presence of HX. The nature of halogen atoms strongly affects the equilibrium constant of the reversible reaction (A). The more reactive the halogen atom, the greater the probability of the reverse reaction occurring and the less the inhibition effect that can be expected from HX.

A comparison of the equilibrium constants $K = k_+/k_-$ in reaction (A) for HBr and HCl shows that for HBr, the equilibrium constant $K_{HBr(A)}$ increases when the temperature of the flame decreases, and is equal to 75, 1605 and 419 000 at 1580°K, 100°K and 600°K respectively.

For HCl, $K_{HCl(A)}$ is 2.5, 2.9 and 3.7 respectively [169]. Thus HCl is not as good an inhibitor as HBr. It is interesting that the effectiveness of the inhibitor action of HBr decreases at a progressive rate as the temperature increases. This means that the inhibitor effect of HX ought to be noticeable at an early stage in low-temperature zones of the flame.

The observed effect of inhibition on the ignition and combustion of the model system in the presence of HX cannot be described by participation in reaction (A). According to the authors of reference [169], HX may be considered as the same kind of homogeneous recombination catalyst for H atoms which inhibits the participation of H in the branching of the oxidation chain:

$$X + X + M \rightleftarrows X_2 + M \qquad (D)$$

$$H + X + M \rightleftarrows HX + M \qquad (E)$$

$$X + H\dot{O}_2 \rightleftarrows HX + O_2 \qquad (F)$$

$$X_2 + \dot{H} \rightleftarrows HX + X \qquad (G)$$

The formed HX molecules again react with hydrogen atoms, which decreases their concentration and retards the whole combustion process.

In general, the effectiveness of the inhibitor action of HX in fuel-rich flames is determined by the values of the equilibrium constants of reactions (A) and (G) and also by the constants of the reaction rates for (D), (E) and (F). Kinetic parameters of the forward reactions and equilibrium constants of the reversible ones are given in [169]. Active $\dot{O}$ and $\dot{O}H$ particles are present inside the high-temperature zone of a flame, which is why reactions (B) and (C) are of great importance.

The nature of the hydrocarbon radical $\dot{R}$ formed from RX_n is not taken into account in the scheme representing the flame initiation process. The nature of the hydrocarbon radical, however, determines not only the stability of the C-X bond inside the compound and the molecular or radical nature of the dehalogenation process but also the probability of subsequent participation of the radical in the combustion process. For data on the greater effectiveness of halogenated aliphatic substances as compared with derivatives of acyclic and aromatic structures see [170].

<u>Metal-Containing Compounds</u>. Salts, oxides, hydroxides and organic derivatives occupy one of the leading positions among flame retardants in reduced-combustibility polymer materials. Most of them have low vapour

pressures (i.e. they are nonvolatile in the normal combustion condition), which is why their action mechanism is most often related to the processes occurring in the condensed phase. In this particular case we do not mean a simple decrease in the percentage of fuel component, or a cooling of the reaction zone as a result of the endothermic formation of nonfuel products during decomposition of such compounds.

The generally accepted viewpoint is that compounds which affect the pyrolysis of polymers during combustion do so in such a way that the yield of nonvolatile products increases; they are more effective than flame retardants with a gas phase action mechanism. Unfortunately, the information on how most flame retardants affect the processes in the C phase during the combustion of polymers is of a qualitative nature; the criteria for flame retardants with differing action mechanisms need to be defined more accurately.

Studies on the retardation of combustion of hydrogen or carbonic flames contain numerous examples of the use of various metal-containing compounds as combustion inhibitors. A number of metal-containing compounds are used as basic ingredients in effective powder-like agents for fire extinguishing [171]. In the case of the combustion of polymer systems, the transition of metal-containing compounds into the gas phase is possible not only when relatively volatile substances are used but also when these form during the high-temperature transformation of the C phase.

When the situation is favourable for metal-containing compounds to enter the flame zone, then their participation in the chemistry of the flaming processes may follow either by a homogeneous mechanism or by a heterogeneous mechanism.

A homogeneous process (at the molecular level in the gas phase) is most often observed with low concentrations of metal-containing compounds. With higher concentrations the probability of condensed particles forming inside the flame increases. The heterogeneous mechanism of flame process inhibition then prevails. A study of the inhibition of hydrocarbon flames by various compounds showed that some metal-containing compounds retard the development of the combustion process more effectively than halogen-containing, organic compounds [172]. The effectiveness of combustion retardation depends not only on the nature of the metal but also on the nature of the fuel substrate [163].

The inhibition action of metal-containing compounds is related to the participation of the compound in the catalysis of destruction of the active centres (atoms, radicals) which are responsible for the development of the chain combustion process. Whether heterogeneous or homogeneous mechanisms of active centre destruction occur in the presence of metal-containing compounds depends on a large number of factors.

Supporters of the homogeneous inhibitor action mechanism for metal-containing compounds assume that the compounds evaporate inside the flame, decompose and produce intermediate active forms of the compounds. Such active forms are, for instance, hydroxides of metals [173].

Alkali metal hydroxides are easily formed inside a flame when decomposition of metal salts occurs in the presence of water vapour. Inside fuel-rich flames, interaction between alkali metals and water vapour (combustion product) may take place according to the equilibrium reaction:

$$Me + H_2O \rightleftarrows MeOH + \dot{H}$$

In the case of lithium metal, for instance, the rate constant for the forward reaction is very high; according to reference [174] in the temperature interval 1600-2400°K, $k \sim 10^{-9} \exp(40\ 000/RT)$ $mmole^{-1} \cdot sec^{-1}$. According to Kaskan [175], the formation of active intermediate metal-containing particles within a fuel-deficient flame occurs as a result of the direct oxidation of alkali metals, and the metal superoxide is formed:

$$Me + O_2 + M \rightleftarrows MeO_2 + M$$

Rate constants for the forward reaction within the H_2-O_2-N_2 flame at 1400-1600°K have been determined for Na, K and Cs [176]; they are equal to 0.82×10^{-33}, 1.02×10^{-33} and $2.1 \times 10^{-33} cm^6 \cdot particles^2 \cdot sec^{-1}$, respectively. Sodium superoxide dissociation requires an energy of 65 ± 3 kcal/mole [77]. Kaskan assumes that metal superoxides react easily with the active centres of a combustion process, forming more stable but active compounds.

$$MeO_2 + O \rightarrow MeO + O_2$$

$$MeO_2 + OH \rightarrow MeOH + O_2$$

The latter products, in turn, react with the active centres of the combustion process.

In general, the catalysis of active combustion centre recombination by an atom of metal formed during the decomposition process can be expressed by the following reactions:

$$\mathrm{Me + A \rightarrow MeA} \qquad \text{(A)}$$

$$\mathrm{Me + A + M \rightarrow MeA + M} \qquad \text{(B)}$$

$$\mathrm{MeA + A \rightarrow Me + AA} \qquad \text{(C)}$$

$$\mathrm{MeA + A + M \rightarrow Me + AA + M} \qquad \text{(D)}$$

where

Me metal atoms

A active centres ($\dot{H}$, $\dot{O}$, $\dot{O}H$)

M third particle (for instance, H_2)

At atmospheric pressure, a trimolecular collision is less likely than a bimolecular collision, and reactions (A) and (C) are the preferred ones. The probability active centres being destroyed according to reaction (C) increases with an increase in the enthalpy of active centre destruction. From this viewpoint the endothermic reaction

$$\mathrm{KOH} + \dot{\mathrm{O}}\mathrm{H} \rightarrow \mathrm{KO} + \mathrm{H_2O} + \Delta H$$

is less favourable than the exothermal reaction with the participation of hydrogen atoms:

$$\mathrm{KOH + H \rightarrow H_2O + K} - \Delta H$$

The destruction of hydroxyl radicals in this case is preferred as a result of reactions (A) and (B). The acceleration of hydrogen atoms and hydroxyl radical formation inside the flame with the addition of various metal-containing compounds has been confirmed experimentally many times [178, 179].

The results in study [179] confirm the homogeneous mechanism of flame

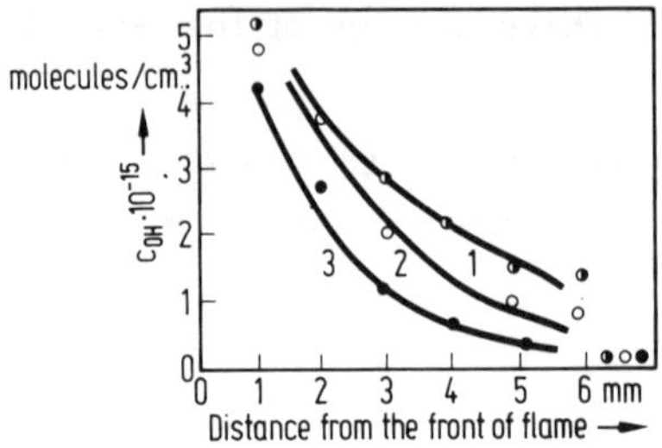

Fig. 3.16 Concentration profiles for OH radicals along the axis of a premixed methane-air flame in the presence of $NaHCO_3$ particles (size 1-10 micrometers). Percentage Na (atoms/cm^3): (1) 0, (2) 1×10^{14}, (3) 5×10^{14}.

inhibition by sodium salts. The authors discovered a direct correlation between the concentration of sodium atoms inside the flame and the change in temperature of the flame and the acceleration of hydroxyl radical recombination (Fig. 3.16). However, calculations showed that sodium is unusually active in the recombination of OH radicals. The major effect observed was explained by the easy polarisation of the alkali metal and its ability to form sufficiently stable complexes with water vapour:

$$Me + H_2O + M \rightarrow Me \cdot H_2O + M \qquad (E)$$

This complex is stabilised by induced dipole-dipole interaction. Its formation is assumed to be favourable because the equilibrium concentration of H_2O in the flame is two orders of magnitude greater than the concentration of the active centres (A).

Thus active centres react with a complex as follows:

$$Me \cdot H_2O + A \rightarrow MeA + H_2O \qquad (F)$$

and react with MeA according to reactions (C) and (D). According to studies, this scheme explains why potassium salts and rubidium salts are more effective in the inhibition of flame spreading than sodium salts.

A study of how the inhibitor action of alkali metal salts and salts of alkali earth metals affects the rate of spreading flames in methane-air mixtures was conducted by Rosser and coworkers [180]. According to calculations, salt particles (carbonates and bicarbonates) with sizes less than 5 microns are able to completely evaporate and decompose. It is assumed that at that time, the atomic metal is an active intermediate form.

The addition of CH_3Cl diminishes the effectiveness of salt action on account of the reaction between the halide and the metal forming a more stable compound.

In terms of the efficiency of the inhibitor effect of salts during combustion, these metals are ranked in the following order [163]: Li < Na < K < Rb. Birchall [181] concluded that the ability to form active intermediates during decomposition of alkali metal salts is a function of the nature of the anion. The effectiveness of flame inhibition decreases in the order: oxalates > cyanides > carbonates > iodides > bromides > chlorides > sulphates > phosphates. Consequently, alkali and alkali earth salts of organic acids are more effective catalysts of active centre recombination during the combustion process.

In the schemes of flame inhibition by metal-containing compounds outlined above, it is assumed that an initial form (atomic metal) develops. If, however, the fuel vapours of sodium or potassium are introduced laterally to the methane-air flame with an oxidising agent, no inhibitor effects are observed [173]. This confirms the importance of hydrogen atom recombination reactions inside the flame of fuel-rich systems.

A study of the formation of condensed particles inside the fuel gas flame during inhibition by certain metal-containing substances [178, 182] demonstrated the presence of such particles even before the formation of the high-temperature reaction zone. Nowadays, the heterogeneous catalysis of recombination of the active combustion centres is illustrated throughout the literature.

Figure 3.17 shows a decrease in the concentration of $\dot{O}H$ radicals inside a propane-air mixture flame with admixtures of metal salt aerosols. The

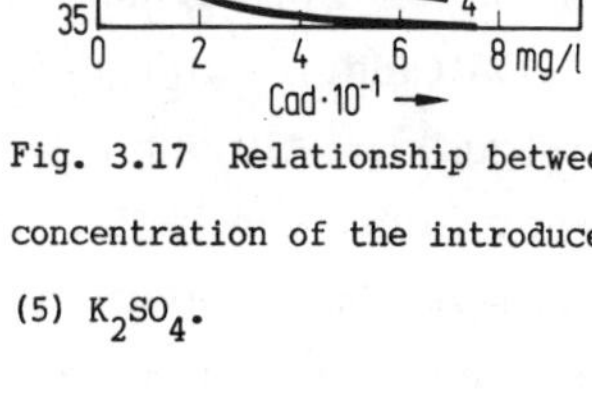

Fig. 3.17 Relationship between concentration of OH radicals (C_{OH}) in a propane-air flame and concentration of the introduced additive (C_{ad}). (1) NaCl, (2) KCl, (3) KBr, (4) KI, (5) K_2SO_4.

decrease in $\dot{O}H$ concentration correlates with the decrease in combustion rate [183]. On the basis of results obtained, conclusions have been drawn as to the heterogeneous mechanism of flame inhibition. It should be mentioned that, in this case, the order of variation of inhibitor efficiency attributed to the anion does not concur with that in reference [181]. At the same time, the relative order of the metal is still K > Na.

In the case of heterogeneous catalysis of inhibition of the fuel gas flame, the nature of the anion affects the "delivery" of the inhibition agent to an appropriate zone of the flame. For the same reason, sodium salts of aromatic acids, for instance, retarded the combustion of a propane-air mixture less effectively than sodium stearate or chloride [184].

Common properties in terms of the recombination of various active centres, the irreversible nature of recombination (unlike the recombination of centres by the homogeneous mechanism) and a relatively weak relationship between recombination and temperature are factors intrinsic to heterogeneous catalysis.

When heterogeneous recombination of oxygen atoms was studied on the surface of alkali and alkali earth metal salts, and also on the surface of anions, a high level of effectiveness of K and Rb halides was discovered. The effectiveness of Ca, Sr and Ba salts, by contrast, is less. Recombination of oxygen atoms is 2.5 times more active on the surface of Fe^{2+} sulphate particles than on the surface of Fe^{3+} sulphate [185].

All experimental data indicate the significant role of the nature of the metals in the recombination of active centres in the combustion process. It was discovered [178] that Mg, Cr, Mn, Sn, Ba and U reinforce the recombination of hydrogen atoms inside the fuel-rich mixtures H_2-O_2-N_2. The metals Na, Co, Ni, Cu, V, Zn, Ga, Th, Ge and La are not as effective [178]. Ca, Sb, Pb, Ti and Si compounds and many others display inhibitor activity [163].

Information on the kinetic parameters of the heterogeneous recombination of active combustion centres is scarce. Reference [186] evaluates the kinetic parameters for the adsorption of hydrogen atoms on the surface of Ag, Au and Cu and also for surface recombination: [2H(ads.) $\rightarrow$ H_2(gas)] and collision recombination: [H(gas) + H(ads.) $\rightarrow$ H_2(gas)]. Although these studies were conducted at low temperatures they indicate a high probability of surface recombination of hydrogen atoms in flame conditions.

Activation energy values in the case of a collision recombination reaction of hydrogen atoms for Ag, Au and Cu are 1.1, 1.4 and 1.3 ($\pm$ 0.2) kcal/mole, and the magnitude of the pre-exponential factor (lg A) is -8.74, -8.70 and -8.82 respectively. The activation energy values for the H surface recombination are 19.8 and 18.4 ($\pm$ 0.5) kcal/mole for Au and Cu respectively. When polymer materials with reduced flammability are developed, antimony oxides are very often used as synergists of halogen-containing flame retardants. Chemical transformations of antimony oxides in the C phase are an example (in this particular case) of the formation of volatile metal-containing compounds during the combustion of polymer materials. The synergistic effect on the system consists in Sb_2O_3-halide derivatives and polymer substrates (or flame retardants) and is responsible, to a large degree, for the formation of volatile halides of antimony ($SbCl_3$ or $SbBr_3$). The halides serve not only as suppliers of halogen atoms into the gas phase, thus increasing the retention time of the latter in the flame zone, but also as suppliers of metal atoms, which in themselves display an inhibitor effect during combustion.

The formation of volatile antimony derivatives when antimony trioxide reacts with hydrogen halide (the product of flame retardants or polymer substrate degradation) occurs through the intermediate stage of antimony oxyhalide formation:

$$Sb_2O_3 + 2HCl \rightleftarrows 2SbOCl + H_2O$$

At temperatures above 200-250°C (depending on the heating conditions) antimony oxychloride decomposes with an activation energy approaching 70 kcal/mole [163]. The thermal degradation of antimony chloride occurs in stages [163, 187]:

$$(1)\ 5[SbOCl] \xrightarrow{240-300°C} [2SbOCl\cdot Sb_2O_3] + SbCl_3$$

$$(2)\ 4[2SbOCl\cdot Sb_2O_3] \xrightarrow{340-380°C} 5[SbOCl\cdot Sb_2O_3] + SbCl_3$$

$$(3)\ 3[SbOCl\cdot Sb_2O_3] \xrightarrow{380-450°C} 4[Sb_2O_3] + SbCl_3$$

$$(4)\ [Sb_2O_3]\ (solid) \rightarrow 2[Sb_4O_6]\ (liquid)$$

Figure 3.18 shows the change in vapour pressure for $SbCl_3$ and Sb_4O_6 when SbOCl is heated under a high vacuum at a rate of 0.2 degrees/min

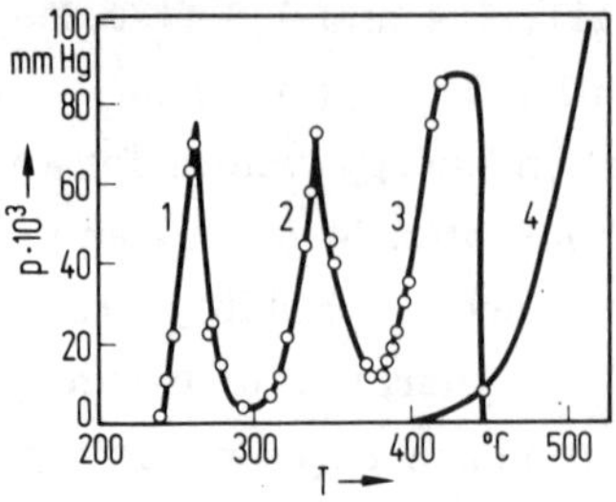

Fig. 3.18 Relationship betwen vapour pressure P for $SbCl_3$ (1-3) and Sb_4O_6 (4) and heating (decomposition) temperature for SbOCl in a vacuum. Heating rate 0.2 degree/min.

[163]. The gaseous products were analysed by means of mass spectroscopy. Sb_4O_6 sublimates in a vacuum at temperatures above 400°C. Under normal conditions of polymeric combustion Sb_4O_6, which boils at 1425°C, does not vapourise. Halides of antimony, SbF_3, $SbCl_3$ and $SbBr_3$, boil at 319°C, 223°C and 288°C, respectively, at atmospheric pressure. When SbX_3 (X = halogen) vapourises, it supplies RX, HX, X and Sb to different zones of the flame:

1. $SbX_3 + \dot{H} \rightarrow HX + SbX_2$
   $SbX_3 \rightarrow SbX_3^* \rightarrow X + SbX_2$
   $SbX_3 + \dot{R}(\dot{C}H_3) \rightarrow RX(CH_3X) + SbX_2$
2. $SbX_2 + \dot{H}(\dot{R}) \rightarrow HX(RX) + SbX$
3. $SbX + \dot{H}(\dot{R}) \rightarrow Sb + HX(RX)$

The participation of HX and RX in the inhibition of flame reactions has already been discussed in the relevant chapter. The participation of antimony atoms in the inhibitor process occurs according to reactions (A)-(D). For example:

4. $Sb + \dot{\underset{\cdot}{O}} + M \rightarrow SbO + M^*$
   $Sb + \dot{O}H + M \rightarrow SbOH + M^*$
   $SbOH + \dot{H} \rightleftarrows SbO + H_2$
   $SbO + \dot{H} \rightarrow SbOH^*$

According to Hastie [163] the reaction $Sb + H_2O \rightleftarrows SbO + H_2$ is also possible.

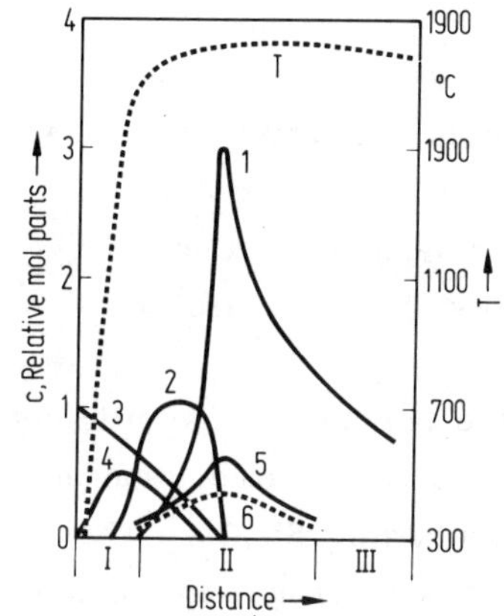

Fig. 3.19 Temperature (T) and concentration (1-6) profiles for products in a methane-oxygen flame in the presence of $SbBr_3$ (pressure 1 atm) (1) Br, (2) HB_r, (3) $SbBr_3$, (4) CH_3Br, (5) Sb, (6) SbO. Zones: (I) preflaming, (II) reaction, (III) postflaming.

A study of the concentration profiles of a methane-air flame with the addition of $SbCl_3$ and $SbBr_3$ demonstrated that RX and HX are formed essentially in a preflaming zone and Sb and SbO in the reaction zone (Fig. 3.19) [188]. Thus inhibition reactions envelop a wide spatial region of the combustion wave.

There is currently a tendency to partially substitute antimony trioxide by metaborates or borates of zinc, barium and oxides of other metals in the development of new, fire resistant polymer materials. Pitts [187] showed that the addition of some metal oxides affects the decomposition rate of SbOCl. Thus it is possible to control the supply of combustion retarding agent to the gas phase. Intensification of the action of the mixture of Sb_2O_3 with Fe_2O_3 and CuO and also with $2ZnO \cdot 3B_2O_3 \cdot 3.5H_2O$ is related to its reduction of the activation energy of SbOCl decomposition [187, 188].

In [189] the effect of different metal oxides as synergists of halogenated flame retardants on polystyrene was discussed. The synergist efficiency of oxides decreases in the order $SCO_2 > Sb_2O_3 > SnO_2 > ZnO_2 > MoO_3$. The energy of the metal-oxygen and metal-halogen bonds increases in the same sequence. Since the intermediate compounds that are formed are not well characterised, correlations with the volatility of existing haloid derivatives are vague. The latter factor is important (from the viewpoint of reagent supply to the flame) when the concentration of reagent and the retention time inside a reaction zone have to be determined. However, in general, the effectiveness of a reagent has to be a function of the rate constants for the reaction with active combustion centres and of the equilibrium constants for the reversible reactions in which the reagent participates. Also the constants, in turn, are determined by the electron structure of the reagents.

A close relationship was discovered between the portion of the antimony compound (per element) that is evaporated during polymer combustion and the change in OI (oxygen index) for a mixture containing Sb_2O_3. The amount of Sb used in the gas phase increases as the Cl component in the polymer increases (Table 3.12) [129].

Table 3.12 Effect of percentage of antimony inside the gaseous phase on the oxygen index of polyolefin

Polymer	Sb:C (in form of Sb_2O_3)	OI	Sb in the gaseous phase, % of initial concentration
$(C_2H_4)_n$	0	0.175	–
	0.005	0.180	~5
$(C_2H_{3.98}Cl_{0.02})_n$	0.013	0.190	~10
	0.005	0.229	31
$(C_2H_{3.94}Cl_{0.06})_n$	0.05	0.261	75

Other examples of how metal-containing compounds act through a gas phase mechanism can be mentioned. Martin and Price [190], for instance, used triphenylic derivatives of antimony, arsenic, bismuth and tributylacetates in order to reduce the combustibility of epoxy polymers. It is known that triphenylstibine evaporates without noticeable decomposition at a temperature close to 360°C. The effectiveness of its action is greater than that of the bromine flame retardant (with 46% Br). Thus when there are 7 Sb atoms per 10 000 atoms of C in the substrate (0.5 mol.% of additive) the oxygen index of the material is higher than in the case of 13.7 mol.% bromine-containing flame retardant, i.e. 350 Br atoms per 10 000 C atoms in the substrate; when the concentrations are equal (2.2 metal atoms per 1000 C atoms), the effectiveness of the compound decreases in the sequence Sb > As > Bi > Sn. Since arsenic compounds are very toxic, they are not of interest in this respect.

The relationship between the ignition index for epoxy polymers in the presence of $(C_6H_5)_3Sb$ and the nature of the oxidiser (O_2 and N_2O) confirms the gas phase mechanism for the effect of this flame retardant. It should, however, be mentioned that no evaluation has been made of the quantity of metal vapourised during the combustion of a polymer. The authors discovered that triphenylstibine is not effective in a mixture with polyethylene. The conditions on the surface of burning polymers, which decompose according to different mechanisms, differ substantially, which is why the conditions of flame retardant evaporation appear to be different.

The possibilities for using metal-containing compounds as flame retardants for polymer materials have hardly been developed. We think that attention should be directed to this class of flame retardants. The available data indicates that their application for the reduction of flammability of polymer materials could be promising. The employment of metals in the form of complex compounds with organic ligands may be one of the ways of inhibiting polymeric flames. Studies involving TGA and gas-chromatographic analysis of different metal-containing volatile complex compounds demonstrate their kinetic stability at sufficiently high temperatures [191].

Phosphorus and its Compounds. Phosphorus and its organic and inorganic compounds are widely used for the production of low-flammability polymeric materials. In spite of the practical importance of this class of flame retardants for polymer materials and the fact that they have been widely used, the chemical transformations of phosphorus-containing compounds and their participation in all stages of the polymer combustion process have not, in fact, been studied. The generally discussed action mechanisms of phosphorus flame-retardants have not been confirmed by experiments. It is still not clear to what degree the oxidation state of phosphorus affects the efficiency of the action of a phosphorus compound during combustion of organic substances. The experimental data is contradictory. In order to reduce the flammability of polymer materials both elemental phosphorus (red modification) and its compounds (three and five-valence state) are used.

An analysis of the percentage of phosphorus in the residue after combustion of polymer materials shows that phosphorus can remain in the condensed phase either almost completely or partially, or that it can completely vapourise ([24] p.64; [192] p. 200). The latter phenomenon depends not only on the chemical nature and initial concentration of the phosphorus flame retardant but also to a high degree on the nature of the polymer substrate and the composition and structure of the material. The relationship between the amount of phosphorus remaining in the C phase and the nature of the polymer indicates the importance of the nature of the reaction between the polymers and the phosphorus compound.

A number of aspects of the action mechanism of phosphorus compounds havè been described in a survey by Lyons ([192] p.1). The function of phosphorus and its compounds as flame retardants is related to the following factors:

1. The specific effect of phosphorus compounds on processes occurring in the C phase during the combustion of polymer materials. Chemical transformations of polymers occurring at this time tend to increase the yield of nonvolatile char residues and decrease the amount of pyrolysis fuel products.
2. The formation of a surface glass or a viscous melted layer of polymetaphosphoric acid. This layer serves as a physical barrier to heat transfer from the flame to the polymer and to diffusion of reagents.
3. The inhibition of gas phase flame reactions.
4. The effect of carbonised polymer pyrolysis product on heterogeneous oxidation.

It was demonstrated with numerous examples that many phosphorus flame retardants of the reactive and additive types contribute to the formation and increase in yield of nonvolatile carbonised residues during the pyrolysis and combustion of polymers. The detailed mechanism of this effect is not clear, as is the case with the mechanism for the carbonisation of polymers.

An analysis of the published data allows one to conclude that phosphorus flame retardants or products of their transformation serve as agents and catalysts for the substituent detachment reactions in the macromolecular chain, for cyclisation reactions and for other reactions of polymers. As has already been mentioned, such reactions favour the formation of the carbon skeleton. Pyrolytic dehydrogenation reactions and dehydration and dehydrohalogenation reactions of organic compounds are aided by the presence of phosphorus compounds. In this case, one of the most important factors to be considered is the relationship between the chemical structure of phosphorus-containing compounds and their reactivity, which determines the specific interaction with a polymer substrate under the conditions of its preliminary treatment and combustion.

The most widely studied reaction involved the effect of phosphorus flame retardants on cellulose materials. In the presence of phosphorus-containing compounds, the decomposition of cellulose occurs in such a way that a decrease is observed in the amount of fuel gases and an increase in the amount of char residue. The amount of char residue can attain a value of almost 80% of a polymer's dehydration, according to the scheme [64, 193]:

$$(C_6H_{10}O_5)_x \rightarrow 6xC + 5xH_2O.$$

This scheme does not reflect the real mechanism of cellulose decomposition. Nevertheless, a tendency can clearly be traced towards a reduction in the combustibility of cellulose materials with an increase in the yield of char residue during combustion. Thus for cotton fabric treated with tetra (oxymethyl) phosphonium hydroxide (30%), the amount of char residue increases from 13.9 to 43.5%, and the oxygen index (OI) almost doubles (from 21.2 to 41.7%). It should be mentioned that the temperature at the start of the degradation of samples treated in this ways drops from 320°C to 274°C [194]. Lyons ([64] p. 165) studied two possible alternative mechanisms of cellulose dehydration in the presence of phosphorus compounds:

1. Preliminary esterification of polymeric hydroxyl groups and subsequent detachment of the connected fragments with formation of double bonds in the macromolecule's section (monomer unit). This esterification of cellulose functional groups reduces the possibility of disruption in the basic chain and the formation of a fuel product such as levoglucosan. This method is often used in practice to obtain modified cellulose materials with reduced flammability.
   Both original compounds and products of their degradation, orthophosphoric and polyphosphoric acids, could serve as esterification phosphorus agents.
   For example,

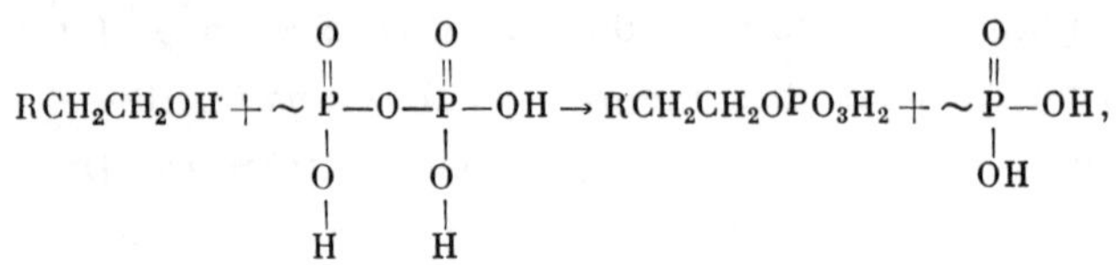

$$RCH_2CH_2OPO_3H_2 \rightarrow RCH{=}CH_2 + H_3PO_4.$$

2. Dehydration occurs according to a carbonation mechanism. As a result of this electrophilic detachment of water, when the oxygen atom forms protonised unshared electron pairs the formation of the carbonic ion is the most important stage of the mechanism:

$$RCH_2CH_2OH + H^+ \rightarrow [RCH_2CH_2\overset{+}{O}H_2] \rightarrow RCH{=}CH_2 + H_2O + H^+.$$

It should be mentioned that phosphorus compounds that can decompose into acids when heated are effective flame retardants. Completely neutralised metal salts of phosphoric acid are not effective. It was discovered, for instance, that with molten polymer based on nonsaturated oligoethers, triphenylphosphate is a less effective flame retardant than methylphosphonic acid. In the presence of the latter, formation of 28% nonvolatile residue was observed during pyrolysis of the polymer at 900°C. In the presence of methylphosphonic acid (by contrast to triphenylphosphate) the reaction of dehydropolycondensation with naphthalene and the formation of a carbon product with a well-ordered structure occurs very readily [195].

As in the case of the synthesis of oligoarylenes from aromatic hydrocarbons [196], it may be assumed that methylphosphonic acid causes the formation of intermediary carbocations, which induce the addition reaction of aromatic rings. Subsequent dehydrogenation produces stable structures.

In polymer chemistry, polyphosphonic acid is known as an activating agent in the electrophilic addition of carbonic acid derivatives to amines, and also in cyclisations (for instance, during production of polyheteroarylenes).

Phosphorus-containing compounds of an acidic nature are able to accelerate the detachment of hydrogen halide from specific halogen derivatives. This catalysis is similar to the one observed during dehydrochlorination of polyvinyl chloride under conditions when the formed HCl does not have enough time to leave the polymer system [53]. It is possible that the surface layer of polyphosphoric acid increases the possibility of a secondary reaction of the polymer pyrolysis products and, as a result, the formation of carbonised residue. From a thermodynamic point of view, the transformation of phosphorus-containing compounds into acids or oxides is a possible process during heating and oxidation.

The study of nonvolatile residues after pyrolysis and combustion of phosphorus-containing polymer systems indicates the presence of acids of phosphorus [24, 193, 195]. Additional possible schemes of phosphoric acid formation during the pyrolysis of polymers containing phosphorus flame retardants will be given below.

Inorganic Derivatives of Phosphoric Acids

$$(NH_4)_2\ HPO_4 \xrightarrow{t^\circ} H_3PO_4 + 2NH_3 \uparrow$$

When orthophosphoric acid is heated, condensed acids of phosphorus are formed [197]:

$$2H_3PO_4 \rightarrow HO-\overset{\overset{O}{\|}}{\underset{\underset{OH}{|}}{P}}-O-\overset{\overset{O}{\|}}{\underset{\underset{OH}{|}}{P}}-OH + H_2O,$$

pyrophosphoric acid

$$3H_3PO_4 \rightarrow HO-\overset{\overset{O}{\|}}{\underset{\underset{OH}{|}}{P}}-O-\overset{\overset{O}{\|}}{\underset{\underset{OH}{|}}{P}}-O-\overset{\overset{O}{\|}}{\underset{\underset{OH}{|}}{P}}-OH + 2H_2O,$$

triphosphoric acid

$$mH_3PO_4 \rightarrow -\left(-\overset{\overset{O}{\|}}{\underset{\underset{OH}{|}}{P}}-O-\right)_m- + (m-1)\,H_2O.$$

polymetaphosphoric acid according to Graham

When the temperature is very high (>1000°C) phosphorus-containing acids are completely split off water and become a phosphorus anhydride [198].

Polyphosphoric acids are characterised by high acidity ($pK \sim 0.5$) and very low volatility. Thus polyphosphoric acid is in itself an equivalent mixture to ortho-, pyro-, tri- and tetraphosphoric acids with a percentage of phosphoric anhydride (83.5 weight%). It has a boiling point of 536°C [199].

Organic Derivatives of Phosphorus-Containing Acids

It is known that, during oxidation, ethers of phosphorus-containing acids (phosphites) are easily converted into ethers of orthophosphoric acid (phosphates) by reacting with radicals according to Arbusov's reaction. In general, the P-O-C bonds are relatively unstable. They break when exposed to heat and are affected by alcohol, hydrogen halide acids and other polar reagents:

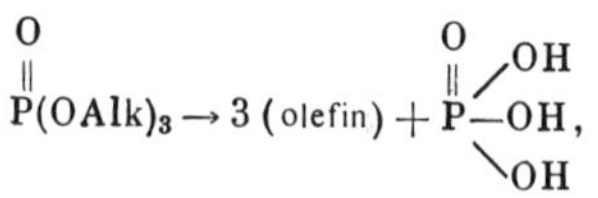

(A)

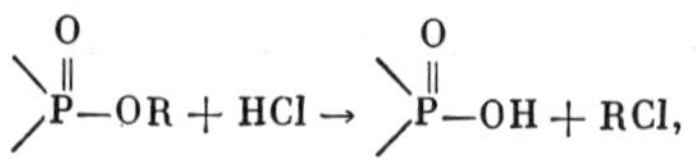

(B)

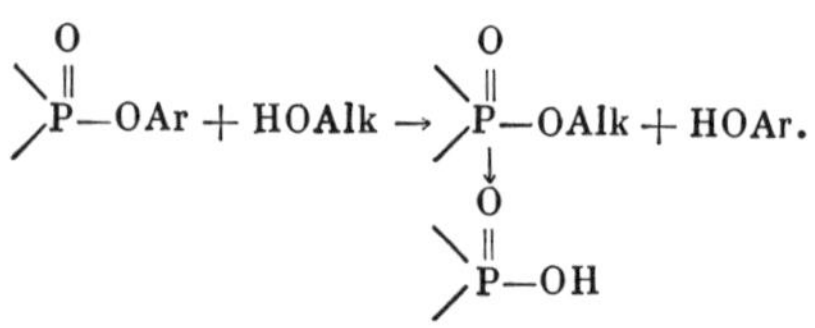

(C)

The increase in effectiveness of aromatic phosphates when combined with halogenated derivatives, as observed in [200, 201], could be caused by the conversion of triarylphosphate into acidic derivatives of phosphorus according to reactions similar to (B). This reaction prevents the conversion of triarylphosphate in the gas phase, reinforces the effect of the phosphorus compound on the processes occurring in the C phase and, at the same time, controls the supply of C_6H_5Cl in the gaseous phase, which is the source of HCl in the flame. Monophosphoric acids formed during reactions (A) - (C) can be converted into polyphosphoric acids by the scheme given above.

A number of patents describe red phosphorus as a flame retardant for polymer materials. Amorphous red phosphorus, P_4, melts at 460°C. Above 416°C it is able to sublime (sublimation heat = 19.7 kcal/mole) [197]. During sublimation, unlike the vapours of the white modification which consist of P_4 molecules, red phosphorus evaporates in the form of

P_2 molecules [202]. At high temperatures, a conversion of the white modification into the red modification is observed.

The mechanism of phosphorus oxidation in the condensed phase has been studied. The basic product of oxidation is phosphorus anhydride P_4O_{10}, which can sublime at ~347°C. It forms acids in the presence of moisture. The reaction between phosphorus vapours and oxygen: P(gas) + O_2(gas) → P_4O_{10}(solid), 730 kcal/mole, can take place in two different ways:

1. Cold flame reaction, which is characterised by luminescence and a small change in temperature. N.N. Semenov presented for the first time a chain-branched free-radical process that served as the basis for the creation and development of the chain reaction theory. However, at present, the mechanism of this process has only been developed as a general outline. The constants of the elementary reactions have not been determined.
2. Hot flame reaction, accompanied by a large increase in temperature and transformation of white phosphorus into the red modification. The characteristics of this reaction have been studied to an even lesser degree than reactions of the cold flame [203].

The following mechanism has been suggested for low-temperature phosphorus oxidation [204]:

Chain initiation:	$P_4 + O_2 \rightarrow P_4\dot{O} + \dot{O}$
Chain growth:	$P_4 + \dot{O} + M \rightarrow P_4\dot{O} + M$
Chain branching:	$P_4O_n + O_2 \rightarrow P_4O_{n+1} + \dot{O}$ (n = 1, 2 ....9)
Chain termination:	$\dot{O} + O_2 + M \rightarrow O_3 + M$
	$\dot{O}$ + X("poison") → stable product
	$\dot{O}$ + wall → stable product

A diatomic radical $P\dot{O}$ was directly identified from among the intermediate products by its radiation spectrum. Continuous background radiation in the visible region has been attributed to the radical $P\dot{O}_2$ [205]. Using the EPR method, Semenov and others [206] directly detected the spectrum of atomic oxygen in a phosphorus flame. This is a basic active centre in the branched-chain process of phosphorus oxidation.

A study of the nature of the radiating species in the chemoluminescence of phosphorus and its spectral characteristics is very important not only for comprehending the mechanism of phosphorus oxidation but also for studying the reaction of flame inhibition by phosphorus compounds. On the basis of a study of white phosphorus vapour oxidation with additional steam (H_2O and D_2O) under atmospheric conditions and varied temperature, the authors of [207] attributed radiation in the visible region (360-800 nm) not to PO_2 (or to $HOP\dot{O}$ [205]) but to the excited dimer particle $(PO^*)_2$. One member is electronically excited in the dimer (PO*...PO). In this way it was discovered that the radiating species during the oxidation of phosphorus are $P\dot{O}$ (groups of γ bands, 228-272.1 nm; β bands, 325-337 nm), HPO or DPO, which appear in the presence of moisture (510-650 nm) and $(PO^*)_2$. New bands, which appeared in the region close to 300 nm in the excitation spectrum at a temperature of 335°C, were assigned to P_2 particles. Inside the hot flames of phosphorus, reactions that involve the participation of P_2 particles are of great significance [203]. In fact, the reaction $P_2 + O_2 \rightarrow [P_2O_2] \rightarrow PO + PO^* \rightarrow 2PO + h\nu$, proposed by Rumpf [208], and the reaction $P_2 + O \rightarrow P + PO$ are possible. The latter reaction is thermally neutral [209].

When polymer materials that contain red phosphorus as a flame retardant burn, it is possible, according to [210] for there to be a direct reaction between the phosphorus and water vapour:

$$P_4 + 6H_2O \rightarrow 2HPO(OH)_2 + 2PH_3$$
$$HPO(OH)_2 + H_2O \rightarrow OP(OH)_3 + H_2$$
$$PH_3 + 4H_2O \rightarrow OP(OH)_3 + 4H_2$$

Condensed acids are formed if there is not enough moisture. Phosphine is highly toxic. Its oxidation occurs with the participation of the active particles $P\dot{H}$, $P\dot{O}$, HPO [205]. If the reagents are well dried, there will not be a reaction between oxygen and phosphine. There are indications of the formation of phosphine and phosphoric acids when polymer materials that contain red phosphorus are heated [211, 212]. To capture the liberated phosphine it was proposed that red phosphorus be used in combination with such substances as MoS_2 PbO_2, $AgNO_3$, CuO, activated coal and others [212].

We have already mentioned that the polymer can vaporise under combustion conditions as a function of the nature of the polymer substrate and the structure of the phosphorus flame retardants. The existence of an organic framework around the element, phosphorus, especially with a small number of carbon atoms in the radical, favourably affects the volatility of phosphorus compounds. The boiling point of aliphatic phosphorus ethers decreases in the following order: phosphates > phosphonates > phosphites. For example, $T_{boil.}$ for $(C_2H_5O)_3PO$, $(C_2H_5O)_2 \cdot P(O)C_2H_5$, and $(C_2H_5O)_3P$ is 215°C, 198°C and 158°C respectively. Aromatic derivatives increase the boiling point (for $(C_2H_5O)_2P(O)CH_2C_6H_5$, $T_{boil.}$ = 270°C).

The volatility of the incompletely substituted phosphorus compounds with free acidic groups drops abruptly. Those compounds are in the associated state and in a hydrogenated form and they can easily react with each other when heated, forming polyphosphorus compounds.

When phosphorus-containing compounds enter the flame zone during the combustion of polymers they are able to react with active centres of the combustion process and retard the rate of the process. The flame reactions of organic compounds with the participation of phosphorus compounds have not been studied sufficiently. A premixed hexane-air (stoichiometric) flame at atmospheric pressure was inhibited by additions of trimethylphosphate, halides and phosphorus thiohalides; it was discovered that the concentration of the admixture is very important [213]. When trimethylphosphate was introduced into a premixed, fuel-rich H_2-O_2 flame, a group of bonds at 510, 525 and 560 nm was discovered in the absorption spectrum. This group has been assigned to HPO particles [209]. By using mass spectroscopy, Hastie [163] discovered P_2, PO and PO_2 (Fig. 3.20) and, at lesser concentrations, HPO, P and PH among the intermediate products of $(C_6H_5)_3PO$ degradation within atmospheric methane and hydrogen flames.

The mechanism of $(C_6H_5)_3PO$ decomposition inside a flame is not clear. The formation of P_2 and P indicates a dissociation of the phosphorus compound and the possibility of equilibrium reactions taking place:

$$PO + \dot{H} \rightleftarrows P + \dot{O}H \qquad (D)$$

$$P + PO \rightleftarrows P_2 + \dot{\underset{\bullet}{O}} \qquad (E)$$

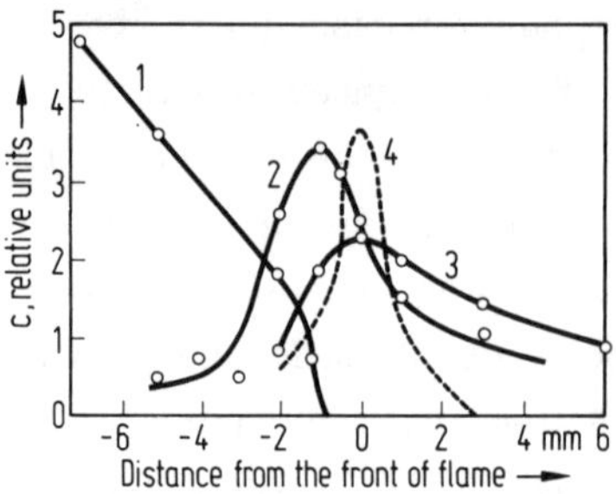

Fig. 3.20 Concentration profiles of products inside a methane-oxygen flame in the presence of $(C_6H_5)_3PO$ (pressure, 1 atm). (1) $(C_6H_5)_3PO$, (2) $P_2(0.5)$, (3) PO, (4) C_4H_2.

The direct equilibrium reaction (E) is endothermic, which is why the existence of PO in the presence of P is possible.

According to Hastie, the inhibition of a flame occurs as a result of catalysis of the destruction of hydrogen atoms by the reactions

$$\dot{H} + PO + M \rightarrow HPO + M$$

$$HPO + \dot{H} \rightarrow H_2 + PO$$

In the flame front, where the concentration of $\dot{O}H$ and $\dot{\underset{\bullet}{O}}$ is comparable to that of H atoms, the reaction

$$\dot{O}H + PO \rightleftarrows HPO + \dot{\underset{\bullet}{O}}$$

is possible. This reaction, which is the same as reaction (D), causes an interchange, not the destruction, of active combustion centres. It decreases the effectiveness of phosphorus compounds in inhibiting a flame. It is assumed that bi- and trimolecular recombination reactions of active combustion centres are possible [205, 214]:

$$PO + \underset{\bullet}{O}(+M) \rightarrow PO_2(+M)$$

$$PO + \dot{O}H(+M) \rightarrow HOPO(+M)$$

The identification of PO_2 and HOPO particles in the continuous part of a flame's emission spectrum (3600-8000 Å) has been criticised, as mentioned before. This does not, however, mean that these reactions are impossible. It should be mentioned that, according to an evaluation of the emission spectrum of a phosphorus vapour/oxygen flame, energy is released in the region of the β bands of PO particles; the energy release corresponds to an emission of one quantum of energy per (2 to 2.5) x 10^4 molecules of P_4. The energy emission corresponding to the continuous part of the spectrum corresponds to one quantum per approximately 130 molecules of P_4 [215].

In the form of triphenylphosphine, phosphorus occupies an intermediate position between Sb and As [191], and in the form of trimethylphosphate it is less effective than $SbCl_3$ [163] in terms of its effectiveness in flame inhibition. Low values of the limiting oxygen index (18-26%) for volatile aliphatic phosphorus-containing ethers (phosphites, phosphonates and phosphates) also indicate a poor level of effectiveness. In terms of action in the gaseous phase [192], phosphorus compounds such as the volatile $POCl_3$, PCl_3 and PBr_3 are more efficient in a flame than trimethylphosphate (by approximately 1.4-2 times) [212]. In turn, they are more effective (by 1.5-3 times) than Cl_2 and Br_2 when inhibiting a hexane flame [163].

If one considers the oxidation of oxyhalides in a flame [216]:

$$4POX_3 + 3O_2 \xrightarrow{500\text{-}1000°C} P_4O_{10} + 6X_2$$

then the contribution of hexane to the general effect of flame inhibition is substantial.

Synergism of phosphorus and halogen-containing substances is often attributed to the formation of oxyhalides and halides of phosphorus [217]. The deformation of oxyhalides, for instance, can occur according to the reaction

$$(RO)_3P = O + 3R'X \rightleftarrows POX_3 + 3\ ROR'$$

However, the generally positive effect that is observed with the joint use of phosphorus and halogen-containing flame retardants may be largely due to the change in direction of the chemical reactions in the C phase. The ratio of halogen to phosphorus and the concentration effects on the efficiency of flame retardants become even more apparent when the effect of the structure of inhibitor compounds is traced.

Examples of synergism and antagonism of phosphorus flame retardants with halogen and nitrogen-containing compounds in fireproofing materials have been discussed in a survey [218].

An important feature of phosphorus flame retardants is their unique ability to extinguish the incomplete combustion reactions of the carbonised residues of polymeric materials. As has been demonstrated in a previous chapter, heterogeneous oxidation of carbon is usually accompanied by the formation of carbon monoxide and carbon dioxide under the normal conditions of the process.

It was discovered that carbon monoxide does not oxidise phosphorus or reduce its oxide. In a mixture of phosphorus and carbon monoxide, phosphorus is oxidised selectively by air without the simultaneous formation of CO_2. In temperatures above 650°C, carbon dioxide oxidises phosphorus [197]. When char containing phosphorus undergoes controlled oxidation, the CO_2:CO ratio of the gaseous products decreases as a result of the enrichment of CO [219]. Carbon is also able to reduce phosphorus compounds into elementary phosphorus.

In view of the concepts regarding the heterogeneous oxidation of carbon, the most satisfactory explanation for the specific influence that phosphorus-containing compounds have on the quenching of the smouldering process is the partial inactivation of active centres on the surface. A shift occurs in the reaction of carbon oxidation, leading to CO formation. This process is followed by weak heat-liberation and promotes the retardation of the self-acceleration of exothermic carbon oxidation reactions.

The inactivation of carbon oxidation active centres by phosphorus compounds can occur through a physical shielding of the carbon surface. However, the effect of small quantities of phosphorus and the suppression of carbon smouldering by other compounds that form protective layers (by silicates, for instance) point to a specificity of the inactivation of carbon centres by phosphorus-containing compounds. We assume that a chemo-inactivation of carbon centres takes place.

As a result, an energetic uniqueness of the remaining centres occurs that is attributed to oxygen. The number of active centres of hydrogen oxidation decreases; these centres are the ones that participate simultaneously in the reaction with oxygen molecules. Thus the reactions

$$3C + 2O_2 \rightarrow 2CO + CO_2$$

$$2C + O_2 \rightarrow 2CO$$

are favoured.

Chemoinactivation is possible not only on account of the reaction between carbon centres and acidic groups of the phosphorus compound but also through the formation of acceptor complexes between the carbon system and phosphorus-containing compounds with vacant *p*-*d* orbitals. However, the geometric correspondence of chemosorption inactivators with the disposition of active centres of carbon oxidation and the stability of the forming bonds is important.

Retardation of Smoke Formation

It is logical to assume that the retardation and quenching of smoke formation during the combustion of organic substances is a function of reactions that could prevent or substantially retard the nucleation and condensation of the combustion products and, in the case of formation of condensed particles, would cause their destruction.

The formation of sooty smoke is typical of the combustion of polymers; it indicates an incomplete combustion of the organic component and retardation of the combustion rate for the system. The basic problems in the reduction of the flammability of polymer materials are linked with soot formation. Flame retardants that diminish the flammability of polymer materials cause an increase in their soot-forming capacity most of the time. It is still not clear how a synchronised reduction of both combustibility and soot formation is to be achieved during the combustion of polymer materials, which is why a balanced approach should be adopted to the problem. The quenching of soot formation during the combustion of gaseous systems occurs through a change in the fuel-oxidiser ratio and in the aerodynamic conditions of flow - to be specific through an increase in the rate of oxidiser flow. At the same time, there is a more complete burning of the fuel. The situation is similar for polymer materials in the majority of cases.

An electric field can be applied to a flame system; this field has a destructive effect on the charged particles inside the flame, retarding the nucleation of soot particles [220]. The most effective method is to use chemical additives that aid the destruction and reduction of charged particle concentration inside the flame and the catalysis of the oxidation reaction for the soot and its hydrocarbon predecessors.

The influence of different metal-containing systems on soot formation inside hydrocarbon flames has been the object of intensive study. Cotton and coworkers [221] studied the effect of adding nearly 40 different metals, introduced in the form of an aqueous salt solution, on soot formation inside a diffusion propane flame. It was discovered that Ba, Mo, W and Sr substantially reduced the rate of soot formation. The authors found that there is an increase in the rate of hydrogen atom destruction in the presence of the additives mentioned.

The quenching mechanism of soot liberation inside a flame has been explained by the participation of metal-containing particles in the gas phase catalysis of H_2O vapour decomposition, with the formation of $\dot{O}H$ radicals:

$$\dot{H} + H_2O \rightarrow \dot{O}H + H_2 \qquad (A)$$

$\dot{O}H$ radicals react directly with soot particles:

$$C(solid) + \dot{O}H \rightarrow CO + \dot{H} \qquad (B)$$

Participation of metal atoms in the catalysis occurs through a cycle of homogeneous gas phase reactions where metal oxides and hydroxides are the active particles:

$$MeO + H_2 \rightleftarrows MeOH + \dot{H} \qquad (C)$$

$$MeOH + H_2O \rightleftarrows Me(OH)_2 + \dot{H} \qquad (D)$$

$$Me(OH)_2 + (M) \rightleftarrows MeO + H_2O + (M) \qquad (E)$$

Such a reaction cycle occurs in a zone where the concentration of fuel is high but the concentrations of active centres of the combustion process ($\dot{H}$, $\ddot{O}$, $\dot{O}H$) are lower than the equilibrium ones. The origination of soot particles is in this zone. The reaction with metal-containing particles leads to activation of the (A) and (B) reactions. Since reaction (D) is endothermic, when the temperature drops the efficiency of the metal in quenching soot formation is reduced.

When a study was made of the effect of metal-containing compounds on soot formation in different zones, it was discovered that metal has a twofold action, namely an "anti-soot" and a "pro-soot" effect [222, 223]. The anti-soot effect on the flame zone of cracking fuel was explained as the destruction of positively charged hydrocarbon ions, which are the nuclei of soot particles when the ions interact with electrons. The source of the latter ions are metal compounds (for instance, $Me \rightleftarrows Me^+ + e^- - I$, where I is the ionisation potential, or $MeO \rightleftarrows MeO^+ + e^-$).

Inside the higher temperature flame zone, where charged particles already exist, a reduction in the positive charge of particles leads to a decrease of their repulsive forces. At the same time, the aggregation of particles leading to the formation of large and less rapidly burning soot particles, is promoted. An antibatic relationship exists between the ionisation potential of metals, I, and their capacity to emit electrons. As a result, alkali and alkali earth metals form ions more easily within a flame. A chemoionisation of the metal is possible along with thermo-ionisation [223]:

$$Me + \dot{O}H \rightarrow MeOH^+ + e^-$$

and

$$R^+ + Me \rightarrow Me^+ + \dot{R}$$

The last reaction reflects the dual role of metal in the soot formation process.

A change in conditions (concentrations of additives and oxidiser) may lead to a change in the nature of the participation of metal ions in soot formation. It was noticed that in the presence of excess oxygen, the "anti-soot" effect of the metal becomes more pronounced. At a relatively high concentration of alkali metal, the formation of soot decreases, and at low concentrations, it increases [224].

Bulewicz and coauthors [224] believe that the major effect of metal addition is a change in the rate of formation of hydrocarbon ions, i.e. the rate of soot nuclei initiation. The effect of metal on the coagulation of particles is a secondary one. The prevailing "anti-soot" effect involves metals that easily form hydroxides [223, 224]. In this case, regardless of the physical effect of the nucleation of soot particles, the oxidation of soot is catalysed according to the scheme described above. Thus the acceleration of oxidation (e.g. the acceleration of extinction) of soot particles inside a flame, combined with a physical factor for the destruction of the initial nuclei, is the most practicable way of quenching the formation of soot during the combustion of organic substances and polymers.

We have already mentioned that the formation of soot inside a hydrocarbon gas flame is quenched by additives such as SO_2 and NO. The mechanism of their action is also linked to the catalysis of water vapour decomposition [221]:

$$NO + H_2O \rightarrow HNO + \dot{O}H \text{ or } SO_2 + H_2O \rightarrow HSO_2 + \dot{O}H$$

$$NO + H_2 \rightarrow HNO + \dot{H} \text{ or } SO_2 + H_2 \rightarrow HSO_2 + \dot{H}$$

$$\dot{H} + H_2O \rightarrow H_2 + \dot{O}H \text{ or } \dot{H} + H_2O \rightarrow H_2 + \dot{O}H$$

SO_2 and NO are less effective at soot quenching than metal compounds. In the case of polymer systems, the formation of smoke depends first of all on conditions favourable to the liberation of volatile products and on their composition, i.e. it is important either in the general pyrolysis of polymers ("flameless" combustion) or in the pyrolysis and combustion of volatile products (flaming combustion). The effect of different factors (ventilation conditions, geometric characteristics of samples, intensity of thermal flow) on the smoke formation of polymers is given in a number of studies [225].

There are recommendations for the production of polymer materials (polyvinyl chloride, vinyl chloride, copolymers, foam polyurethanes) with the use of different metal-compound additives. For example, in the case of polymer materials based on vinyl chloride, copper cyanide and copper thiocyanate [226], iron-oxide mixtures of powdered iron with copper oxide or molybdenum oxide, mixtures of copper oxide, molybdenum oxide and vanadium oxide [227] are recommended. Dicyclopentadienyl iron is recommended for materials based on polyurethanes [228].

The use of oxygen-containing acid-type additives was also proposed. There are, for instance, recommendations for the use of adipic acid [229], fumaric acid [230] and isophthalic acid [231, 232]. The quenching mechanisms of soot formation by these recommended substances have not been studied.

We can assume that the enrichment of volatile products by oxygen-containing substances reduces the soot-forming capacity of volatile products as a result not only of a source of additional oxygen inside the flame but also of the tendency for oxygen-containing compounds to capture electrons and form negatively charged ions. The recombination of the latter with positively charged hydrocarbon ions (the nuclei of soot particles) decreases the probability of soot nucleation.

When metal-containing substances are used, alternative effects are possible:

1. Additions affect the pyrolysis of polymers in such a way that the composition of volatile products changes or the yield of char residues increases.
2. As a result of transformations in the C phase, additives form volatile products during combustion, then vapourise and quench soot formation according to the above scheme.

In the case of polyvinyl chloride (PVC), however, the substances serve as acceptors of hydrogen chloride. The metal chlorides formed are able to vapourise under the conditions of polymer combustion. In fact, $FeCl_2$ and $FeCl_3$ sublime at temperatures near 300°C and $MoCl_5$ at 268°C. Dicyclopentadienyl iron sublimes at about 400°C.

While studying the effect of ferrocene on the quenching of smoke formation during combustion of PVC, the authors of [233] disovered that the addition of 0.1-1.5 mol.% ferrocene causes a reduction in the amount of benzene during combustion of the polymer. A linear relationship between the formation of smoke during PVC combustion and the amount of benzene liberated emphasises the importance of this product as a predecessor of soot particles.

It was clearly shown by gamma-resonance spectroscopy that ferrocenes react readily in an inert medium with gaseous HCl, forming $FeCl_3$, small quantities of $FeCl_2$ and traces of ferrocene chloride. In the presence of oxygen in air at 500°C, it is principally Fe_2O_3 that is formed. If ferrocene is present then the yield of carbonised residue from the combustion of PVC decreases. It should be mentioned that in the carbonised residue, the percentage of iron reached 25% of the theoretically possible amount. The remaining 75% passed into the gaseous phase.

References

1. Kontratiev V. N., Nikitin E. E., Kinetika i mekhanizm gosophaznih reaktsyi. M.: Nauka, 1975.
2. Geydon A, Spectroskopiya plamen. V. N. Kondratiev, tr. M.: Izd-vo Inostr. lit., 1959.
3. Lewis B., Elbe G., Gorenie, plamia i vzriva v gazah. Frank-Kámenetskii, tr. from English, M.: Izd-vo Inostr. lit., 1948.
4. Kondratiev V. N., Emanuel N. M., Usp. chim., 1956, vol. 25, no. 2 pp. 393-431.
5. Wall L. A., Fire Research Abstracts and Reviews, 1971, vol. 13, no. 3, pp. 204-219.
6. Blazowski W. S., Cole R. B., McAlevy R. F. III, 14th Symposium (Intern.) on Combustion. Pittsburgh: Combust. Inst., 1973, pp. 1177-1186; AIAA Journ., 1968, vol. 6, No. 6, pp. 1137-1145.
7. Schultz R. D., Dekker A. O., 5th Symposium (Intern.) on Combustion. N.Y.: Reinhold Publ. Corp., 1955, pp. 261-271.
8. Chaiken R. F., Andersen W. H. et al, Combust. amd Flame, 1959, vol. 3, no. 3, pp. 227-235; J. Chem. Phys., 1960, vol. 32, no. 1, pp. 141-145.
9. Cantrell R. H. et al, AIAA Journ., 1963, vol. 1, no. 7, pp. 1544-1551.
10. Shteynberg A. S., Sokolova I. A., Dokl. AN USSR 1964, vol. 158, no. 2, pp. 448-451.
11. Shteynberg A. S., Ulihbin V. B. Phys. goreniya i vzriva, 1969, vol. 5, no. 1, pp. 31-36.
12. Tkachenko E. V., Ulihbin V. B., Shteynberg A. S., Phys. goreniya i vzriva, 1969, vol. 5, no. 1, pp. 16-20.
13. Merzhanov A. G., Dubovitskyi F. I., Dokl. AN USSR, 1960, vol. 135, no. 6. pp. 1439-1443.
14. Kowts M., Raketnaya tekhnica i cosmonautica, 1965, vol. 3, no. 7, pp. 55-61.
15. Shteynberg A. S., Ulihbin V. B., Dolgov E. I., Manelis G. B., Goreniye i vzriv. M.: Nauka, 1972, pp. 124-127.
16. Madorsky S., Termicheskoe razlozheniye organicheskih polymerov. M.: Mir, 1967.

17. Brauman S. K., J. Polymer Sci., Polymer Chem. Ed., 1977, vol. 15, no. 6, pp. 1507-1510.

18. Burge S. J., Tipper C. F. H., Chem. Ind. London, 1967, no. 2, pp. 362-367; Combust. and Flame, 1969, vol. 13, no. 5, pp. 495-505.

19. Stuetz D. E., Diedwardo A. H., Zitomer F., Barnes B. P., Polym. Sci., Polymer Chem. Ed. 1975, vol. 13, no. 3, pp. 585-621.

20. Kiryushkin S. G., Yakimchenko O. E., Shlyapnikov U. A. et al, Visokomol. soed., 1975, vol. 176, no. 5, pp. 385-387.

21. Grassi N., Khimiya protesessov destruktsee. M.: Izd-vo Inostr. lit., 1959.

22. Aseeva R. M., Smutkina Z. S., Berlyn A. A., Kasatochkin V. I., Structurnaya chimiya ugleroda i ugley. M.: Nauka, 1969, pp. 161-200.

23. Korshak V. V., Visokomolec. soed., 1977, vol. 19A, no. 6, pp. 1179-1187.

24. Kodolov V. I., Goruchest i ognestoykost polymernih materialov. M.: Khimiya, 1976.

25. Wall L. A. SPE J., 1960 no. 8, pp. 810-821.

26. Plate N. A., Kinetika i mekhanica obrazovaniya i prevrashcheniya macromolecule: Dokl. on simposium. Erevan. M.: Nauka, 1968, pp. 250-275.

27. Pudov V. S., Buchackenko A. L., Usp. Khim., 1970, vol. 39, no. 1, pp. 130-171.

28. Shilov U. B., Denisov E. T., Dokl. AN USSR, 1976, vol. 226, no. 4, pp. 887-890.

29. Bugdasar'yan Kh.S., Teoriya radikalmoy polymerizatsee. M.: Nauka, 1966.

30. Aseeva R. M., Zelenetskaya T. V., Selskaya O. G., Berlin A. A., Visokomol. soed., 1972, vol. 14A, no. 9, pp. 1573-1579.

31. Aseeva R. M., Berlin A. A., Ushkov V. A. et al, Croatica chem. acta 1974, vol. 46, no. 3, pp. 183-186.

32. Zelentskaya T. V., Dublinskaya A. M., Aseeva R. M., Berlin A. A., Izv. AN USSR. Ser. Khim., 1972, no. 9, pp. 1965-1969.

33. Aleksandrov U. A., Nikitina G. S., Pravednikov A. N., Visokomolecul. soed., 1968, vol. 10A, no. 5, pp. 1078-1085.

34. Dudina L. A., Karmilova L. V., Enikolopian N. S., Dokladi k yubileynoy sessee po visokomoleculiarnim soedineniyam. M.: IKhF AN

USSR, 1970, pp. 243-268; J. Polym. Sci., 1967, vol. 16C, pp. 2277-2291.

35. Wall L. A., Analyticheskaya khimiya polymerov, A. A. Arest-Yakubovich, tr. from Eng. M.: Mir, 1965, vol. 2, pp. 152-208.
36. Berlin A. A., Enikolopian N. S., Visokolec. soed., 1968, vol. 10A, no. 9. pp. 1475-1490.
37. Boucher E. A., J. Polymer Sci., Polymer Phys. Ed., 1977, vol. 15, no. 6, pp. 117-1127; J. Chem. Soc. Faraday Trans. II, 1976, vol. 72, no. 7, pp. 1697-1701.
38. Barlow A., Lehrle R. I., Robb J. C., Sunderland F., Polymer, 1967, vol. 8, no. 10, pp. 537-549.
39. David C., Comprehensive chemical kinetics, C. H. Bamford, C. F. H. Tipper, eds., Amsterdam, etc.: Elsevier Sci. Publ. Co., 1975, vol. 14, chap. 1, pp. 1-173.
40. Tsuchiya Y., Sumi K., J. Polymer Sci. A1, 1969, vol. 7, no. 5, pp. 1599-1607.
41. Wall L. A., J. Elastoplast., 1973, vol. 5, no. 1, pp. 36-65.
42. Bresler S. E., Kazbekov E. N., Usp. khim., 1968, vol. 36, no. 4, pp. 720-736.
43. Szöes F., Plaĉek J., J. Appl. Polymer Sci., 1970, vol. 14, pp. 2629-2637; Eur. Polymer J., 1972, vol. 8, no. 4, pp. 525-532; 1974, vol. 10, no. 8, pp. 725-734.
44. Petrov A. I., Betekhin V. I., Zakrevskyi V. A., Mekhanica polymerov, 1976, vol. 2, no. 2, pp. 207-212.
45. Maro S., Shiwada S., Kashiwabara H., Sohma J., J. Polymer Sci., A2, 1968, vol. 6, no. 3, pp. 1435-1447.
46. Katsuhida M., Tadakiko M., Nippon Kagaku Kaishi, 1975, no. 7, pp. 1241-1245.
47. Matui S., Aida H., Polymer, 1976, vol. 17, no. 3, pp. 199-204.
48. Smith S., J. Polymer Sci., 1958, vol. 30, no. 2, pp. 459-471.
49. Katorzhnov N. D., Strepikheev A. S., ZhPKh, 1959, vol. 32, no. 2, pp. 625-630; no. 4, pp. 1363-1369.
50. Straus S., Wall L., J. Res., 1959, vol. 63A, no. 2, pp. 269-280.
51. Ritchie P. D., J. Chem. soc., 1957, no. 1, pp. 525-530; no. 4, pp. 2107-2111; Thermal degradation of polymers. L.: Soc. Chem. Ind. Monogr. 1961, no. 13.
52. Minsker K. S., Berlin A. A., Lisitskyi V. V., Visokomolec. soed.

1977, vol. 19A, no. 1, pp. 32-39; 1976, vol. 18A, no. 1.
53. Pupko R. A., Pudov V. S., Viskomolec. soed., 1976, vol. 18B, no. 11, pp. 865-868; 1970, vol. 12B, no. 3, pp. 218-221.
54. O'Mara M. M., Pure and Appl. Chem., 1977, vol. 49, no. 5, pp. 649-654.
55. Kasatochkin I. I., Berlin A. A., Smutkina Z. S. et al., Izv. AN USSR. Ser khim., 1965, no. 6 pp. 1003-1009.
56. Bersch C. F. et al., J. Res. NBS, 1958, vol. 60, no. 5, pp. 2863-2871.
57. Kipling J. J., Shrotes P. B., Carbon, 1966, vol. 4, no. 1, pp. 1-16.
58. Holve D. J., Sawyer R. W., in 15th Symposium (Intern.) on Combustion. Pittsburgh, Combust. Inst., 1974, pp. 351-361.
59. Yavorsky I. A., Elohina V. I., Gurjiants V. M., in 5th Intern. Symposium on Combustion Processes. Abstrs. Krakov: Pol. Acad. Sci., 1977, pp. 26-27.
60. Geyderikh M. A., Izucheniye termicheskogo prevrashcheniya polyacrylonitrila. Autoref. dis., Ph.D. M.: INKhS AN USSR, 1965.
61. Hay J. N., J. Polymer Sci., A1, 1968, vol. 8. no. 6, pp. 2127-2137.
62. Watt W., Green J., in Intern. Conference on Carbon Fibres: Their Composites and Applications. L.: Plast. Inst., 1971, Pap. no 4.
63. De Winter E., Macomol. Chem. Rev., 1966, no. 1, pp. 336-341.
64. Lyons J. W., The chemistry and uses of fire retardants N.Y.: Wiley Intersci., 1970.
65. Rogovin Z. A., Khimiya i tekhnologiya polymerov, 1960, no. 7/8, pp. 174-186.
66. Poliakov A. M., Derevitskaya V. A., Rogovin Z. A., Visokomolec soed., 1960, vol. 2., no. 2, pp. 386-392; 1963, vol. 5, no. 2, pp. 161-169.
67. Tang M., Bacon R., Carbon, 1964, vol. 2, no. 3, pp. 211-231.
68. Siling M. I. in Itogy nauky i tehniky. Ser. khim i. tekhnolog. VMS M.: VINITI, 1977, vol. 11, pp. 119-162.
69. Berlin A. A., Tsvelikhovsky G. I., Aseeva R. M. et al., Plastmassi, 1969, no. 1, pp. 23-25.
70. Aseeva, R. M., Almanbetov K. in IUPAC Intern. Symposium on Macromolecular Chemistry Bp.: Acad. Kiadó, 1969, vol. 5, pp. 379-382.

71. Berlin A. A., Yarkina V. V., Firsov A. P., Viskomolec. soed., 1968, vol. 10A, no. 8, pp. 1913-1921; no. 9, pp. 2197-2205.
72. Friedman H. L., J. Poly. Sci. C6, 1964, vol. 1, no. 1, pp. 183-187.
73. Shulman G. P., Lochte H. W., J. Appl. Polymer Sci., 1966, vol. 10, no. 2, pp. 619-628.
74. Conley R. T., Bicron J. F., J. Appl. Polymer Sci., 1963, vol. 7, no. 1, pp. 103-107; pp. 171-176.
75. Jackson W. M., Conley R. T., J. Appl. Polymer Sci., 1964, vol. 8, no. 4, pp. 2163-2171.
76. Ouchi K., Honda H., Fuel, 1959, vol. 38, no. 1, pp. 429-443; Carbon, 1964, vol. 4, no. 1, pp. 59-77.
77. Moyseev V. D., Neyman M. B., Raspopova E. I., Plastmassi, 1960, no. 6, pp. 11-13.
78. Learmouth G. S., Osborn P., J. Appl. Polymer Sci., 1968, vol. 12, no. 8, pp. 1815-1827.
79. Winkler E. L., Parker J. A., J. Macromol. Sci. Revs. Macromol. Chem., 1971, vol. C5, no. 2, pp. 245-251.
80. Adorova I. V., Issledovaniye kinetiky i mekhanizma polycondensatsee phenola s formaldegidom v kisloy srede: Autoref. Ph.D. dis., M.: MKhTI, 1974.
81. Mitchell S. J., Pickering R. S., Thomas C. R., J. Appl. Polymer Sci., 1970, vol. 14, no. 1, pp. 175-183.
82. Gudings E. P., Khim. i tekhnolog polymerov, 1961, no. 3, pp. 104-107.
83. Kardash I. E., Pravednikov A. N., Medvedev S. S., Dokl. AN USSR, 1964, vol. 156, no. 3, pp. 658-661.
84. Mashall G., Todd A., Faraday Soc., tr. 1953, vol. 49, no. 1, pp. 67-70.
85. Kovarskaya B. M., ZhVKhO D. I., Mendeleev, 1966, no. 3, pp. 267-271.
86. Avondo R., Vovelle G., Delbourgo P., in 16th Symposium (Intern.) on Combustion. Pittsburgh: Combust. Inst., 1976, pp. 301-310.
87. Rafikov S. R., Roda V. V., Zhuravlyova I. V., in Stareniye i stabilizatsiya polymerov. M.: Khimiya, 1966, pp. 67-103.
88. Lee L. H., J. Polymer Sci. A2, 1964, vol. 2, no. 11, pp. 2859-2867.
89. Davis A., Golden J. H., J. Chem. Soc., 1968, vol. 13, no. 1, pp. 40-47.

90. Mikheev U. A., Lednyova O. A., Toptigin D. Y., Visokomolec. soed., 1971, vol. 13A, no. 4, pp. 931-937.

91. Davidov E. Y., Lednyova O. A., Mikheev U. A. et al., Dokl. AN USSR, 1970, vol. 195, no. 4 pp. 875-878.

92. Kosobutskaya A. A., Issledovaniye vliyaniya khimicheskogo stroeniya aromaticheskih polymidov na ikh termostoykost: Autoref. ph.D. dis., Kalinin, KGU, 1973.

93. Krasnov E. P., Savinov V. M. et al., Visokomol. soed., 1966, vol. 8, no. 2., pp. 380-387.

94. Gribkova P. N., Roda V. V., Vigodsky Y. S. et al., Visokomol. soed., 1973, vol. 15B, no. 2, pp. 243-247.

96. Kosobutsky V. A., Dokl. AN USSR, 1975, vol. 223, no. 3, pp. 636-639.

97. Valetskaya N. Y., Issledovaniye v oblasty ulutsheniya tekhnologicheskh i ekspluatatsionnih svoystv polyamidov no osnove aromatikcheskih diaminov i izophtalevoy kisloti.: Autoref. Ph.D. dis., M.: MKhTI, 1973.

98. Day M., Wiles D. M., Textile Research J., 1974, vol. 44, no. 11, pp. 888-900.

99. Korshak V. V., Khimicheskoe stoeniye i temperaturniye kharacterisktiky polymerov. M.: Nauka, 1970,

100. Bishop D. P., Smith D. A., J. Appl. Polymer Sci., 1970, vol. 14, no. 2, pp. 345-357.

101. Kamzolkina E. V., Nechaev P. P., Markin V. S. et al., Dokl. AN USSR, 1974, vol. 219, no. 3, pp. 650-652; 1977, vol. 233, no. 1, pp. 156-159.

102. Ehlers G. et al, J. Polymer Sci., 1970, vol. 8, no. 9, p. 3511-3520.

103. Gay F. P., Berr C. E., J. Polymer Sci., 1968, vol. A16, no. 5, pp. 1935-1947.

104. Hirsch S. S., J. Appl. Polymer Sci., 1967, vol. 11, no. 1, pp. 305-309.

105. Kovarskaya B. M., Levantovskaya I. I., Blumenfield A. B. et al., Plastmassi, 1971, no. 8, pp. 54-56; 1973, no. 1, pp. 64-66; Visokomol. soed., 1971, vol. 13B, no. 1, pp. 210-203.

106. Kronguaz E. S., Bochvar D. A., Stankevich I. V., Korshak V. V., Kokl. AN USSR, 1968, vol. 179, no. 1, pp. 94-97.

107. Rode V. V., Bondarenko E. M., Korshak V. V. et al., J. Polymer

Sci., A1, 1968, vol. 6, no. 3, pp. 1351-1357.
108. Moos E. K., Skinner D. L., J. Cell. Plast., 1977, vol. 13, no. 4, pp. 276-280.
109. Grigorovskaya V. A., Lapina N. A., Ostrovsky V. S. et al., Khim. tvyordogo topliva, 1975, no. 3, pp. 121-126; 1977, no. 1, pp. 126-131.
110. Grigorovskaya V. A., Basin V. E., Khakimova D. K., J. Polymer Sci. Polymer Phys. Ed., 1977, vol. 15, pp. 2075-2085.
111. Fish R. H., Parker J. A., J. Appl. Polymer Sci. Appl. Polymer Symposia, 1973, no. 2, pp. 22-24.
112. VanKrevelen D. W., Polymer, 1975, vol. 16., no. 8, pp. 615-621.
113. Matsumoto T., Fujiwara T., Kondo J., in 12th Symposium (Intern.) on Combustion. Pittsburgh: Combust. Inst., 1969, pp. 515-524.
114. Lavrov N. V., Physico-khimicheskiye osnovi protessa goreniya topliva. M.: Nauka, 1971.
115. Khimicheskiye i physicheskiy svoistva ugleroda. Woker F., ed. M.: Mir, 1969.
116. Petrenko I. G., in Gazifikatsiya tvyordogo topliva. M.: Gostoptekhizdat, 1957, pp. 312-318.
117. Shylov N. A. et al, Z. Phys. Chim., 1928, vol. 133, pp. 188-198; 1929, vol. 143, pp. 41-49; 1030, vol. 149, pp. 211-221.
118. Garten V., Weiss, D. Proc. of 3rd Conference on Carbon. N.Y.: Pergammon Press, 1959, pp. 295-311; Revs. Pure and Appl. Chem., 1957, vol. 7, no. 1, pp. 69-81.
119. Jackson C., Wynne-Jones W. F. K., Carbon 1964, vol. 2, no. 3, pp. 227-238.
120. Armstrong J. W., Jackson C., Marsh H., Carbon 1964, vol. 2, no. 3, pp. 239-250.
121. Wright F. J., in 15th Syposium (Intern.) on Combustion. Pittsburgh: Combust. Inst., 1974, pp. 1449-1460.
122. Fenimore C.P., Jones G. W., Combust. and Flame, 1966, vol. 10, no. 3, pp. 295-301.
123. Zhobanov B.E., Davlichin T. Kh., Gihbov K. M., Visokomolec. soed., 1975, vol. 17B, no. 10, pp. 746-749.
124. Mararil P. Kh., Mekhanizm i kinetika gomogehhih termicheskih prevrashchenyi uglevodorodov. M.: Khimiya, 1970.
125. Richard J. R., Vovelle C., Delbourgo R., in 15th Symposium

(Intern.) on Combustion. Pittsburgh: Combust. Inst., 1974, pp. 205-216.

126. Brown W. P., Tipper C. F. H., in Combustion Institute European Symposium. L.: Acad. Press, 1973, pp. 137-141.

127. Richard J. R., Vovelle C., Delbourgo R., ibid., pp. 131-136.

128. Fenimore C. P. in Flame Retardants Polymeric Materials, Levvin M., Allas S. M., Pierce C. M., eds. N.Y.-L.: Plenum Press, 1975, ch. 9, pp. 371-390.

129. Ephremov V. L., Kolyesnikov B. Y., Shapirin V. A. et al., in Khimisheskaya physika protsessov goreniya i Tchernogolovka: Ed.-publ. division of OIKhF AN USSR, 1977, pp. 14-17.

130. Stepukhovich A. D., Kinetika i mekhaniszm termicheskogo crekinga alkanov Saratov: Saratov Univ. 1965.

131. Maltsev V. M., Maltsev M. I., Kasporov L. Y., Osnovniye Kharacteristiky goreniya. M.: Khimiya, 1977.

132. Robertson A. F., Fire Eng., 1973, vol. 126, no. 9, pp. 97-104.

133. Street J. C., Thomas A., Fuel, 1955, vol. 34, no. 1, pp. 4-15.

134. Fabris H. J., Sommer J. G., Flame retardancy of polymeric materials, Kuryla W. C., Papa A. J., eds., N.Y. : Marcel Dekker, 1973, vol. 2, ch. 5, pp. 130-199.

135. Geydon A. G., Volfgurd K. G., Plamya yego structura, izlucheniye i temperature. Goldenberg S. A., tr., M.: Metallurgizdut, 1959.

136. Kokurin A. D., ZhPKh, 1969, vol. 52, no. 7, pp. 1952-1958.

137. Rabinovich E. A., in Pererabotka prirodnogo gasa. M.: VNIIGAS, 1961, pp. 61-67 (works of VNIIGAS).

138. Hommann K. H., Wagner H. G., 11th Symposium (Intern.) on Combustion. Pittsburgh: Combust. Inst., 1967, pp. 371-379.

139. Frazee J., Anderson R., in Proc. of 3rd Conference on Carbon. Oxford; London: Pergamon Press, 1956, pp. 405-415.

140. Tesner P. A., in Pererabotka prirodnogo gasa. M.: Gostoptekhizdut, 1959, pp. 3-12 (works of VINIIGAS).

141. Sweitzer G., Heller G., Rubber World, 1956, vol. 134., pp. 855-867.

142. Stehling F. C., Frazee J. D., Anderson R. C., in 6th Symposium (Intern.) on Combustion. N.Y.: Reinhold Publ. Corp., 1956, pp. 247-254.

143. Parker W., Wolfhard H., J. Chem. Soc., 1950, pp. 2038-2041.

144. Wersborg B. L., Yeung A. C., Howard J. B., in 15th Symposium

(Intern.) on Combustion: Abstrs. of Pap. Pittsburgh: Combust. Inst., 1974, pp. 282-284.

145. Jinno H., Fukutani S., Takaya A., in 16th Symposium (Intern.) on Combustion: Abstrs. of Pap. Pittsburgh: Combust. Inst., 1976, pp. 131-132.

146. Schaefer B. A., Combust. and Flame, 1969, vol. 13, pp. 208-217.

147. Knewstubb P. F., Sugden T. M., in 7th Symposium (Intern.) on Combustion. L.: Butterworths Sci. Publ., 1959, pp. 247-253.

148. Karachevtsev G. V., Talroza V. L., in Goreniye i vzriv. M.: Nauka, 1972, pp. 726-736.

149. Howard J. B., in 12th Symposium (Intern.) on Combustion. Pittsburgh: Combust. Inst., 1969, pp. 877-887.

150. Berlin A. A., Geyderikh M. A., Davidov B. E. et al., Khimiya polysopryazhyonnikh system. M.: Khimiya, 1972.

151. Lavrov N. V., Evlanov S. F., Dokl. AN USSR, 1967, vol. 173, no. 1, p. 59.

152. Rice F. O., Dooly M. D., J. Amer. Chem. Soc., 1934, vol. 56, pp. 2747-2751.

153. Kassel L. S., J. Amer. Chem. Soc., 1932, vol. 54, p. 3949.

154. Cullis C. F., Hucknall D. J., Shepherd J. V. in Combustion Institute European Symposium. L. N.-Y.: Acad. Press, 1973, pp. 111-116.

155. D'Alessio A., Di Lorenco A., Sarofim A. F. et al., in 15th Symposium (Intern.) on Combustion. Pittsburgh: Combust. Inst., 1974, p. 1427-1438.

156. Berlin A. A., Grigorovskaya V. A., Skurut V. E., Visikomolec. soyed., 1966, vol. 8, no. 11, pp. 1976-1982.

157. Kondratiev V. N., in Comprehensive Chemical Kinetics, Bomford C. H., Tipper C. F. H., eds, Amsterdam, etc.: Elsevier Sci. Publ. Co., 1969, vol. 2, ch. 2, pp. 81-188.

158. Gontkovskaya V. T., Merzhanov, Ozerkovskaya N. I., in Khimicheskaya phyzika protsessov goreniya i vzriva. Kinetika khimicheskih reaktsyi. Tchernolgolovka. Ed.-publ. div. of OIKhF AN USSR, 1977, pp. 30-32.

159. Fristrom V. M., Vestenverg A. A., Structura plamenyi. M.: Metallurgizdut, 1969.

160. Antonova I. N., Kuzmin V. A. et al. Izv. AN USSR, Ser. Khim. 1955, no. 6, pp. 789-794.
161. Basevich V. Y., Kogarko S. M., Furman G. A., Izv. AN USSR, Ser. Khim., 1971, no. 7, pp. 1406-1410, 2191-2197; 1972, no. 10, pp. 2193-2143.
162. Basevich V. Y., Kogarko S. M., Neygau M. G., IZV. AN USSR, Ser. Khim., 1976, no. 1, pp. 43-47.
163. Hastie J. W. J. Res. NBS, 1973, vol. 77A, no. 6, pp. 733-754.
164. Arthur J. R., Bowring J. R., J. Chem. Soc. (London), 1949, no. 1, pp. 1-10.
165. Creitz E., J. Res. NBS, 1961, vol. 65A, no. 4, pp. 389-400.
166. Creitz E. C., J. Res. NBS, 1970, vol. 74A, no. 4, pp. 521-530.
167. Mills R. M., Combust. and Flame, 1968, vol. 12, no. 6, pp. 513-521.
168. Tolles Z. E., Tolles C. J., Plast. and Polymers (London), 1972, pp. 319-328.
169. Dixon-Lewis G., Simpson R. J., in 16th Symposium (Intern.) on Combustion. Pittsburgh: Combust. Inst., 1976, pp. 1111-1119.
170. Green J., Plast. Technol., 1975, vol. 21, no. 8, pp. 44-49.
171. Baratov A. A., ZHVKhO D. I. Mendeleev, 1974, vol. 19, no. 5, pp. 531-537.
172. Vorob'yov V. A., Andrianov R. A., Ushkov V. A. Goryuchest polymernih striotelnih materialov. M.: Stroyizdut, 1978.
173. Friedman R., Levy I. B., Combust. and Flame, 1963, vol. 7, no. 2, pp. 195-203.
174. Bulewicz E. M., James C. G., Sugden T. M., Proc. Roy. Soc. (London), 1956, vol. 235, pp. 89-97.
175. Kaskan W., in 10th Symposium (Intern.) on Combustion. Pittsburgh: Combust. Inst., 1965, pp. 41-46.
176. Carabetta R., Kaskau W. E., J. Phys. Chem., 1968, vol. 72, no. 7, pp. 2483-2491.
177. Evans M. C., Phillips L. F., Trans. Faraday Soc., 1976, vol. 62, pp. 1717-1720.
178. Bulewicz E. M., Padley P. J., in 13th Symposium (Intern.) on Combustion. Pittsburgh: Combust. Inst., 1971, pp. 73-80; Proc. Roy. Soc. (London), 1971, vol. 323A, pp. 377-385.
179. Iya K. S., Wollowitz S., Kashan W. E., in 15th Symposium (Intern.) on Combust. Pittsburgh: Combust. Inst., 1974, pp. 329-336.

180. Rosser W. A. Jr., Inami S. H., Wise H., Combust. Flame, 1963, vol. 7, no 1., pp. 107-118.
181. Birchall J. D., Combust. and Flame, 1970, vol. 14, no. 1, pp. 85-93.
182. Cotton D. J., Jenkins D. R., Trans. Faraday Soc., 1971, vol. 67, pp. 730-733.
183. Gogol L. A., Kononenko K. M., Odnorog D. S. et al., Ingibirovaniye tsepnih gazovih reaktsiy. Alma-Ata. Publ. by KuzGU, 1971.
184. Glazkova A. P., Nikolenko A. F., Sevrikov V. V., Dokl. AN USSR, 1975, vol. 222, no. 4, pp. 869-873.
185. Petviashvily D. I., Museridze M. D., Dzoyenidze Z. G., Baratov A. N., in Goreniye i problemi tusheniya pozharov. M.: VNIIPO, 1977, pp. 13-17.
186. Kislyuk M. U., Tret'yakov I. I., Kinetika i Kataliz, 1973, vol. 14, no. 5, pp. 1497-1502; 1974, vol. 15, no. 3, pp. 710-715; Dokl. AN USSR, 1973, T. 208, no. 5, pp. 1134-1138.
187. Pitts J. J., Fire and Flammability, 1972, vol. 3, no. 1, pp. 51-56.
188. Hastie J. W. Combust and Flame, 1973, vol. 21, no. 1, pp. 49-61.
189. Hasimoto S., Furukawa J., Sci. and Ind., 1971, no. 4, pp. 211-215.
190. Martin F. T., Price K. R., J. Appl. Polymer Sci., 1968, vol. 12, no. 1, pp. 143-151.
191. Sokolov D. N., Usp. khim., 1977, vol. 56, no. 4, pp. 740-776.
192. Flame Retardants. Hilado C. J., ed. N.Y.: Technomic Publ. Co., 1973.
193. Perfect J. R. W., J. Soc. Dyers Colour, 1958, vol. 74, no. 12, pp. 829-832.
194. Flammability of Fabric. Hilado C. J., ed., N. Y.: Technomic Publ. Co., 1974.
195. Kodolov V. I., Povstugur V. I., Alyamovsky S. T. et al., Vysokomolec. soyed., 1977, vol. 19B, no. 4, pp. 283-287.
196. Kovačic P., Kyriakis A., J. Amer. Chem. Soc., 1963, vol. 85, no. 2, pp. 454-461.
197. Van Wazer. Phosphor i ego soeyedineniya, M.: Izd-vo Inostr. lit., 1962.
198. Brown E. H., Whitt C. D., Ind. Eng. Chem., 1942, vol. 44, pp. 615-617.

199. Mikhailin A. D., Postnykov N. N., Klenitsky A. I., in Trudi/NIUIF, 1969, ed. 211, pp. 36-42.

200. Benbow A. W., Cullis C. F., Combustion Institute European Syposium. L.,N.-Y.: Acad. Press, 1973, pp. 183-188.

201. Kim L. B., Plastmassi, 1974, no. 4, pp. 52-54.

202. Melville H. W., Gray S. C., Tr. Faraday Soc., 1936, vol. 32, pp. 271-273.

203. Dainton F. S., Kimberly H. M., Tr. Faraday Soc., 1950, vol. 46, no. 332, pp. 629-633.

204. Dainton F. S., Tr. Faraday Soc.,1946, vol. 42, pp. 377-382.

205. Davies P. B., Thrush B. A., Proc. Roy. Soc. 1968, vol. A302, no. 1469 pp. 243-248.

206. Semenov N. N., Soroka L. B., Azatyan V. Z., Kinetika i Kataliz, 1975, vol. 16, no. 3, pp. 811-820.

207. Vanzee R. J., Khun A. U., J. Chem. Phys., 1976, vol. 65, no. 5, pp. 1764-1771.

208. Rumpf K., Z. Phys. Chim., 1938, vol. 38B, pp. 469-473.

209. Fenimore C. P., Jones G. W., Comb. and Flame, 1964, vol. 8, no. 1, pp. 133-144.

210. Brunauer S., Schultz J. F., Ind. Eng. Chem., 1941, vol. 33, no. 6, pp. 828-837.

211. Piechota H., Kunstst.-Rdsch., 1965, vol. 12, no. 4, pp. 191-198.

212. Pat. 1435446 (Gr. Brit.).

213. Lask G., Wagner H. G., in 8th Symposium (Intern.) on Combustion. Baltimore : Williams-Wilkins Co., 1962, pp. 432-438.

214. Cordes H., Witshel W., Z. Phys. Chem., 1965, vol. 46, no. 1, pp. 35-41.

215. Nevrovsky V. A., Soroka L. B., ZhFKh, 1977, vol. 41, no. 2, pp. 376-379.

216. Vorob'yov N. I., Pechkovsky V. V., Ptashkova G. V., ZhPKh, 1971, vol. 44, no. 7, pp. 1445-1451.

217. Hilado C. J., Flammability Handbook for Plastic. Stamford: Technomic Publ. Co., 1969.

218. Weil E. D., in Flame Retardancy of Polymeric Materials, Kuryla W. C., Paps A. J., eds., N.Y.: Marcel Dekker, 1975, vol. 3, pp. 185-244.

219. Coppie S., Church J., Lietle R., Ind. Eng. Chem., 1950, vol. 42, pp. 415-419.
220. Hardesty D. R., Weinberg F. J., in 14th Symposium (Intern.) on Combustion. Pittsburgh: Combust. Inst., 1973, pp. 907-918.
221. Cotton D. H., Frieswell N. J., Jenkins D. K., Combust. and Flame, 1971, vol. 17, no. 1, pp. 87-93.
222. Salooja K. C., Nature, 1972, vol. 240, no. 5380, pp. 350-353.
223. Feugier, A. in 15th Symposium (Intern.) on Combustion. Abstrs. of Pap. Pittsburgh: Combust. Inst., 1974, pp. 285-287.
224. Bulewicz E. M., Evans D. G., Padley P. J., in 15th Symposium (Intern.) on Combustion. Pittsburgh: Combust. Inst., 1974, pp. 1461-1470.
225. Smoke and Products of Combustion. Hilado C. J., ed., N.Y.: Technomic Publ. Co., 1973, vol. 2. p. 405.
226. Pat. 3819574 (USA)
227. Pat. 2209793 (France)
228. Pat. 2307387 (Fed. Rep. Germany)
229. Pat. 3637542 (USA)
230. Pat. 3639307 (USA)
231. Pat. 1325134 (Gr. Brit.)
232. Pat. 2128428 (France)
233. Lecomte L., Bert M., Michel A., Guyot A., J. Macromol. Sci. Chem., 1977, vol. 11A, no. 8, pp. 1467-1489.

Addendum

Since this book was written, a number of new results have been obtained. The following section discusses these developments.

The Effect of Oxygen on Polymer Decomposition

The question as to how oxygen affects the degradation of polymers during combustion still attracts the attention of many scientists. The accumulation of information on the high temperature degradation of polymers under different conditions is useful for a definitive solution of this problem. Gibhov [1] studied the linear pyrolysis of polymers under the effect of a highly heated flow of gas with various concentrations of oxygen (0-21 vol.%). It was discovered that when the temperature of the flow is 700°C, oxygen does not substantially affect the linear rate of degradation for polystyrene, has a slight effect on it in the case of PMMA and reduces it (concentration above 3 vol.%) in the case of epoxy polymer on account of the formation of a carbonised layer on the surface. The temperature at the contact area between the carbonised layer and the degrading epoxy polymer was 100°C lower than the surface temperature of a polymer heated in a stream of inert gas. It was concluded that oxygen participates in the degradation of the condensed phase but that the nature of the oxidising reactions is determined by the nature of the polymer.

According to reference [2] the replacement of air by nitrogen or helium does not substantially affect the change in the vaporisation constant of PMMA during the heating of spherical polymer samples in the 400-900°C temperature range.

Jakes and Drews [3] studied the concentration of peroxide and carbonyl groups in the superficial layers of polypropylene, sectioned in the region of the flame edge and from the influx before the flame front. The latter method was used to obtain direct proof of a heterogeneous polymer oxidation reaction in the case of flame diffusion over the horizontal surface of the samples. It was discovered that the percentage of the above-mentioned groups was small; the percentage was greater inside the layer in front of the flame than in the front part of the flame edge. It initially drops to a minimum ($Y_{ox.} = 0.23$) with an increase in oxygen concentration, but then increases again. The authors believe that the data proves what Stuetz and coworkers previously stated about the participation of oxygen in the degradation of the condensed phase during combustion. In order to verify his idea, Stuetz conducted a series of experiments [4]. He analysed the relationship between the combustibility of a large number of polymers and their thermal oxidising stability in an atmosphere containing 1% O_2 and 99% N_2 (conditions close to those on the surface of burning samples) (see Fig. 3A.1). He used the temperature at which the polymer is degraded at a rate of 3.3% per minute as a characteristic of thermal stability. A correlation was determined between the limit oxygen concentration during upwards combustion of the sample and the degradation temperature. Polymers that have decomposition temperatures of above 525°C belong to the group of materials that do not burn in an atmosphere of air, according to accepted standard tests.

It becomes apparent that the pure thermal stability of polymers changes in the same way as $T_{degrad.}$ in an atmosphere with 1% O_2. Das [5] determined the mass and linear rate of combustion for cross-linked polymers based on unsaturated polyethers, reinforced by styrene, at a different oxygen pressure. He observed a carbonisation of the surface with low rates of oxidiser flow and an essential increase in the rate of thorough combustion with an increase in pressure. He assumed that these facts indicate that an impórtant role in the combustion of these polymers is

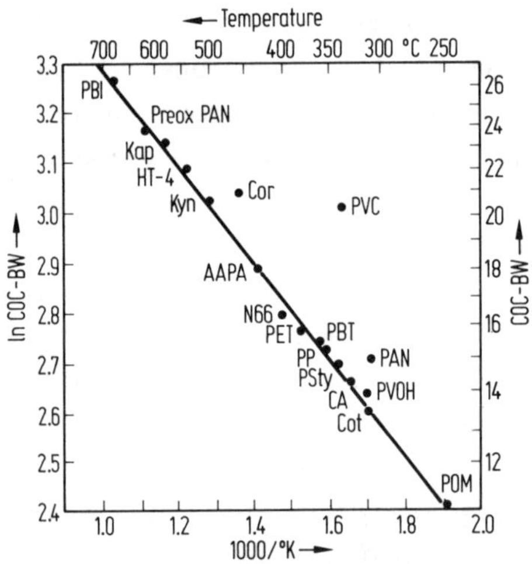

Fig. 3A.1 Correlation between combustibility of polymers and their temperature of degradation in an atmosphere of 1% O_2 and 99% N_2.

attributable to atoms of the oxidiser coming from the flaming zone and attacking the surface. Usually, very high rates of heating are observed in the case of polymer combustion. These rates do not coincide with the rates that are used in thermal analysis (TGA). An extrapolation of TGA data was proposed [6] in order to predict the thermal stability of substances under high heating conditions. Since the thermal degradation process has a relaxation nature, the characteristic parameters, the temperature of degradation origination $T_{d.o.}$ of the substance and the rate of heating dT/dt, must be related via a simple relationship:

$$1/T_{d.o.} = c_1 - c_2 \left(\frac{dT}{dt} \right)$$

Constants c_1 and c_2 can be readily determined from this linear equation using experimental $T_{o.d.}$ values, obtained by the TGA method with two preset values for the heating rate. The authors of reference [6] thus determined constants c_1 and c_2 for three polymers and later computed $T_{o.d.}$ for high rates of heating (10-500°K/sec). A method using a laser probe was developed in order to check the equations given above. Figure 3A.2 shows a good correlation between the experimental and computed values of $T_{o.d.}$. The disagreement does not exceed 30°K.

The conditions for heating greatly affect the rate of vaporisation (Fig. 3A.3) and, correspondingly, affect the ignition temperature of a polymer (Fig. 3A.4). As Fig. 3A.4 shows, the character of the change in ignition temperature due to the heating rate depends on the nature of the polymer

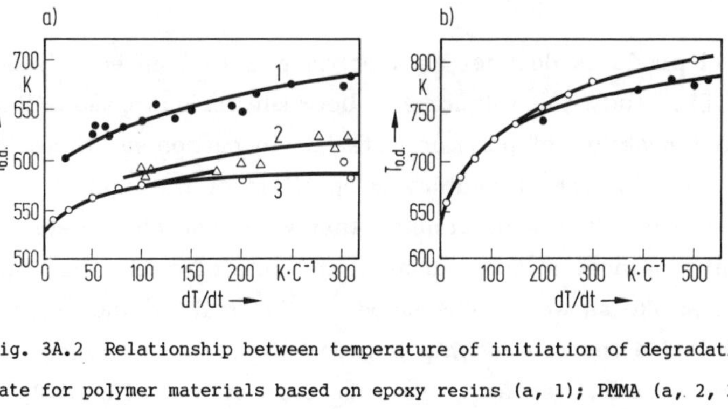

Fig. 3A.2 Relationship between temperature of initiation of degradation $T_{o.d.}$ and heating-up rate for polymer materials based on epoxy resins (a, 1); PMMA (a, 2, 3); polyoxodiazole (b), o = experimental data obtained with a laser probe, X = TGA data, o = computed data.

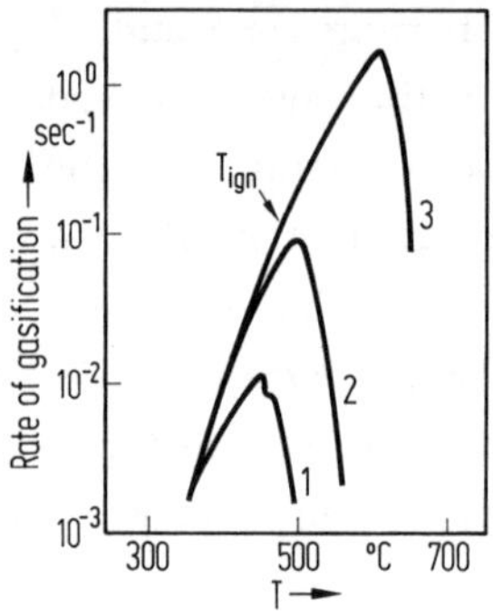

Fig. 3A.3 Relationship between rate of vaporisation of polymer materials and temperature at a constant heating rate: (1) 1°C/sec, (2) 10°C/sec, (3) 100°C/sec. The arrow indicates ignition of the polymer.

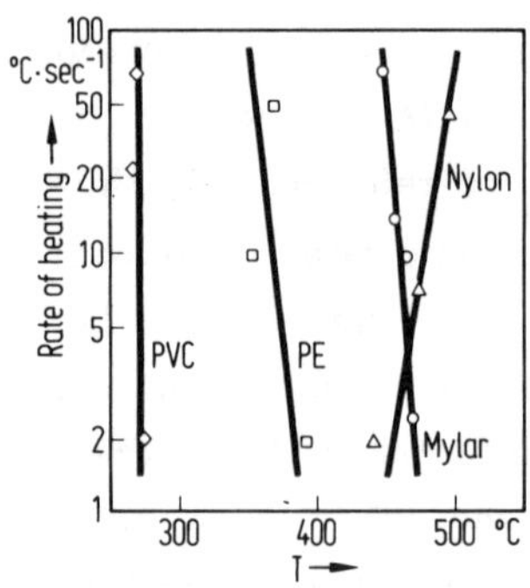

Fig. 3A.4 Relationship between ignition temperature of polymers and the rate of heating in an air atmosphere.

[7]. The heating rate, however, also affects the mechanism of degradation.

The yield of fuel products declines as a polymer is carbonised. The works of Brauman [8] study the relationship between various combustibility characteristics and the ability of polymer materials to carbonise, the effect of the morphology and also the elementary composition of the char layer formed during pyrolysis. It was determined that when the char yield exceeds 5%, not only is there a sharp drop in the rate of combustion but in the liberation of smoke as well. The slope of the straight lines (Fig. 3A.4) which characterise the relationship between the limiting oxygen index and the char number, may serve as a measure of how effectively the char layer decreases the combustibility of polymer materials. Thus the modification of aromatic polycarbonates by siloxanes increases the efficiency of the char by comparison with the char efficiency of polymers that do not contain silicon. When char residues of modified polymers were studied with the aid of electron microscopy, the presence of layers of a different nature were discovered on the carbon surface [9]. It should not be thought, however, that carbonisation by itself reduces all the fire characteristics of polymers. Hilado [10] showed examples where carbonisation during combustion of polymers did not reduce but increased the toxicity of gases, and the combustibility of the degradation products increased (for the same char numbers). The determining factors are the mechanism of polymer decomposition and the nature of the fuel that enters the flame. This is why the fundamental problem of the interrelationship between the chemical structure of polymer materials and their combustibility still remains in the forefront of the study [11].

When discussing the question of the efficiency of the char layer as a heat barrier that retards the heat transfer towards the virgin polymer, an analogy can be made with the properties of swelling coatings. The analogy results from the analysis of coating models [12, 13] and stems from the fact that the effectiveness of the char layer depends on such factors as heat conductivity of the carbon foam, layer thickness, degree of exothermal and endothermal char formation, convectional heat transfer by the products of degradation, radiation or reflection of heat by the surface.

One of the objective characteristics that reflects the structural properties of the substances in combustion reactions is the heat of combustion. Johnson [20] showed a simple empirical ratio that links OI to the specific heat of combustion for polymers, OI = $1.9/\Delta H_{com.}$, where $\Delta H_{com.}$ is expressed in cal/g. However, in many cases this ratio is not accurate. Another relationship is given in [15]:

$$OI = 0.01286 \left(\frac{\Delta H_{com} \, MW}{m_p + m_n} \right) - 0.044\%$$

where

$\Delta H_{com.}$ is also expressed in cal/g

MW molecular weight of a polymer chain

m_p, m_n number of moles of combustion products for complete stoichiometric combustion of polymer and nitrogen per mole of polymer chain

The equation was derived on the basis of the assumption that when the limiting oxygen index is calculated, the heat of combustion is used for heating the sample to the flame temperature of an inert gas, which comes from the surrounding atmosphere, and heat is also absorbed for the increase in enthalpy of the combustion products. The larger the factor in parentheses, the higher the limiting oxygen index should be for polymer materials. The nature of the relationship between the derived equation and the value of the reciprocal in parenthesis is reminiscent of the VanKrevelen ratio.

To show more clearly how structural factors affect the combustibility of polymers and also to obtain additional information, the authors decided to investigate the relationship between the limiting concentration of oxygen and the specific combustion heat in terms of stoichiometric coefficients of oxidiser-fuel, r, in the combustion reaction [16]. In the case of polymers that form a carbonised residue, it is necessary to consider the quantity Y_{ch} and the heat of combustion $\Delta H_{e,ch.}$ The graphic representation of the relationship between $Y^*_{ox.}/r$ and $1/\Delta H_{com.}$ for polymers that nearly totally decompose during pyrolysis in an inert atmosphere is shown in Fig. 3A.5. The slope of the derived straight lines corresponds to the amount of heat consumed for the liberation of a fuel mass unit into the gaseous phase and for the increase in enthalpy of the products.

The nature and number of end groups, the type of bonds, the presence of aromatic rings and carbonate groups in the molecule of an unsaturated ether and other specific properties of the structure affect the flammability of polymer materials. The tangent of the slope increases during the transition from methacrylic polymers to acrylic polymers. Thus for polymers of alkylene glycol dimethacrylates, this is equal to 2.82 kJ/g; for polymers with carbonate bonds, 2.86-2.94 kJ/g; for acrylic polymers it increases to 3.27 kJ/g. Linearly structured polymers such as PMMA, PE and others have a smaller slope, corresponding to the value of 2.68 kJ/g. The homologous series of polymers based on methacrylic and acrylic ethers of various glycols show a tendency towards increased flammability with an increase in the number of recurring groups in the initial ether; in other words OI decreases in the order of the decreasing thermodynamic stability of unsaturated ethers and the polymers based on them.

The calculation of the combustion heat and the consumption of oxygen during combustion enables one to show how certain structural similarities affect the flammability characteristics of polymer materials.

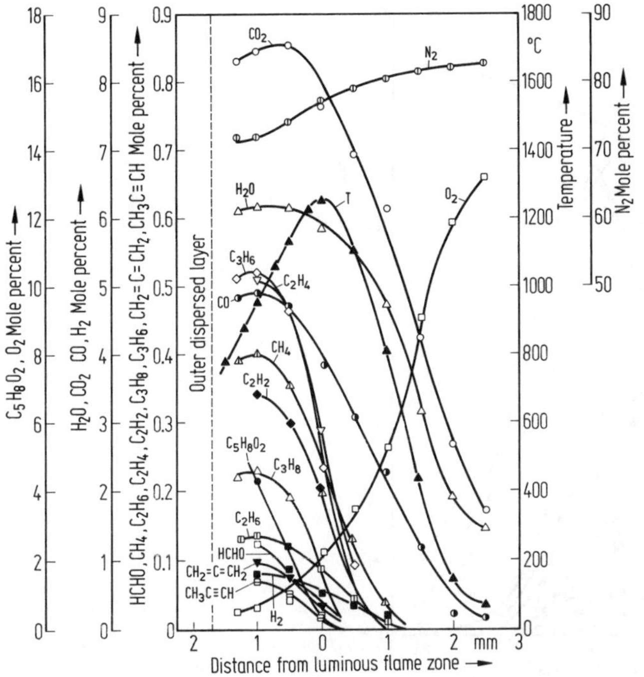

Fig. 3A.5 Reationship between $Y^*_{ox.}/r$ and specific combustion heat for linear and crosslinked polymers.

Chemical Processes in the Gaseous Phase

In reference [17] a comparative analysis was conducted of the concentration and temperature profiles of a plane diffusion flame of PMMA and of its monomer in a configuration involving a counterflow of oxidising agent (O_2/N_2). The analysis was performed for almost equal rates of flow and concentrations of oxygen. An analysis of the temperature profiles of the PMMA combustion wave, recorded with the aid of a thermocouple inserted from the side of fuel and flame, led the authors to the conclusion that there is a two-phase dispersion layer on the surface of the polymer. Figures 3A.6a and 3A.6b show the concentration profiles for degradation products of PMMA and MMA in the zone above the dispersion layer. The concentration of oxygen penetrating the diffusion flame towards the surface of the dispersion layer (foamed melt) is nearly identical for PMMA and MMA. It is small (~0.5 mol.%) and, according to the authors, the oxidising decomposition of polymer during combustion does not play an essen-

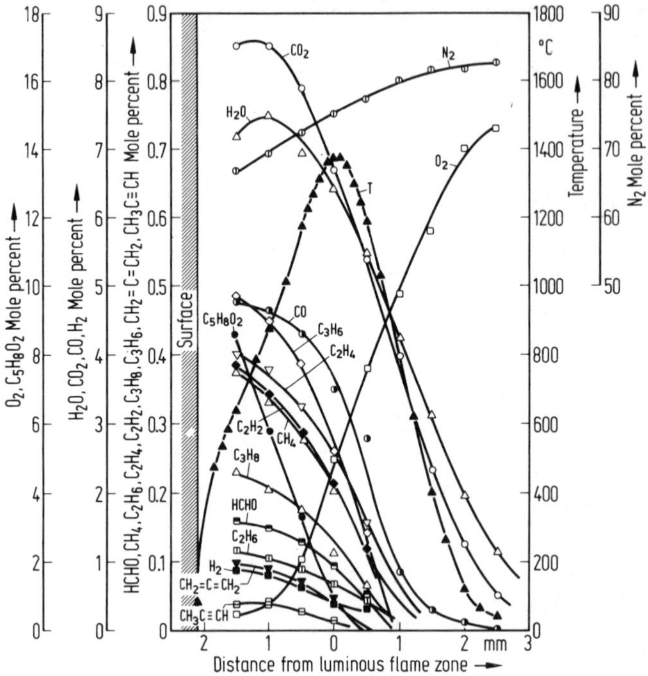

Fig. 3A.6 Temperature and concentration profiles of molecular products inside the flame: (a) PMMA (at oxidising agent flow rate of 0.315 m/sec, $Y_{ox.}$ = 0.178); (b) MMA (at oxidising agent flow rate of 0.327 m/sec, $Y_{ox.}$ = 0.18).

tial røle. It was, however, mentioned that with nearly equal ratios of O_2 and MMA in the gaseous phase during polymer combustion, larger concentrations of CO_2, CO, H_2 and C_2H_4 and smaller concentrations of H_2O, O_2 and MMA were observed; with the aid of gas chromatography, nearly 16 molecular products of decomposition were discovered. The majority of these are formed as a result of the decomposition of monomer; they are also produced by induced molecular oxygen or by active flame particles. Assuming that CO is a product of thermal degradation and that CO_2 is a product of gas phase oxidation, according to the analysis of the CO_2 profile, the authors of reference [17] concluded that the oxidation of fuel only occurs in the gaseous phase. It should be mentioned that as the quenching limit is approached, which corresponds to a concentration of oxygen inside the flow, a clear shift was noticed in the location of the maxima of the principal products of oxidation towards the surface of burning MMA. The concentration of oxygen near the surface increased.

Chemical Aspects of the Reduction in Combustibility of Polymer Materials and Smoke Formation during their Combustion

Recent literature consists of an increasing number of studies that deal with the determination of the relationship between composition and concentration of combustion products (with the products formed inside the flame) and the formation of smoke and its toxicity. Critical surveys of the mechanism of soot formation are given in references [18, 20]. Using PVC as a sample, the link between the yield of benzene and other aromatic compounds (precursors of soot nuclei) and the density of smoke was demonstrated. It was concluded on the basis of an analysis of how metal-containing additives affect the yield of benzene, CO and CO_2, that the additives induce smouldering (incomplete combustion) of linear polyenes formed during the dehydrochlorination of PVC, and then quench the formation of benzene and smoke [21].

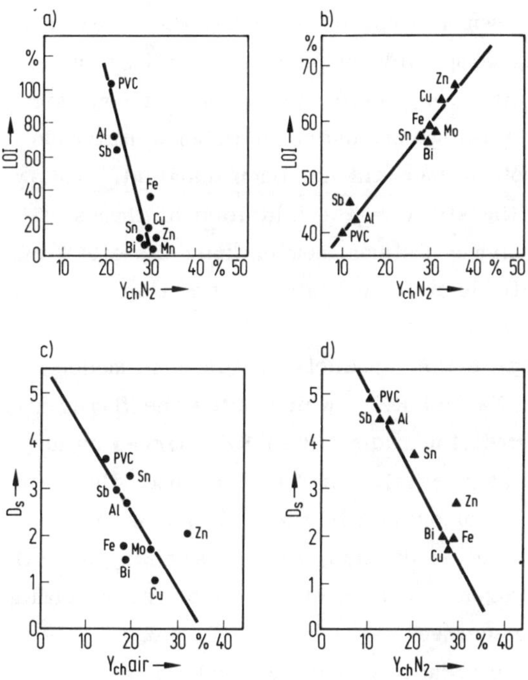

Fig. 3A.7 Effect of metal oxides on the formation of benzene in the degradation of PVC (700°C); (a) oxygen index, (b) density of smoke.

It was discovered that many of the metal oxides reduce the yield of benzene and the density of smoke, and increase the limiting oxygen index of polymers [22]. These changes are proportional to the yield of char residue in the case of PVC degradation at 470°C in an atmosphere of air ($Y_{ch.}$, air) or of nitrogen ($Y_{ch.}$, N_2) (Figs. 3A.7a, b). Since some additives (MoO_3, ZnO) retarded the liberation of CO_2 during incomplete combustion, it was concluded that the formation of smoke was caused by the pyrolysis of PVC. Those metal oxides that actively reduce the formation of smoke catalyse the crosslinking of the polyenes and thereby quench the yield of aromatic compounds.

Oxidation of the char layer, catalysed by certain metals (Fe, Cu, Bi) reduces the yield of char, which is why there is a deviation from the correlation of Ds vs. $Y_{ch.}$.

Halogen-Containing Flame Retardants

In accordance with the well-known macroscopic criteria, the overwhelming majority of halogen-containing compounds behave as gas phase flame retardants in polymer systems. However, there can be various concrete mechanisms for the quenching of flaming combustion of polymer materials by means of a halogen. This problem has still not been clarified. Following the same methodology as for the study of the inhibition of flames of gaseous systems in the case of polymer material combustion, the effect of flame retardants is studied by introducing them into the flame on the fuel side or the oxidiser side.

Flames of monomer-polymer degradation products are used as models. The quenching mechanism of HX, X_2 and RX_n, which enter the flame, may have a physical and chemical nature (the radical R in RX_n serves as an additional source of combustion in any case). In the first case, the flame retardants reduce the concentration of oxygen in the fuel mixture within the flaming reaction zone through simple dilution, in the same way as with dilution by carbon dioxide or nitrogen. The specific heat of the resulting mixture affects the flow of heat consumed.

In the second case, the flame retardant is directly involved in the elementary flaming reactions and affects the kinetics of the complex combustion process. Dixon-Lewis [23] discovered that inhibition reactions of the hydrogen-air flame by HBr lead to an essential decrease in the rate of heat liberation, and that there is no basis for reconsidering the approach to the chemical inhibition mechanism. The reverse reaction in the balance $H + HX \rightleftarrows H_2 + X$ decreases the effectiveness of flame inhibition by HX but its exclusion from the kinetic scheme predicts a situation pointing to a complete absence of inhibition of the high orders of magnitude for the transformation of fuel into combustion products [24].

One of the main objections to the chemical mechanism of inhibition by halogens is the necessity of introducing large quantities of flame retardant (12-15% Br and ~20% Cl) in order to noticeably reduce the flammability of a polymer [25].

Using the examples of the combustion of halogenated benzenes and other hydrocarbon systems and with the introduction of halogen-containing compounds from within the fuel, Larsen showed that the minimum weight percentage of halogen at the end of combustion is 75-80%. The efficiency of all the halogens is the same when compared on a gravimetric basis but that does not apply to the volumetric quantities. In the latter case, the efficiency is proportional to the atomic weight of the different halogens.

The role that the halogens play in the reduction of polymer flammability is simply to increase the rate of material supply into the flame, without making an essential contribution to the heat of combustion or, accordingly, to the heat flow from flame to fuel. When CF_3Br enters the flame of polymethylmethacrylate with the oxidiser the halogen-containing compounds display a higher efficiency. The effectiveness depends on the direction of combustion (candlelike combustion vertically, downwards or upwards, full coverage by the flame). For example, the minimum concentration of CH_3Br added to an "airy" (O_2/N_2) atmosphere at the end of PMMA quenching is 5% and 10% of total weight during candlelike and upward combustions respectively. This is almost half as much as the quenching concentration of nitrogen. If the concentration of oxygen stays constant, the mass rate of thorough polymer combustion in an atmosphere with an admixture of CF_3Br increases in the atmosphere of O_2/N_2. Quenching does not occur. The data obtained indicated that CF_3Br "works" in a flame according to both a physical and a chemical mechanism [26]. The problem is how to distinguish between these effects. Calculations [26] show that the addition of the inhibitor CF_3Br to the flame of PMMA, which reduces the rate of the gas phase reaction (there is a change in the kinetic parameters of the initial gas phase reaction), should not affect the mass rate of thorough polymer combustion up to the complete quenching of the flame. However, quenching will occur with a higher concentration of oxygen in the atmosphere. As attempt has recently been made to distinguish the physical and chemical effects [27, 28]. It was discovered that the value of the oxygen index does not depend on how the flame retardant is introduced into the flame zone [28]. The inhibiting effect of the chlorine-containing compounds on the combustion of polymer materials decreases on dilution. The contribution of $C_2F_4Br_2$ chemical inhibition is small, just as with CF_3Br [27].

Phosphorus-Containing Flame Retardants

Phosphorus-containing compounds may affect the combustion process in both the condensed and the gaseous phase by physical and chemical mechanisms. The most common effect of P flame retardants appears in the C phase [29]. It was noticed in many cases that, in the presence of phosphorus flame retardants, the degradation of polymers occurs at a faster rate. In fact, such behaviour has been observed in the case of the combustion of copolymers of MMA together with diallylmethylphosphate (DAMP) and in the combustion of PMMA, which contains various additive-type flame retardants. A more detailed study of high-temperature pyrolysis of these polymer systems determined the mechanisms of the process [30]. It was discovered that, for a constant temperature of the surrounding atmosphere, the temperature of the samples increases with a rise in the percentage of phosphorus (Fig. 3A.8a).

If the constants of thorough combustion of polymers are compared, not for a constant surrounding atmosphere temperature but for a constant stable regime on the surface, it appears that the constants become smaller in the presence of phosphorus (Fig. 3A.8b).

At present, several assumptions have been verified regarding the mechanism of phosphorus flame retardant action in the C phase. A direct analysis of the carbonised residues after the pyrolysis and combustion of polymers using X-ray electronic spectroscopy established that there is formation of PO_4 fragments, called polyphosphoric acid. The latter is localised in the presurface layer; on the surface itself, carbon accumulated. Migration of phosphorus occurs under certain conditions [31].

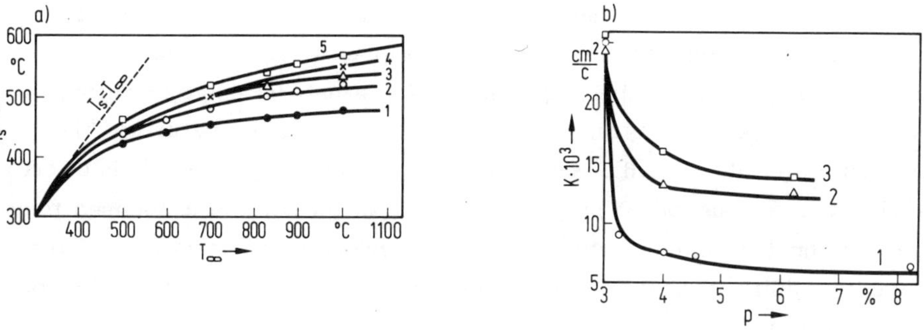

Fig. 3A.8a Relationship between the surface temperature T_s and the temperature of the surrounding atmosphere: (1) PMMA; (2, 3, 4, 5) copolymers containing 0.31, 1.49, 2.88, and 11.23% phosphorus.

Fig. 3A.8b Relationship between vaporisation constant and percentage of phosphorus in MMA copolymers for equal temperature regimes on the surface. (1) Copolymer of MMA with DAMP, (2, 3) copolymer of MMA with nonreactive flame retardants. T_s = 490°C (1, 2) and 485°C (3).

The treatment of char by phosphorus-containing compounds causes a reduction in the surface tension and in the rate of the liquid passing into the char layer, and it also retards the rate of combustion. However, the highly foamed carbonised layer does not show the expected insulation properties: oxygen easily penetrates the inner surface of the initial polymer [32].

Using the method given in reference [28], it was disovered that a few volatile phosphorus flame retardants (trimethylphosphate) react according to a physical mechanism.

In a series of studies using resonance fluoresence spectroscopy, the kinetic parameters of a series of elementary reactions of atoms and active particles were determined for the first time for simple phosphorus compounds [33].

Metallic Compounds

The way in which metallic compounds affect the combustion of polymers depends on a number of factors. A large role is played by the nature of the metal, the degree of its oxidation, the type of compounds, the manner in which the metal is introduced into the polymer, its concentration, the order W > Mo > V, the combustibility of wool falls. In terms of macroscopic criteria, titanium and zirconium complexes react in the condensed phase. They catalyse the degradation of the natural polymer during conditions of polymer pyrolysis and combustion and the nature of the polymer material. Metallic compounds can affect the processes that occur during the combustion of polymers in the gas phase, the condensed phase and also on the contact area phase. The velocity and reactivity of the metal-containing compound determines the probability of its participation in the flame reaction.

At present it has been shown that there is a means by which an active reduction in combustibility can be achieved for a number of polymer systems by using metallic compounds at very low concentrations. Metals and their organic derivatives can be introduced into the polymer macrochain, can be linked by coordination bonds to the polymer's functional groups, or can be used as additives.

The combustibility of wool articles is reduced, for instance, by modification with K_2ZrF_6 and K_2ToF_6, combining them with isopolymolybdate, vanadates, tungstates of alkali metals, their complex compounds, and organic acids. The oxygen index of the wool articles increases from 24.5 to 31.5 as a result of such treatment. It was discovered that as the atomic radius of Me ions, and accordingly $(MeF_6)^{-2}$ ions, decrease in the combustion (see Friedman [29]). When zinc pyromellitate (1-5 pph) is added to PVC, the yield of char increases almost twofold, the yield of volatile fuel drops by 39% and the oxygen index increases from 50 to 72-75. The enhancement effect of char formation correlates better with the boiling point of the formed metal halides than with their Lewis acid strength. In other words, the less volatile metal halides promote cross-linking on account of the dehalogenation of polymers [34] by remaining in the C phase.

Small additions of Co^{2+}, Ni^{2+}, Mn^{2+} and Fe^{2+} salicylates (as was shown in [35]) accelerate the thermo-oxidative degradation of polystyrene and reduce the time delay of its self-ignition with an increase in the redox potential of Me^{2+} (Fig. 3A.9). Some shifting occurs in the case of Ni^{2+}. It was disovered that at the high self-ignition temperature of polystyrene, Co salicylate is converted to Co_3O_4. The authors assume that the complexes of metals with variable valence accelerate the processes that occur according to the mechanism of electron transfer:

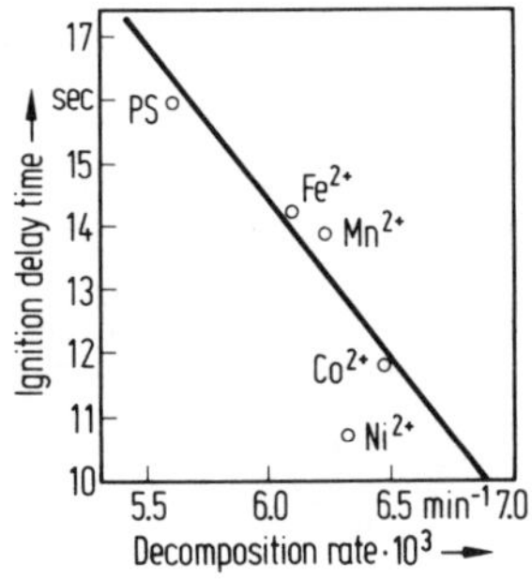

Fig. 3A.9 The effect of salicylates of transitional metals on ignition time delay and rate of polystyrene decomposition.

$$Me^{2+} \rightarrow Me^{3} + e^{-}$$

$$\sim R + e^{-} \rightarrow \sim\dot{R}$$

$$Me^{3+} + \sim\dot{R} \rightarrow Me^{2+} + \sim R'$$

Volatile metal halides, which have a relatively low strength of Me-X bonds, are particularly effective flame retardants. Halides of two and three-valence iron in very low concentrations (1-5 pph) essentially decrease the combustibility of ABS plastics (OI increases from 18.5 to 30). In combination with halogen-containing compounds, oxides and salts of Mn^{2+}, Sn^{4+}, Mo^{6+} and other metals also reduce the flammability of ABS. The effectiveness of $FeCl_3$ is many times greater than that of $SbCl_3$ [36].

All the data together shows extensive possibilities for employing metal compounds in order to selectively regulate the thermal properties of polymer materials. This field is just at the outset of its development. There is much that remains unclear. The analysis is often difficult on account of the dual function of metal compounds (especially *d* and *f* types) in the case of redox reactions. Their ability not only to inhibit but also to accelerate the process of polymer destruction is very clear and appears in concentration effects. A specific influence of certain metals on the ignition and on the combustion rate of polymer materials indicates a specific role of heterogeneous oxidation reactions. It is known that ions of the metals with variable valence accelerate the decomposition of hydroperoxides according to a radical and molecular mechanism through intermediate complexes [37]:

$$nROOH + mMe^{(n+1)+} \rightarrow [nROOH\ldots mMe^{(n+1)+}] \rightarrow ROO\cdot/H^{+} + Me^{n+} + (n-1)ROOH + (m-1)Me^{(n+1)+}$$

$$nROOH + mMe^{n+} \rightarrow [nROOH\ldots mMe^{n+}] \rightarrow RO\cdot + OH^{-} + Me^{(n+1+} + (n-1)ROOH + (m-1)Me^{n+}$$

$$\downarrow$$

$$[Me^{(n+1)+}, OH^{-}, RO\cdot] \rightarrow Me^{n+} + R'C(O)R'' + H_2O$$

In the higher valence state, $Me^{(n+1)+}$ ions are able to react with the radicals $\dot{R}$, $R\dot{O}$ and $R\dot{O}_2$ and are also able to accelerate the decomposition of oxygen-containing compounds: aldehydes, ketones, acids and alcohols. A positive effect of metal compounds on the retardation of flaming combustion of polymers as a result of solid phase reactions contributes first of all to the reduction in the yield of volatile fuel products during the pyrolysis of polymers.

There are two alternate directions:

1. Catalysis of the formation of nonflammable products (CO_2, H_2O).
2. Retardation of the rate of polymer degradation at the high temperatures on the surface in the case of stationary combustion.

References

1. Ghibov K. M., Abdikarimov M. P., Zhubanov B. A. Vysokomolec soyed., 1978, vol. 20A, pp. 2688-2693.
2. Galchenko A. G., Khalturinsky N. A., Berlin A. A., Vysokomolec. soyed., 1980, vol. 22A, no. I, pp. 16-21.
3. Jakes K. A., Drews M. J., J. Polymer Sci., Pol. Chem. Ed., 1981, vol. 19, p. 1921.
4. Stuetz D. E., et al, J. Polymer Sci., Pol. Chem. Ed., 1980, vol. 18, pp. 967, 987.
5. Das A. N., Combust. and Flame, 1981, vol. 42, p. 35.
6. Vishnevsky G. E., et al., Dokl. AN SSSR, 1981, vol. 261, no. 5, p. 1119.
7. Hedges J. H., Baer A. D., Ryan N. W., 17th Symposium (Intern.) on Combustion. Pittsburgh: Combust. Inst., 1979, p. 1173.

8. Brauman S. K., J. Fire Retardant Chem. 1979, vol. 6, pp. 249, 266.
9. Kambour R. P., Klopfer H. J., Smith S. A., J. Appl. Pol. Sci. 1981, vol. 26, p. 847.
10. Hilado C. J., et al., J. Fire and Flammability, 1978, vol. 9, pp. 367, 553.
11. Zaks Y., et al., J. Appl. Polym. Sci., 1982, vol. 27, p. 913.
12. Ellard J. A., Am. Chem. Soc. Div. Org., Coatings and Plastics, 1973, vol. 33, p. 531.
13. Cagliostro D. E., Riccitiello S. R., J. Fire and Flammability, 1975, vol. 6, p. 205.
14. Johnson P. R., J. Appl. Pol. Sci., 1974, vol. 18, p. 491.
15. Ohe H., Matsuura K., Textile Research J., 1975, vol. 45, p. 778.
16. Aseeva, et al., Plastmassy, 1982.
17. Sechadri K., Williams F. A., J. Pol. Sci. Pol. Chem. Ed., 1978, vol. 16, pp. 1755-1778.
18. Tesner T. A., Physika goreniya i vzriva, 1980, vol. 16, p. 1.
19. Bankston C. P., et al., Combust. and Flame, 1981, vol. 41, pp. 273-292.
20. Calcotte H. F., Combust. and Flame, 1981, vol. 42, p. 215-242.
21. Descamps J. M., Delfosse L., Lucquin M., Fire and Materials, 1980, vol. 4, p. 37.
22. Ballistreri A., et al., J. Pol. Sci. Pol. Chem. Ed. 1981, vol. 19, p. 1397.
23. Dixon-Lewis G., Combust. and Flame, 1979, vol. 36, p. 1.
24. Lavachev L. A., Combust. Sci. and Technology, 1981, vol. 25, p. 49.
25. Larsen E. R., Ludwig R. B., J. Fire and Flammability, 1979, vol. 10, p. 69.
26. Sibulkin M. et al., Fire and Flammability, 1979, vol. 10, p. 263; Combust. and Flame, 1982, vol. 44, p. 187.
27. Tucker D. M., Drysdale D. D., Rasbach D. J., Combust. and Flame, 1981, vol. 41, p. 293.
28. Akopyan S. V., et al., Dokl. AN USSR, vol. 264, no. 3, p. 939.
29. Flame Retardant Polymeric Materials, M. Lewin, S. M. Atlas, C. M. Pierce eds., N.Y.-L.: Plenum Press, 1978, vol. 2.
30. Galchenko A. G., et al., Vysokomol. soyed., 1982, vol. 24A, p. 63.
31. Lihpanov A. M., et al. in Khimicheskaya Physika goreniya i vzriva.

Goerniye Kondensirovannih i geterogennih system. Tchernogolovka: OlkhF AN USSR, 1980, p. 85

32. Ghibov K. M., Zhubanov B. A., Shapovalova L. N., in Thesis of the Symposium, October 19-21, 1981 Tallinm Institute of Chemistry AN USSR 1981, p. 93.
33. Alekandrov E. N., et al., Dokl. AN USSR, vol. 260, p. 113; Izvestiya AN USSR, Ser. Khim. 1982, no. 1.
34. Brauman S. K., J. Fire Retardant Chem. 1980, vol. 7, pp. 119, 154, 161.
35. Kishore K., Prasad G., Nagarajan R., J. Fire and Flammability, 1979, vol. 10, p. 296.
36. Whelan W. P. Jr., J. Fire Retardant Chem., 1979, vol. 6, p. 206.
37. Blyumberg F. Y., Dokl. AN SSSR, 1978, vol. 242, p. 358.

Chapter 4

Methods for the Study of the Combustion Process and Estimation of the Flammability of Polymer Materials

In this chapter we shall discuss the chief methods available for analysing the different characteristics of the combustion of polymer materials and also the methods developed for fire modelling (simulation) and estimation of the flammability of polymer materials. It should be emphasised that a general theory of combustion has still not been formulated. Consequently, it is vital to have information characterising the combustion of polymer materials under different conditions and to be able to simulate combustion in small-scale experiments, thereby allowing the behaviour of polymer materials in critical situations to be predicted.

To formulate a scientific solution to the problem of of polymer flammability it is important to have information on the action mechanism of the different flame retardants.

Experimental Methods for the Study of the Combustion of Polymer Materials. Determination of Combustion Characteristics.

In the overwhelming majority of cases, the instrumental techniques and methods for the study of polymeric materials are similar to those previously developed for the analysis of various condensed fuel systems. A methodical approach has been adopted to the study of the combustion of these materials.

Usually attention is focussed either on general macroscopic combustion characteristics (speed, limitations of stable combustion, temperature of the combustion products and their composition, etc.), such as the standard testing of the combustibility of a material, or on a more detailed analysis of flame structure and burning surface. In the latter case, one has to determine the hydrodynamic characteristics of the fuel reagent and oxidising agent flow, the temperature and concentration range of the combustion wave and the correlation between these and the molecular and physical structure of the material and its physico-chemical properties.

An approach of this type permits an elucidation of the mechanism and the kinetic principles that characterise the combustion of a polymeric material and its components. The following types of combustion, characterised by different geometric configurations of diffusion flames, are the types most frequently examined:

1. Candlelike combustion;
2. Formation of a "flat" flame as a result of the counterflow effect of the oxidising agent on the sample's surface;
3. Combustion of spheres;
4. Flame spread over the surface.

The method of a moving substrate or a moving wire has been developed for studying the ignitability and combustibility of polymers [1, 2]. This technique enables one to examine the interaction between the polymer and the flame used as a high-temperature reaction medium with a known chemistry and properties (for instance, an $H_2 + O_2 + N_2$ flame or a $CH_4 + O_2 + N_2$ flame is used).

Let us look briefly into the experimental techniques for determining the most important characteristics of the combustion of polymer materials.

Rate of Combustion

Compared with gaseous and liquid fuel systems, polymer materials have relatively low values for their linear rate of combustion. This simplifies the determination of the linear rate of combustion. During the standard testing of the combustibility of polymer materials, the combustion rate is determined visually by recording the time and distance of the shift in the combustion front.

When visual observation does not give sufficient accuracy in measurement or is too difficult, there are various other techniques applied to monitor the shift in the combustion front: photography, or temperature recording at specific distances by means of two thermocouples. In addition, it is possible to use the method of burning wires, which was developed in order to determine the normal rate of combustion of miscible solid fuels [3].

For small-scale laboratory research it is feasible to weigh the burning sample continuously during the stationary combustion process and thereby directly determine the rate of mass change with time. This method has been used, for instance, in the case of combustion of polymer spheres [4].

Various types of electromechanical transducers are useful for rapid measurement of weight [5]. The mechanical signals picked up by the transducer when there is a change in the linear combustion front or in the force or the pressures of the tested system are transformed into an electric current. The electric signal is amplified and recorded.

The knowledge of the shape and size of the burning surface and the density of the material permits calculation of the mass $\dot{m}$ and linear rates v for the polymer sample, using the relationship $\dot{m} = \rho \cdot v$ (g/cm²·sec).

The shift in the combustion front creates difficulties in the determination of local flame characteristics. It is thus understandable that there should be a tendency in research to create a set-up which will allow sample input at a rate that will compensate for the rate of the flame's combustion and maintain a steady burning of the surface. References [6, 7] may be cited as examples of such an approach. The configuration of a set-up for the study of the flat flame of a polymer material with a counterflow of oxidising agent is shown in Fig. 4.1 [8].

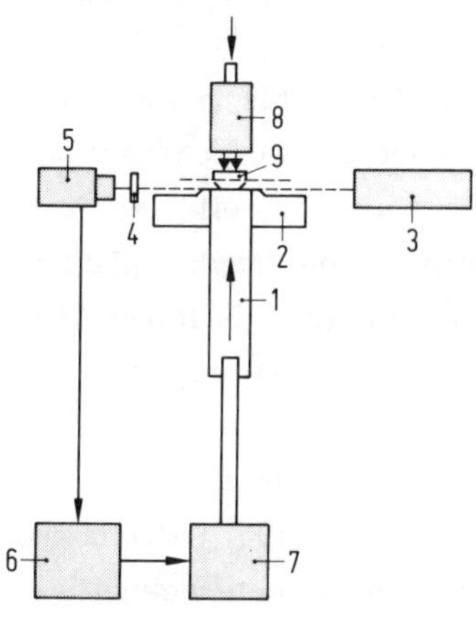

Fig. 4.1 A device for the generation of a flat flame. (1) Polymer sample, (2) cooling device, (3) He-laser, (4) red filter, (5) photodiode, (6) electronic feedback system, (7) servomotor, (8) nozzle for oxidiser, (9) flame.

An optical system for controlling the burning surface level is combined with an electronic feedback system which regulates the operation of the servo motor. According to the authors, the whole system maintains the burning surface of the polymer samples at a preset level with an accuracy of $\pm$ 0.01 mm. In this case the linear combustion rate can be recorded and estimated from the rate of linear shift in the samples (if there is no physical transport of the material as a result of the melting).

Shape and Structure of Flame and Burning Surface

Optical methods have been the most widely used for these purposes. Examples that should be cited are:

1. Photographing or filming the burning object in its own light;
2. Interference methods, including those that enable reproduction of a volumetric (holographic) picture;
3. Schlieren-photography or Tepler's method;
4. Shadow method.

The photographing of an object by flame light gives useful information about the shape and size of the flame and the burning surface. However, it is not always possible to determine or to estimate the heterogeneities in the medium (swirling, frothing of the superficial layer) in this way. The method based on the interaction between reflected and transmitted light and the medium is more promising.

The holographic method is even more important [9]. This method involves the constructive and destructive interference of two coherent light beams generated from a common laser source. One beam is reflected from an object and recombined with the second beam on a photographic plate in order to produce an interferogram. When the interferogram is illuminated with a coherent light source a three-dimensional image of the object is produced.

This method is especially useful for the study of condensed phase dispersion phenomena during the combustion of polymers [10]. It enables the position coordinates and the size and concentration of heterogeneities in the medium to be determined.

Common interference methods involve the passage of coherent rays through the medium. Where there are regions containing optical heterogeneities (regions with different refraction indexes) there is a deflection of the light rays and a separation due to phase and interference.

The so-called interference method is characterised by the fact that the size of heterogeneities is determined by the time delay needed for rays to reach the screen. This time delay is proportional to the variations in the index of refraction. The equipment and the experimental technique are described in [11]. Complicated and expensive equipment limits the application of the interference method. It should, however, be noted that the high accuracy and good steric resolution of the holographic method are particularly valuable for the determination of other flame characteristics, such as temperature profile.

Schlieren-photography (Tepler's) method also involves the interference of light beams as the rays pass through a flame. The name of the method comes from the word "Schlieren", which means "optical inhomogeneities". With this method, the angle of deflection of light rays passing through the optical heterogeneity in the medium is determined in relation to the initial position of incidence. By contrast to the holographic method, Schlieren-photography is simpler in terms of the equipment used and its convenience. Since the angle of deflection is proportional to the density gradient and the magnitude $(1/T)(dT/dz)$, the method can also be applied to determine the temperature field within a flame. Reference [12] may be cited as an example of how Schlieren-photography is used to analyse the shape of a flame and to estimate the temperature profile during the combustion of polymer materials. In this study, the combustion of paper at different angles was analysed.

The simplest method to determine whether or not there are heterogeneities inside a flame is the shadow method. The shadow method system usually consists of a point source of light projected on a screen. The flame under study is placed at a specific distance between the screen and the light source. When light rays pass through the heterogeneous medium, dark spots appear on the screen due to the change in screen illumination that is caused by the deflection and interference of the light rays.

The shadow method permits the location of the heterogeneity inside the flame to be established. A more detailed interpretation of the interference image is difficult. In principle, the shifting of dots on a screen on which light rays impinge (in cases where the rays of light are disordered by the heterogeneity) is proportional to the second derivative of the refractive indexes at different heterogeneous points. Quantitative determination of this shifting, however, is impossible with the cited system. The shadow method clearly indicates the presence of turbulence inside the flame.

Temperature Profiles during Combustion

Contact and noncontact methods are used for the determination of temperature within the combustion wave of a polymer material. Contact methods are used mostly for the determination of temperature fields within the condensed phase. Noncontact methods are used for measurement of the temperature of the burning surface and the temperature profile within the flame. The contact method used most frequently for the study of polymers is that of thermocouple probes.

The active junction of the thermocouple can be placed inside the sample, on its surface or at a fixed distance from the point of ignition, or can be introduced into the flame. The shifting of the junction in order to measure the temperature inside the combustion wave takes place as the combustion front shifts. If a special device is used to keep the position of the burning polymer surface at a pre-set level, then the temperature profile of the flame can be determined as a function of the shifting of the probe through the use of a manipulator. In this case, it is necessary to measure the position of the probe inside the flame relative to the burning surface. The latter operation is carried out with a cathetometer or by using optical methods of flame imaging as described in the previous section.

The method of thermocouple probing is simple and practical. When the size of the junction is small (a few microns) a high steric resolution can be achieved. It should, however, be mentioned that when the thermocouple probe is introduced into the reacting medium it may cause turbulence, react with the medium, affect the dynamics of the reactions or be a source of heat loss from the flame. If there are large temperature gradients within the combustion wave, the thermal response of the thermocouple and the response of the recording device cause an essential distortion in the shape of the temperature curve. Devices with a fast response (loop oscillograph) are preferable for recording thermocouple signals.

The source of errors during determination of the temperature characteristics of the combustion of condensed substances (including polymers) and the means of eliminating them with allowance for the influence of different factors, are discussed at length in references [13-17]. The technique of thermocouple preparation and the method of applying a silicate coating to the thermocouple to prevent a reaction with the medium were described by Fristrom and Vestenberg [13].

The use of a thermocouple probe is limited by temperature. By using a platinarhodium-platinum thermocouple, temperatures of up to 1600°C can be determined; with a tungsten-rhenium thermocouple, temperatures of up to 2500°C can be determined in an inert medium. At temperatures greater than 1000°C in an oxidising environment, the tungsten-rhenium thermocouple begins to oxidise appreciably and produces readings that are too high. With an increase in the temperature of the medium there is an increase in the error caused by the rise in thermal losses due to the heat transfer and radiation from the probe.

The noncontact methods, which in the overwhelming majority of cases have no limitations due to an upper temperature limit, are considerably more promising in respect of thermal losses. There are a large number of noncontact methods for determining temperature. For the study of polymer combustion, however, it has been the optical methods that have been used in practice. The possibility of applying the optical interference method and Schlieren-photography for determining the temperature of polymer flames has already been mentioned.

The comparison of various gas phase temperature profiles during the spreading of a laminar flame over the surface of polymethylmethacrylate, determined by thermocouples and by interference methods, is given in reference [18]. Another technique of optical temperature measurement is based on the absorption of electromagnetic waves by a hot surface or flame. An entire group of radiation pyrometry methods was established on the basis of this principle [19]. The most simple radiation-pyrometry methods consist in measurement of the intensity of a continuous spectrum of radiation from the object in the IR or visible region as compared with the intensity of the luminous source at a certain temperature. IR pyrometry has a number of advantages as compared with pyrometry which uses radiation from the visible part of the spectrum: the extension of the measurement range towards lower temperatures makes it possible to examine flames with a pronounced formation of soot.

On account of steric resolution, the methods of radiation pyrometry are not as good as the interference methods. In addition, the temperature values derived in the first case are averaged out along the whole cross-section of the flame.

The temperature distribution within the condensed phase in the case of a constant linear rate of thorough polymeric combustion may be represented by Michelson's equation:

$$T_x = T_{\text{in}} + (T_{\text{s}} - T_{\text{in}})\exp(-vx/a)$$

where

x distance from the surface

T_{in} initial temperature of polymer sample, °K

T_x temperature at point x

T_{s} temperature of the burning surface

a temperature coefficient of polymer (assumed to be constant)

During combustion the temperature gradient changes on the phase contact area. At this time a deviation should be observed from the linear nature of the semilogarithmic relationship between (T_x - T_{in}) and x, from which the temperature of the burning surface T_{s} can be determined with a good degree of accuracy.

Concentration Profiles of Products inside a Flame

Two approaches have been adopted to the determining of the composition and concentration of compounds in various zones of the flames of polymer-containing systems:

1. Removing samples from the flame with a probe, followed by analysis of the sample by different methods
2. Direct analysis of the flame by spectral methods.

The first method has been used primarily for the analysis of compounds that are stable during the probing process (see Chapter 3). Sampling could be done either periodically or continuously.

The application of flame probing always raises the question as to the extent to which the introduction of the probe and the sampling leads to disturbance of the flame. In essence, the probing of the flame in order to obtain samples of products causes the same negative effect as thermocouple probing. In this sense, the noncontact research methods are the more promising ones.

A direct analysis of the flame by spectroscopic methods allows one to identify and determine the local concentration of both stable and short-lived particles - atoms and radicals. The analysis can be carried out by measuring the absorption and radiation spectra of the flames.

The research involving analysis of the nitrocellulose flame [20] may be cited as an example of determining the concentration profile for molecular products and radicals with respect to the radiation spectra. The authors determined the concentration profile for radicals ($\dot{O}H$, $\dot{C}N$, $\dot{N}H$, $\dot{C}_2$) and molecular particles (CO_2, CO, NO, H_2, O_2, N_2). The determination was conducted on the basis of the intensity of certain spectral lines in visible UV bands of the radiation spectrum, which were analysed along the height of the flame of the polymer substance. A similar analysis was also conducted for the IR spectra of the flame radiation in the 0.8-9.0 micrometer range.

Despite the apparent usefulness of the nonperturbing methods of flame analysis, difficulties arose in the identification and quantitative analysis of flame spectra of multicomponent mixtures or organic compounds.

The method of sampling by probing allows this difficult problem to be solved by separation of the sample (preliminary or in the analyser) into individual components. Samples from the flame are analysed by the usual methods of gas chromatography and mass spectroscopy. We shall not discuss these methods here, since they are well known and widely applied by chemists. Each of them has its advantages and disadvantages in respect of the analysis of the concentration and distribution of particles within the flames of condensed systems. It should be mentioned that the bulk of experimental problems is connected with probe sampling rather than with the analysis.

In order to obtain sufficiently high steric resolution to reduce the disturbing effects and diminish the catalytic effect of the probe, it was recommended that conically-shaped quartz microprobes with an aperture angle of 15-45° and small inlets be used. The technique for the fabrication of quartz microprobes was established by Fristrom [21]. The stretching of capillary tubes is accomplished by the use of small loads, as shown in Fig. 4.2.

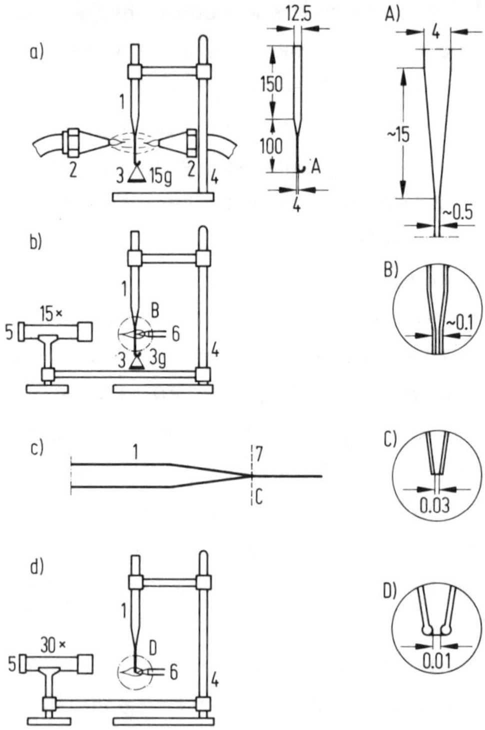

Fig. 4.2 Preparation and set-up of quartz probe. (a) Coarse stretching of capillary, (b) fine stretching, (c) cutting, (d) fire-polishing. (1) Quartz tube, (2) oxygen, (3) load, (4) stand, (5) microscope, (6) microburner, (7) knife for glass. All sizes are in mm.

The methods of mass-spectroscopy probing of flames of condensed systems are given in reference [22]. The methods of mass spectroscopy make it possible, in principle, to determine not only the presence of molecular particles but also radical particles and ions. The sampling of such particles from the flame with a probe, however, requires special techniques and is not easy. In addition, the presence of short-lived particles complicates the problem of analysis of the local composition within the flame.

Gas-Dynamic Characteristics

In order to determine the mechanism of flame spreading during the combustion of polymers it is important to know the values of the gas-dynamic characteristics of the flows of oxidising gas and fuel reagents. In the case of diffusion combustion, a matter of special interest is the nature of motion and the local rates of flow of a gaseous oxidising agent near the leading edge of the flame.

The simplest and most frequently applied method in the study of polymer combustion for determining local rates of a gaseous flow is the modelling of the flow using suspended microparticles and the compilation of a photographic record of particle trajectories.

Finely dispersed powders (particle diameter, 1-5 micrometers) of magnesium, titanium, zirconium and aluminium oxides are used as injection particles. The system includes a device for injecting the particles into the oxidising agent stream, an illuminating system for the customary inspection of the particles and an optical system for photographing the flame. The lighting system provides intermittent illumination of the object at pre-set time intervals. Figure 4.3 shows the set-up of a lighting system used for measurement of the gas rate near the leading edge of the flame in the case of a flame spreading over a paper sheet [23]. A beam of light from a high-intensity lamp (750 Watt) passes through an aperture and is intercepted by a rotating disk. There are 16 rectangular notches around the disk circumference. The resultant pulsed beam of light is reflected by two

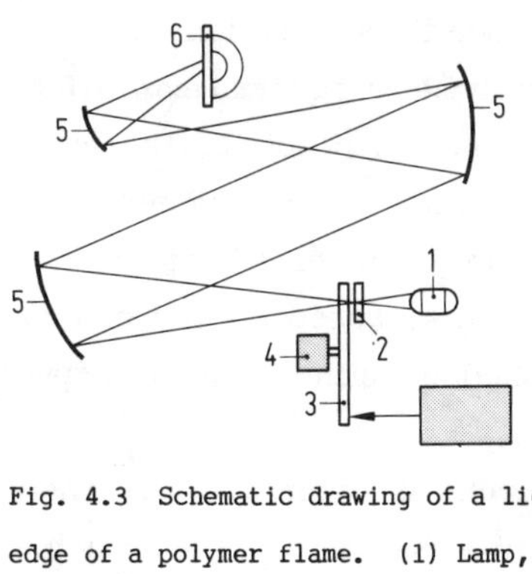

Fig. 4.3 Schematic drawing of a lighting system for determining a gaseous flow rate near the edge of a polymer flame. (1) Lamp, (2) aperture, (3) rotating disk, (4) motor, (5) concave mirror, (6) testing sample.

concave mirrors and focussed by a third concave mirror on the gaseous stream near the leading edge of the flame.

The local rate of gas flow is determined by measuring the length of an interrupted trajectory of the particle on a photographic enlargement (see Figure 2.13). The length and time scales are known.

Errors in trajectory studies of gas-dynamic characteristics may be linked to the effect of particle inertia, particle response time, where there is a rapid change in the rate of gas flow, and other factors. Many of these errors can be discounted [13].

In the case of experimental studies of the combustion of polymer materials, the most complicated problem is the determination of the physical and chemical processes that occur both in the condensed phase and on the surface of the phase contact area, and a quantitative estimation of the changes detected.

At present, it is common practice to quench and cool the burning surface quickly with liquid nitrogen. Then either microtome cuts of the condensed phase or the surface can be analysed by various methods used in the study of condensed systems (NMR and EPR, spectroscopy, electron microscopy, IR spectroscopy, elemental analysis etc.). Electron spectroscopy and a number of other methods can be considered for the study of the processes that occur within the surface layers during combustion of polymeric materials [24, 25].

Estimation Methods for the Flammability of Polymer Materials

When polymer materials are produced for a specific application, an important question is the extent to which the polymeric material can affect the safety, health and well-being of the consumer. In practice, the rules and standards of fire prevention are an inherent part of more general standards and regulations that guarantee the safety and health of the population. In addition, certain corresponding laws, based on the criteria of flammability of materials, have been established by either international or national standards [25a].

In general, the flammability of polymer materials is determined by their tendency to ignite and spread the combustion process and by the possible after-effects that injure the health and welfare of people. In order to obtain a precise estimate of the degree of flammability of a material it is necessary to have quantitative criteria that objectively characterise the above-mentioned phenomena.

Many difficulties arise, however, because these characteristics are not stable and depend not only on the chemical nature of the material but also on the size and geometry of the samples, the nature of the surrounding environment, the aerodynamics of motion, the direction of reagent flow, the source of ignition, the response time of the effect, and other factors. Consequently, a large number of methods have been created for the evaluation of potential flammability of polymer materials. However, as each of the existing methods allows only the influence of just a few factors on the level of flammability to be estimated, none of the methods used in isolation can adequately describe the real flammability of polymer materials.

It is understandable that, in many cases, the results of evaluations by different methods do not correlate. Standard methods of testing for the flammability characteristics of polymer materials accepted in different countries are frequently based on different quantitative criteria. In addition, the situation is complicated by the fact that many accepted methods are strictly research tests. Thus the results as to the degree of flammability of materials are not recognised by law.

International organisations that coordinate research in the field of fire prevention (the International Standardisation Organisation, ISO; the European Standardisation Committee, CEN; and the Standardisation Committee of the Council for Mutual Economic Assistance, CMEA) are responsible for such problems as the determination of the principal differences between national standards and the development of international standards and of efficient programmes for testing and controlling polymer materials.

The adoption of a single set of standardised methods and programmes for the testing of material flammability, compiled on a strictly scientific basis, would be a guarantee of an objective evaluation of the degree of flammability of polymer materials for any application.

According to the existing norms for fire prevention, the combustibility of materials is taken as a major characteristic of a potential fire hazard (combustibility is the capacity of a material to ignite and perpetuate the combustion process). However, the physical destruction of structures based on polymeric materials, smoke formation, and the formation of toxic pyrolysis and combustion products are very hazardous during fires. This is why there is a need for a polymeric material to be identified as a potential fire hazard on the basis of an evaluation of data obtained during a special testing programme. The testing programme should consist in establishing the characteristics that determine the inclination of materials to ignite, their capacity to sustain the combustion process, their fire resistance, their ability to form smoke and the toxicity of their pyrolysis and combustion products.

In practice, the programme is only carried out at the final stage of materials testing when recommendations are made on how to use the developed material. The initial testing stage is confined to the study of the material's flammability characteristics. Depending on the purpose and the material requirements, the programme could be complemented by other experiments.

We have already indicated that the combustion characteristics of polymeric materials depend to a great extent on the size and geometric shape of the sample, which is why an appropriate method for characterising the flammability of a given type of material is recommended for each type of polymer material (films, threads, rigid and flexible plastics, foams). Thus according to a subcommittee for the testing of the flammability of floor coverings (American Society for Testing Materials, ASTM) more than 40 different testing methods are applied in order to determine ignitability and combustibility [26].

Depending on sample size, the tests are divided into small-scale (up to 30 cm), medium-scale (30-120 cm), large-scale (>120 cm), and full-scale categories. The most complete description of a material's flammability is given by the full-scale and large-scale testing programmes, which closely simulate real fire conditions. Even such tests, however, cannot reflect all the varied situations leading to the initiation, propagation and spreading of fire and its consequences. Full-scale and large-scale tests are expensive and difficult to control. Furthermore, such tests do not provide a detailed study of the combustion process. They are usually carried out at the final stage of the testing programme and are designed mainly for estimates of materials with limited flammability.

The tunnel method, ASTM E84, is the most well-established test among the large-scale methods and has been recommended by the ASTM for the evaluation of the degree of flammability of polymer materials used in the building trade. Nowadays, this method is also used for other purposes.

The measured characteristics are: the rate of flame propagation over the surface, the heat liberation during combustion and the density of smoke. A sample of 7.6 m length and 50 cm width placed horizontally serves as a ceiling for a tunnel chamber. Two gaseous jets placed at a distance of 30 cm from the end of the sample and 19 cm below the sample surface are used as sources of ignition. The height of the flame is 135 cm.

The flammability estimate is made on a relative scale. According to this scale, the degree of flame propagation over the surface of an asbestos-cement plate is assumed to be 0, and over a surface of oak, 100%. Materials are sub-divided into the following groups: nonignitable with a flame spreading index of 0 to 25 units, flame retarding (25-50 units), slow-combustible (50-75 units), ignitable (75-100 units) and readily burning (more than 100 units). This method has been modified by the Canadian Standardisation Association (CSA) in order to test polymer floor coverings. In particular, Standard CSA B54.9 specifies the placement of material on the floor of the testing chamber and the spreading of flame along the top side of the horizontal surface. Method ASTM E286 is used for the determination of the flammability of cellular plastic materials. The sample is placed at a small angle.

There are a number of large-scale tests for the determination of fire-resistance and flammability of polymer materials used in the building trade, and various ignition sources are employed in these experiments. Fire resistance is defined as the length of time for which the effective properties of the materials are retained under the action of fire.

Fire-prevention norms and regulations assume the determination of fire-resistance properties of construction materials on the basis of international standard (ISO R843) and national standards (BS 476, p. 8, DIN 4102, NF 92.201, ASTM E119). In the USSR, fire chambers are used for large-scale fire-resistance tests (size of sample >6 m) for building structures containing polymer materials. The degree of flame action and the load regime can be regulated in these chambers. The deformation changes, the destruction (loss of strength) of the structures by the flame, and also the rate of the spreading of the combustion process [27] are registered in these tests. Medium-scale and small-scale testing methods are more practical.

Determination of the Ignitability of Polymeric Material

The relative tendency of polymer materials to ignite is determined by such characteristics as self-ignition and ignition temperature, oxygen and temperature indexes, and the influence of ignition sources with different powers and different lifetimes.

In order to determine the self-ignition and ignition temperature, a testing method is used that is based on a procedure established in 1949 [28]. According to the method set out in ASTM D 1922, a sample weighing 3 ± 0.5 g is used. With the aid of thermocouples, the temperature of air is maintained constant inside the furnace and the temperature increases near the surface of the sample as a result of the acceleration of the oxidising reaction, which promotes ignition (in the presence of an additional ignition source) or leads to self-ignition of the fuel gas-products of pyrolysis of polymer materials.

In the USSR, a vertical tubular furnace constructed in the All-Union Scientific Research Institute of Fire Prevention, ASRIP, is used for the determination of ignition temperature ($T_{ign.}$) and self-ignition temperature ($T_{s\text{-}ign.}$). The size of the reaction chamber is $d = 1 \times 10^{-1}$ m, $h = 2.5 \times 10^{-1}$ m [29]. Preheated air enters the chamber at a rate of 45×10^{-3} m/min. A holder with a sample of $d = 3.5 \times 10^{-2}$ m, $h = 5 \times 10^{-2}$ is placed in the chamber at a pre-set temperature. When the temperature on the surface of the samples rises sharply, either a filament or a jet flame is brought up to the sample and the initiation of flame is observed. A minimum air temperature is fixed in the series of experiments, at which ignition develops into stable combustion of the substance inside the furnace. In the same way, the self-ignition temperature without application of an additional ignition source is determined at the same time.

Naturally, the lower the self-ignition or ignition temperature, the higher the tendency of polymer materials to initiate the combustion process. This criterion is not always useful, however. Flame retardants can affect $T_{ign.}$ and $T_{s\text{-}ign.}$ in a complex way. There are a large number of instances when phosphorus-containing and halogen-containing flame retardants reduce $T_{ign.}$ and $T_{s\text{-}ign.}$ appreciably on account of the decrease in thermal stability of the composite material [30].

In accordance with the recommendations of the ISO, standard CMEA 382-76 was developed and ratified for the testing of construction materials in order to determine their flammability. The method consists of the determination of flammability criteria for materials in air when the material is heated to a temperature of 825 ± 5°C over a period of 20 minutes.

For the test, five cylindrical samples are used each time (diameter, 45 ± 2 mm; height, 50 ± 3 mm; volume 80 ± 5 cm^3). Samples are prepared from materials with thicknesses less than 50 mm by piling up an appropriate number of material layers. The weight of each sample is determined prior to testing. The testing takes place in a tubular furnace with a chamber of size d = 75 ± 1 mm, h = 150 ± 1 mm. Three thermocouples monitor the temperature inside the furnace (the hot junction is placed at a distance of 10 mm from the wall), the temperature on the surface of the central portion of the sample and the temperature inside the sample (the hot junction is placed in the centre of the sample). The hot junction of all three thermocouples is set at the same horizontal level. The holder with the sample is lowered quickly into the furnace with an equilibrated temperature. The sample must be heated for 20 minutes after its initial placement in the hot zone. During that time, the temperature is monitored by the thermocouples and the position and time of the sample's ignition and the duration of the flame are recorded. Ignition is considered to be stable if the flame exists for a period of 10 seconds and more. After testing, the weight of each sample is determined.

The material belongs to the noncombustible or fire-proof groups if the mean values of the maximum thermocouple readings both inside the furnace and near the surface of the sample do not exceed the initially established temperature by more than 50°C. In addition, the average loss of the sample's weight after testing should not exceed 50% of the initial weight, and the maximum duration of the flame should not be more than 10 sec.

The oxygen index (OI) method has been widely used of late to test the relative ignitability of polymer materials. This method was approved by standards ASTM D2863 and GOST 21793-76 (GOST: All-Union State Committee for Standards, USSR). The OI represents a minimal volumetric concentration of oxygen within the flow [(4 ± 1) x 10^{-2}m/sec] of oxygen-nitrogen mixture. This concentration is needed for the ignition and maintenance of stable, candlelike combustion of a sample over a period of 180 ± 3 sec or for propagation over a length of 50 mm, depending on which condition is achieved first. Three to 10 samples are used for testing. The method is characterised by high reproducibility. According to GOST 21793-76, the method is not practical for determining the flammability of plastics and cellular materials. This method is only used for a relative evaluation of the ability of polymer materials to burn during research experiments.

In other countries, however, there is a tendency for the method to be used for all kinds of polymer materials in order to test their ignitability as a criterion for potential fire hazard. Materials with an OI greater than 21% are difficult to ignite, and those with an OI greater than 27% are self-quenching in air.

The English firm Stenton Redcroft developed equipment for determining the OI. The equipment was adapted so that more extensive information could be obtained on the ignitability, combustibility, smoke-forming capacity and other characteristics of polymer materials. In particular, a device was proposed for the determination of the ignitability of polymer materials at high temperatures (up to 400°C) and for the determination of the so-called temperature index. It is at a minimum temperature that ignition occurs and the stable candlelike combustion of samples in an atmosphere of air is maintained. Determination of the temperature index is carried out for polymer materials with OI >21%.

The method for determining the limiting concentration of oxygen (LCO) with more stable burning conditions is the same method as the oxygen index method. According to this method, introduced by the All-Union Research Institute of Fire Prevention, the ignition of samples is carried out from below and the spreading of the combustion wave takes place in an upwards direction. The LCO method is preferable for the testing of polymer materials with OI > 30 to 40% [31]. LCO values are always smaller than OI values.

The most objective quantitative criteria for the ignitability of polymer materials, however, should be the minimum values of energy required for their ignition. These characteristics may be represented as the time delay before ignition, multiplied by the magnitude of the thermal pulse. Practical determination of the latter, however, is difficult.

Determination of the Capacity of Materials to Propagate Combustion

A close relationship exists between the ignitability of a material and its capacity to propagate the combustion process. The spreading of a flame over a surface is the result of a constant initiation of ignition in fresh adjacent sections. The classification of materials on the basis of their degree of flammability is based mostly on quantitative criteria of the propagation of combustion under different conditions. The existing methods of testing the combustibility of polymer materials are very varied.

Figure 4.4 shows a scheme that indicates some of the technical aspects of the testing of materials by the most widely used standard methods. The rate of flame diffusion or the rate of thorough combustion of a sample is the basic quantitative standard.

Many methods involve the simultaneous determination of a material's capacity to ignite under the effect of an ignition source during a certain period of the ignition process, and of a number of the effects after the removal of the ignition source.

USSR Standard GOST 17088-71 advises determination of the flame diffusion rate in the case of ignition of vertically placed samples of plastic materials in order to classify the readily flammable materials. The flame of a flowing gas is used as the ignition source with a lifetime of 2 minutes. If just one out the six samples tested is capable of spreading a flame all along its length for this period of time then the material is categorised as a readily flammable material.

GOST 21207-75 was developed for determination of the flammability of blocks and films of polymeric materials. The method comes under the class of small-scale methods. In fact, it is similar to standards ISO R1210 and ISO R1326. Ignition in this test is achieved from the flame of a Bunsen burner during a 60 second period. After the ignition source is removed incomplete combustion and the time for self-combustion are recorded. Materials are classified in the following categories:

1. The sample does not smoulder and does not burn after the flame is removed
2. The sample burns or smoulders for less than 15 seconds and at that time a portion of the sample remains unburned
3. The sample burns or smoulders for more than 15 seconds but less than 120 seconds and burns incompletely
4. The sample burns completely or burns for a period of longer than 120 seconds.

The method is recommended for research purposes and is not designated for the determination of flammability of cellular materials (foams).

However, according to American standard ASTM D1692, which is similar in its technical specifications to USSR GOST 21207-75, the relative flammability of not only rigid sheet materials but also of cellular plastics, may be studied in the same way. The testing method ASTM D1692 also designates that the flash time of the flame is equal to 60 seconds. After the flame is removed, the rate of combustion (with respect to 10 samples) or quenching time is registered and additional characteristics (melting, formation of drops, carbonisation, formation of smoke) are recorded.

The ASTM method specifies a 30-second interaction of the flame with the sample. According to the results of the tester, the materials are subdivided into incombustible (samples do not ignite after 30 seconds' interaction with the flame, temperature of the flame ~960°C), self-quenching and combustible groups. For self-quenching, (according to this classification for plastics), it was recommended that method ASTM D757 be used for obtaining additional information. Ignition in the latter case is carried out by using a Globar element with a temperature of 950° $\pm$ 50°C for a period of 30 minutes. The British standard BS 476 p. 7, specifies a 60-second exposure time for a propane flame in order to ignite the sample. According to this standard, materials are subdivided into four classes:

1. Samples burn out during a 60-second period at a distance of less than 76 mm
2. at a distance of less than 150 mm
3. at a distance of less than 230 mm
4. at a distance of more than 230 mm from the ignited end.

The testing of flammability for vertically placed samples ignited from below involves more stringent conditions for the development of the combustion process. ASTM D568 belongs to the group of similar medium-scale methods of testing. Samples are exposed to a Bunsen burner flame for a 15-second period. If, during that time the material does not ignite, it is classified as incombustible. If the flame does not reach a point situated 76 mm away from the fixed end, then the material is classified as self-quenching; if the flame reaches this point the material is classified as combustible. This method is used for flammability testing of plastic materials and foam plastics.

Standard ASTM D626 is recommended for the testing of the flammability of fabrics. Ignition is performed for 12-second periods using the flame of a hydrogen-methane-carbon monoxide-ethane mixture. Method BS 4735 designates the exposure of a sample to the flame for a period of 1 minute in order to determine the flammability of plastic materials and rubbers. The rate of flame distribution and the quenching time are recorded.

In the US, Underwriters' Laboratories Inc. developed a method (UL 94) for determining the flammability of polymer materials used in electronics and instrument fabrication. Samples are placed vertically. After a double exposure of the sample to a flame within a 10-second period, the quenching time, occurrence of smouldering and formation of burning drops are recorded. Materials are subdivided into three self-quenching groups: SE-0 (quenching time less than 10 sec.) SE-1 (less than 30 sec., no burning drops), and SE-2 (less than 30 sec., formation of burning drops).

In a number of methods, one factor that is regarded as a characteristic of flammability is a loss in sample weight following interaction with the ignition source; for instance, exposure to a flame during either a 2-minute period (GOST 17088-71, "flame tube"; and GOST 16363-76, "ceramic tube") or during 10-second periods (ASTM D3014). Combustible materials are characterised either by a weight loss of more than 20% and by a duration of combustion (after removal of the ignition source) of more than 1 minute (according to GOST 17088-71) or by a 30% loss of weight (according to GOST 16363-76).

In the latter case, the relative heat liberation during combustion of the tested samples (according to the change of the temperature of the exhaust gases over time) is determined in addition to the weight loss.

Testing methods are represented in Figure 4.4; the sample is placed at an angle of 45° or 30° to the horizontal and in the direction of the combustion wave, which travels upwards. The test methods differ in sample size and in the sources that initiate the ignition. Foam plastics are often tested according to ASTM C209. Ignition is carried out by a flame which forms as a result of the combustion of 1 ml of ethyl alcohol. The ASTM D1430 method was designated for the testing of film and elastic sheet of polymer materials; ASTM D1230, for the testing of fabrics. In both cases, the flame of a butane jet is used.

Standard ISO E155 is similar to ASTM D1692 in its technical specifications and designation. The ASTM E162 method in which ignition is carried out by the flame of a gas jet and by simultaneous thermal radiation from a porous refractory material panel with a temperature of ~670°C, that affects

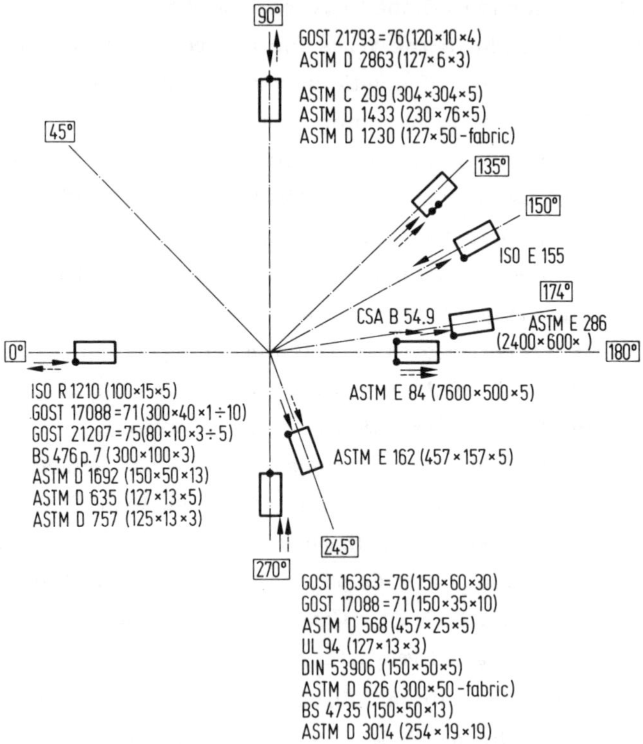

Fig. 4.4 Diagram of a three-dimensional layout of samples for the standard methods of testing flammability of polymer materials. Direction of the combustion wave (→) and flow of oxidising agent (--→). Numbers in parentheses show size of samples in mm.

the sample, is widely used for the testing of floor coverings (rugs, carpets). This method determines flame diffusion, heat liberation, and the formation of smoke. The direction of a combustion wave affects the development of the combustion process.

A method for testing the flammability of fibrous materials has been proposed and is based on determination of the angle of a sample's position at which spontaneous quenching of the flame occurs in an atmosphere of air [30].

The determination of relative heat liberation during the combustion of a polymer material enables an opinion to be formed about the material's potential ability to spread a flame, and its flammability. In the USSR a calorimetric method was developed (GOST 17088-71) for determining the combustibility class of polymer materials in terms of the heat liberated during combustion. This method involves determining the quantity of heat (Q_{HCS}) liberated during combustion of the sample [70 x 35 x (1 to 10)mm] in an atmosphere of air, and the minimum quantity of heat consumed during ignition of the sample ($Q_{ign.}$). Parameter $Q_{ign.}$ is determined experimentally by comparing the quantity of heat liberated to the quantity of heat absorbed from the source, Q_S. The rate of flammability, $K = Q_{HCS}/Q_S$, serves as a criterion for the classification of materials. Materials are divided into incombustible ($K < 0.1$), combustible with difficulty ($K = 0.1$ to 0.5) and combustible ($K > 0.5$). Within the combustible groups there are materials that do not ignite readily ($K = 0.5$ to 2.1) and materials that ignite readily ($K > 2.1$).

The drawback to this standard method, which is considered to be an arbitrary one, is the possibility of erroneous conclusions connected with the determination of Q_S. The thickness of the material affects Q_{HCS} during the combustion of samples in air. The evaluation of possible errors and a description of the means of eliminating them are given in reference [32].

A generalised criterion, called the "flame diffusion index" I was proposed for the evaluation of the flame hazard for polymer materials [33]. I is determined in terms of the quantity of heat liberated during combustion $Q_{com.}$, the rate of flame diffusion $v_{fl.\ diff.}$ over the surface and the distance of flame diffusion $l_{fl.\ diff.}$. The criterion is applicable in the case of homogeneous and multilayer materials with different thermophysical properties.

The analytical expression of I is in general:

$$I = \left[\frac{Q_{com}}{Q_{ign.}}(1 + i^2 v_{fl.diff.})\right]^{1/2},$$

where

$i = l_{fl.\ diff.}/\Delta l$

It is assumed that on small sections over which the flames diffuse, Δl, the thermal effect and the rate of flame diffusion are constant.

For an analytical solution of the equation based on the thermal unidimensional combustion model, it is necessary to have information on ignition temperature, the effective kinetic parameters of pyrolysis for polymer materials during combustion, the thermophysical properties of polymers at different temperatures and thermochemical and other characteristics.

Experimentally, the flame diffusion index I is determined from the results of ignition and diffusion of the flame over a sample (320 x 140 x δ mm) when the sample has been ignited by the flame of a gas jet and subjected to the effect of heat from a radiation panel with a temperature of ~900°C. According to technical data, this method is similar to ASTM E162. During the course of the experiment the following quantities are determined: ignition time, time of stable combustion initiation (t_{isc}), time required for the flame front to pass over the surface zone ($t_{s.c.}$), distance of flame diffusion ($l_{fl.\ diff.}$) maximum temperature of fuel gases (T_{max}) and the time required to attain T_{max}.

The flame diffusion index is calculated from the simplified formula:

$$I = \left[\frac{\beta\theta\Delta T_{max}\Delta t_{max}}{t_{isc}}\left(1 + \mu l_{fl.diff.}\sum_{1}^{i}\frac{1}{t_{s.c.}}\right)\right]^{1/2},$$

where

β thermal coefficient of the assembly in Watt/degree

$\Delta T_{max} = T_{max} - T_0$

T_0 temperature of fuel gases during the testing of a standard sample of incombustible material, °C

$\Delta t_{max} = t_{max} - t_{isc}$

μ and Θ dimensional coefficients

According to this testing method, combustible materials are subdivided into materials with a slow ($I \leqq 20$) and fast ($I > 20$) surface flame diffusion. Materials with $I < 20$ burn and distribute a flame only in the zone of flame activity and high temperatures.

Determination of the Smoke-Formation Capacity of Polymer Materials

The smoke-formation capacity of polymer materials during combustion and the effect of high temperatures is determined with the aid of optical methods on the basis of various characteristics. All the methods used are based on measurement of the optical density of fuel gases, i.e. products of pyrolysis and combustion of polymer materials. For this purpose, either special optical combustion chambers are used, such as the smoke chamber issued by the All-Union Scientific Research Institute for Fire Prevention [34], NBS or XP-2 in accordance with standard ASTM D2843 [35], or existing equipment is adapted.

The specific optical density of smoke $D_{sp.}$ is determined mainly with respect to the volume of the chamber V within which the combustion products are distributed, and the area of the sample surface S from which the liberation of smoke occurs: $D_{sp.} = DV/S\delta$, where D is the integral optical density (proportional to the logarithm of the ratio of incident light to the light that passes through an absorbing medium) and δ, the thickness of the absorbing medium's layer through which a ray of light is passing.

The specific optical density of smoke could be related not only to the area of the sample's smoky surface S but also to its weight.

The magnitudes used as criteria for the smoke-forming capacity of polymer materials during combustion are the maximum specific optical density $D_{sp.max}$ and its attainment time t_{max}, the maximum rate of increase in the specific optical density $R_{sp.\ max} = d(D_{sp.max})/dt$, and the time of its attainment t'_{max}, and time t_{16}, the time required to attain the critical value of smoke density $D_{sp.} = 16$, above which visibility inside the medium drops sharply.

The complex property called the index of smoke opacity, ISO, takes into account $D_{sp.max}$, R_m and t_{16} [35]:

$$ISO = D_{sp.max} R_m / t_{16}$$

where

R_m mean rate of change in specific optical density of smoke in the range 10-90% $D_{sp.max}$.

In the USSR, a smoke-formation coefficient K_S is used to determine the smoke-formation capacity of materials in accordance with the method of the All-Union Scientific Research Institute of Fire Prevention [34]. This coefficient allows for the maximum values of the optical density increase rate in cases where a flow of fuel gases is at its maximum speed. K_S also correlates with the rate of thorough combustion of a material.

Determination of the Toxicity of Polymer Combustion Products

A great hazard to human health and life during a fire is presented by noxious toxic combustion products that are inhaled or absorbed through the skin. It has been proposed that materials should be assessed on the basis of their specific index of toxicity ($IT_{sp.}$).

$IT_{sp.}$ takes into account the concentration of substances formed under the actual standard conditions of combustion (or of decomposition at a certain temperature), the permissible concentration (PC) of these substances in the surrounding medium, the life-important concentration of oxygen ($[O_2]_{l.i.}$) and the actually observed concentration ($[O_2]_t$) [34]:

$$IT_{sp.} = K\left[\frac{[O_2]_{l.i.}}{[O_2]_t} + \frac{[CO]_t}{[CO]_{PC}} + \frac{[X]_t}{[X]_{PC}} + \ldots\right],$$

where

X combustion product formed during the combustion of 1 g of material and distributed inside 1 m³ of the surrounding medium per million,

K constant that accounts for the liberation of substances during combustion from a unit of surface into a unit of volume of the surrounding medium.

Indexes of toxicity depend on combustion conditions. The maximum values of $IT_{sp.}$ characterise a material's potential flammability. In practice, an integral criterion W_{CL-50} (in mg or g), i.e. the amount of material that creates a concentration of combustion products in a volume such that 50% of animals die ("lethal concentration") [34, 36], is used to characterise the toxicity of products. In the same way, the experimental results obtained during determination of the correlation between the toxic effect of combustion products and the weight of burned material is calculated from a statistical analysis.

An index of a material's toxicity could also be expressed by the parameter H_{CL-50} in units of the material's weight per unit volume (kg/m³) that causes the death of 50% of test animals during combustion [37].

The fire prevention research centre at Ghent University (Belgium) has developed a method for calculating complex criteria of a material's flammability in terms of the fire index R_a. This index takes into account the most important factors that characterise a fire hazard. The index R_a is calculated according to the formula:

$$Ra = \frac{k_1Q + k_2I + k_3P + k_4E + k_5F}{\sum k_n},$$

where

Q index of heat liberation

I index of ignitability under the influence of a flame

P index of the flame's diffusion over the surface of the material

E index of ignitability of the whole surface under the effect of thermal radiation

F index of smoke danger which accounts for the density, toxicity and corrosion capacity of the combustion products

k_n correction factor accounting for the environment and the way in which the materials are used ($0 \leq k \leq 1$; $n = 1$ to 5).

Each of the indexes mentioned is represented by a dimensionless value ranging from 0 to 100 units. Index R_a changes within these limits. Thus the R_a of a material based on phenol-formaldehyde resin and sawdust as a filler (thickness 4 mm and density 1375 kg/m³) is 1.54; for sheets of polystyrene (thickness 3 mm, density 100 kg/m³) it is 76.1; for foam polyurethane R_a is 65 to 88.

In order to quantitatively evaluate the fire hazard of objects made of combustible and polymer materials, it is recommended that use be made of the parameter for the fuel content of the object (C_i) per 1 m³ of volume, which accounts for the efficiency of each fuel material and its fire index [38]:

$$C_i = \frac{\sum \text{Ra}\, m Q_{\text{com.}} + \sum \text{Ra}'\, m' Q'_{\text{com.}}}{100Sh} \ \text{megajoule/m}^3$$

where

m and m' the weight of combustible building materials and the weight of the fuel in the material in kg, respectively

$Q_{\text{com.}}$ and $Q'_{\text{com.}}$ the heat of combustion for these materials respectively, megajoule/kg

S and h surface and height of the object respectively, m² and m

The fire hazard of objects with a fuel content, C_i, of less than 20 megajoule/m³ is negligible; for $C_i \geq 2000$ megajoule/m³ the fire hazard is very great.

Modelling Methods for the Full-Scale Combustion of Polymer Materials

Two methods of modelling full-scale combustion of polymer materials have recently been developed:

1. Modelling at atmospheric pressure
2. Modelling at elevated pressure.

Both methods are based on the general assumption that fires are diffusion flames and conform to general hydrodynamic rules. The modelling methods are based on the application of the similarity theory. This permits various combustion characteristics of polymer materials to be determined using available means and allows the prediction of their behaviour under conditions of full-scale combustion (i.e. a prediction of the materials' potential flammability).

Modelling at Atmospheric Pressure

This method involves the conduction of small-scale combustion tests of materials at atmospheric pressure. When geometric similarity is retained but the scale of the tested sample is decreased, the rate of the observed process decreases. This change may occur if the conditions imposed by the Froude number are satisfied both in the model and in the full-scale prototype. The Froude number (Fr) represents a correlation between the forces of dynamic pressure and the forces of gravity for convectional movement of a gas:

$$\mathrm{Fr} = U^2/gL$$

where

U rate of gas flow by means of convection

g acceleration of gravity

L characteristic size

The Froude number can be represented as

$$\mathrm{Fr} = \rho U^2/\Delta\rho g L$$

where

ρ and $\Delta\rho$ density of gas and its change due to a change in temperature, respectively

According to Froude, modelling is used for the analysis of turbulent diffusion flames, for which the effect of viscosity forces is insignificant, and heat and mass transfer inside the flame are controlled in general by turbulent natural convection.

Analysis shows that when proceeding from a model to full-scale combustion, the rate of the process under study changes in proportion to $L^{1/2}$ and the heat flow that occurs during combustion is proportional to $L^{5/2}$ [39].

Modelling at Elevated Pressure

Full-scale combustion of materials under normal atmospheric conditions may be simulated by small-scale tests at elevated pressure. The theoretical development of the model and its experimental proof are given in reference [40]. Further development of the pressure model for the application of studying polymer combustion is given in reference [41].

Modelling at elevated pressure eliminates many of the difficulties that occur during modelling according to Froude, i.e. it permits allowance to be made for viscous forces in the moving medium and thus allows laminar and turbulent flames to be analysed. In order to conduct such small-scale tests, however, it is necessary to have equipment that can sustain a sufficiently high pressure.

The method of modelling at elevated pressure is also based on the invariance of a number of dimensionless criteria or numbers that characterise different forces that act during motion of the medium. It is assumed that geometric similarity is maintained between the full-scale model and the prototype.

Hydrodynamic similarity of diffusion combustion is described by such parameters as the Grashof number (Gr), Reynold's number (Re), Fourier's number (Fo), Nusselt's number (Nu) and others. During modelling, the change in model dimensions and in the pressure required to keep the hydrodynamic similarity constant is established. For instance, the Grashof number determines the nature of motion in the case of free convection:

$$\mathrm{Gr} = \frac{gL^3\rho_0^2}{\mu^2}\gamma\Delta T,$$

where

$\rho_0 = PM/RT$ density of the surrounding medium (P, pressure; M, molecular weight)

$\gamma \sim 1/T$ volumetric expansion coefficient for gas

ΔT difference in temperature due to convection process

μ dynamic viscosity of gas (not dependent on pressure)

As viscosity, molecular weight, flame temperature and surrounding atmosphere temperature do not depend on pressure, the pressure should be increased in the proportion $L \sim P^{-2/3}$ in order to keep Gr invariant for a reduction in the dimensions of the sample tested.

In the case of forced convection calculations, it is necessary to keep the Reynold's number invariant, Re = $\rho UL/\mu$. As the density of the gas P is proportional to pressure, when $L \sim P^{-2/3}$ Re will be preserved if the rate of flow is $U \sim P^{-1/3}$.

In a similar way, processes of flame diffusion in a model and in a full-scale prototype are examined, assuming that Fourier numbers are invariant in gas and solid phases. In particular, the rate of flame spread (or diffusion) over a polymer material surface will vary with pressure in accordance with the law

$$v_{\mathrm{fl.diff.}} \sim P^{2/3}.$$

References

1. Fristrom R. M., Grunfelder G., Hunter L. W., Combust. and Flame, 1976, vol. 27, no. 1, pp. 33-43.
2. Sckacke H., Hunter L. W., Fristrom R. M., Grunfelder G., in 16th Symposium (Intern.) on Combustion. Pittsburgh: Combust. Inst., 1976, pp. 1317-1327.
3. Bakham N. N., Belyaev A. F., Goreniye geterogennih Kondensirovannih system. M.: Nauka, 1967.
4. Waibel R. T., Essenhigh R. T., in 14th Symposium (Intern.) on Combustion, Pittsburgh: Combust. Inst., 1973, pp. 1413-1420.
5. Berlin G. S., Electronniye pribori s mekhanicheskimi upravlyaemimi electrodami. M:: Energiya, 1970.
6. Stuetz D. E., Diedwardo A. H., Zitomer F., Barnes B. P., J. Polymer Sci., Polymer Chem. Ed., 1975, vol. 13, no. 3, pp. 585-621.
7. Holve D. J., Sawyer R. W., in 15th Symposium (Intern.) on Combustion. Pittsburgh: Combust. Inst., 1974, pp. 351-361.
8. Matthews R.D., Sawyer R.F., J. Fire and Flammability, 1976, vol. 7, no. 2, pp. 200-216.

9. Physika goreniya i metodi eyo issledovaniya. Tcheboksary: Published by Tcheboksar University, 1975, ed. 5.
10. Zarko V. E., Kutsenogyi K. P., Ginzburg V. M. et al., Dokl. AN USSR, 1974, vol. 216, no. 1, pp. 120-122.
11. Abrukov S. A., Tenyeviye i interferentsionniye metody issledovaniya opticheskih neonorodnostey. Kazan.: Publ. by KGU, 1962.
12. Hirano T., Noreikis S. E., Waterman T. E., Combust. and Flame, 1974, vol. 22, no. 3, pp. 353-363.
13. Fristrom R. M., Vestenberg A. A., Structura plameny. M.: Metallurgiya, 1969.
14. Maltsev V. M., Maltsev M. I., Kasporov L. Y., Osnovniye Characteristiky goreniya. M.: Khimiya, 1977.
15. Alekseev U. I., Korolyev V. Y., Knyazhitsky V. P., in Goreniye i vzriv. M.: Nauka, 1972, pp. 128-131.
16. Zehnin A. A., Inzh.-phys. zhurn., 1962, vol. 5, no. 5, pp. 68-74.
17. Sato A., Combust. and Flame, 1975, vol. 24, no. 1., pp. 35-41.
18. Fernandez-Pello A., Williams F. A., in 15th Symposium (Intern.) on Combustion. Pittsburgh: Combust. Inst., 1974, pp. 217-231.
19. Pokhil P. F., Maltsev V. M., Zaitsev V. M., Metody issledovaniya protsessov goreniya i detonatsee. M.: Nauka, 1969.
20. Maltsev V. M., Stalenko A. G., Seleznyov V. A., Phys. goreniya i vzriva, 1973, no. 2, pp. 220-226.
21. Fristrom R. M., in The Experimental Methods in Combustion Research. J. Surukue, ed., Oxford: Pergammon Press, 1961; pp. 14-19.
22. Korobeynichev O. P., Tereshchenko A. G., Dokl. AN USSR, 1976, vol. 231, no. 5, pp. 1159-1161.
23. Hirano T., Noreikis S. E., Waterman T. E., Comb. and Flame, 1974, vol. 23, no. 1 pp. 83-96.
24. Siegbahn K., Nordling K., Fulman A. et al., Electronnaya spectroskopiya, Borovsky I. B., tr. from Eng. M.: Mir, 1971.
25. Protopopov O. D., Ozhe-spektroskopiya i. primenence k issledovaniyam poverkhnosty slozhnih emitterov. M.: Institute of Electronics, 1970, ad. 16.
25a. Troitzch J., International Plastics Flammability Handbook. Hanser, 1983 (distributed by Macmillan Publishing Co., New York).
26. Day M., Wiles D. M., J. Fire and Flammability, 1973, vol. 4, no. 4, pp. 165-173.

27. Bushev V. P., Pchelintsev V. S., Yakovlev A. I., Ognestoykost zdaniy. M.: Stroyizdut, 1970.
28. Setchkin N. P., J. Res. NBS, 1949, vol. 43, no. 6, pp. 591-605.
29. Monakhov V. P., Metodi issledoaniya pozharnoy opasnosty veshchstv. M.: Khimiya, 1972.
30. Nehorlavost polymernych materialov. Bratislava: Dom Techn. CSVTS, 1978.
31. Goreniye e problemy tusheniya pozharov. M.: VNIIPO, 1977.
32. Andrianov K. A., Kiselyov V. F., D'yachenko B. M. et al., in Steklyannoe volokno i stekloplastiky. M.: NIITEKhIM, 1976, no. 7, pp. 26-30.
33. Kiselyov Z. F., Marchenko V. M., Rivkind V. I., in Protessi goreniya i protessi tusheniya pozharov. M.: BNIIPO, 1973, pp. 15-20.
34. Metodika opredeleniya dimoobrazuyushchey sposobnosty stroetelnikh materialov. M.: VNIIPO, 1974.
35. Smoke and Products of Combustion. Hilado C. J., ed. Technomic Publ. Co., 1973, vol. 2.
36. Ilichkin V. S., Vasilyev G. A., Romanov E. I., Pozharnaya zashchita sudov. M.: 1976, no. 7, p. 71.
37. Eytingon A. M., Poddubnaya L. T., Naumova L. S., Grybunova G. P. Khim. volokna, 1977, no. 4, pp. 49-51.
38. Herpol G., Bull. tech. suisse rom., 1974, vol. 100, no. 2, pp. 432-437.
39. De Ris J., J. Appl. Polymer Sci. Appl. Polymer Symp., 1973, no. 22, pp. 185-193.
40. De Ris J., Kanury, A. M., Juen M. C., in 14th Symposium (Intern.) on Combustion. Pittsburgh. Combust. Inst., 1973, pp. 1033-1044.
41. Kanury A. M., in 15th Symposium (Intern.) on Combustion. Pittsburgh: Combust. Inst., 1974, pp. 193-202.

Chapter 5

Reduction of the Flammability of Polymeric Materials

The problem of flammability reduction in polymer materials is usually discussed with respect to the multistage character of the materials' diffusion combustion process. Retardation and inhibition of the process can be induced at each stage by physical and chemical means.

The physical methods include:

1. Diminishing the supply of heat to a polymer material (such as by heat insulation or shielding the surface)
2. Cooling the combustion zones by increasing the number of paths available for the flow of heat into the surrounding atmosphere (such as an outflow of heat from a polymer coating through a heat-conducting base layer, losses through the evaporation of components, transport and elimination of heat via melt drops)
3. Making it more difficult for the reagents to transfer to the front of the combustion zone (creation of a physical barrier between the polymer and the oxidising medium, retardation of the diffusion of fuel components within the material)
4. Disruption of the flame by a gas flow
5. The effect of an acoustic and gravitational field.

The chemical methods include:

1. Changes in the composition and structure of the polymer molecule, in the composition of the material and in the material component ratio, leading to a change in the kinetics and mechanism of the chemical reactions responsible for the decomposition of the polymer materials and in the ignition and combustion of the fuel products in order to inhibit these reactions
2. The action of chemical reagents on the flame: such reagents act as inhibitors of gas phase combustion reactions.

Physical and chemical processes that are initiated externally in order to act on the polymer system are used in practice for quenching processes already in progress, i.e. for extinguishing a fire. Strictly speaking, however, the problem of reducing the flammability of a polymer material is connected with the use of these methods as preventive measures to deter the initiation or retard the development of material combustion. In other words, the effect of physical and chemical factors must be manifested in the system itself without the aid of external influences.

At a certain stage of the combustion process the system reacts against the effects of heat, oxidising medium or fire by certain self-preserving physical or chemical factors. As far as the latter are concerned, attention can be focussed on the predominant mechanism for reducing the flammability of polymer materials.

As a rule, chemical factors (which are characterised by rate constants for the specific reactions) are intimately connected and are a function of physical parameters (which are characterised by parameters for heat and mass transfer). Determining the role of each factor and estimating the contribution of each such factor to the net effect is the most important task in establishing whether or not polymer flammability is reduced and in determining the efficiency of the method developed.

General Tendencies in the Design of Polymer Materials with Reduced Flammability

In the development of polymer materials with reduced flammability, recent research can be resolved into the following areas:

1. Synthesis of non-combustible polymers
2. Chemical modification of polymers
3. Use of flame retardants
4. Use of fillers
5. Application of fireproofing coatings
6. Combination of various methods for obtaining low-combustible materials which depend on the proposed application, technical specifications and production cost of the material.

From the viewpoint of the combustibility characteristics of polymer materials, the synthesis approach is the most productive and promising direction. Progress in the field of synthesis of new, nonflammable, thermostable polymers is very significant. However, insufficient raw materials, the uneconomically small scale of production of such polymers, difficulties of synthesis and reprocessing into products and, finally, the high cost, all limit the application of these materials. Such polymer materials are designed to be used in technological areas where problems of fire hazard are particularly important on account of the drastic conditions under which the material is used.

Development in this direction does not diminish the importance of the other factors that permit reduced-flammability materials to be created using traditional polymer materials. These latter materials are manufactured on a significant scale.

The modification of polymers is a natural development in the change of structure and properties of macromolecules and it allows the flammability of the materials to be reduced. The concept of "modification" has a very broad definition. Any change in the chemical structure of a polymer material under the effect of chemical and physical agents may be noticeable in the properties of the final material. The end product is a modified version of the initial material.

The modification of polymers to increase their thermal stability, to reduce the rate of gasification and the rate of fuel gas outflow and to enhance their capacity to form a carbonised residue under the conditions of high-temperature pyrolysis and combustion should be included in this category. This modification is most frequently performed during the process of polymer synthesis. At such time, either different comonomers are used in polyreactions, or polymers with different chemical bonds and new properties are obtained by using the same initial monomers with different catalysts and under different conditions.

A great deal of attention has thus recently been focussed on the chemical modification of polyurethanes. It is known that during the synthesis of polyurethanes different polyreactions occur for the same initial components (diisocyanates and diol). As a result, it is possible to form other groups and bonds in addition to urethane in the molecular structure of the polymer - such as biurethane, allophanate, carbodiimide, isocyanurates and others. The relative content of these groups in the final product affects many of the physical and chemical properties of the polymers, including flammability.

The use of special catalysts for the synthesis permits preparation of polyurethanes with an increased percentage of carbodiimide and isocyanurate groups, which gives the polymer better resistance to the influence of the flame. In this way, the modification is very similar to the first method listed.

It is possible to chemically modify a polymer that has already been prepared by treating it with different chemical agents. Such treatment may be total or superficial. In the latter case, only the surface layers of the polymer material are affected. For instance, there are methods for decreasing the flammability of polymer materials by halogenation, sulfonisation and phosphorylation of the material surface etc. Surface modification of polymer materials is more economical than bulk modification and it is a promising method for reducing the flammability of many types of materials (films, fabrics).

The use of reactive flame retardants locked into the molecular structure of the final product as a result of polyreactions with an initial monomer results in a chemical modification. The application of flame retardants in this way is clearly a separate case from the retardation of combustion being provided by chemical groups or elements.

The use of flame retardants is the most widespread method for reducing the flammability of polymer materials. Apart from reactive flame retardants, use is frequently made of inert flame retardants, or additive-type retardants as they are often called. These blend mechanically with a polymer substrate. The share of inert flame retardants in the total volume of flame retardants used is very large (80-85%). Inert flame retardants can be extracted by water or by detergents; they can migrate readily and can seep out of the material. The flammability of such materials increases during the ageing process. These disadvantages are the reason for the recent tendency to reduce the consumption of inert flame retardants in materials by comparison with reactive retardants. Both the active type of filler (these produce an intensification of the mechanical properties of the materials) and inert fillers are used in the production of many types of polymer materials with reduced flammability.

Fireproof coatings are also used in order to reduce the flammability of fuel materials. Such coatings reduce the probability of protected objects igniting under the action of combustion. Fireproof coatings have applications in industrial and civil engineering, in the manufacture of conveyor devices, in the prevention of flammability in wood articles, in fibrous wood panels, in plastics, in the protection of metal surfaces and in construction.

Let us discuss briefly the general trends in progress on the methods cited. At the same time it should be mentioned that the determination of the flammability of polymer materials is relative. So-called incombustible materials which are able to withstand short periods of flame in a temperature range of 900-1500°C will burn on more intense heating or with increased exposure time.

Synthesis of Incombustible Polymers

Developments in this area are following two main directions:

1. Synthesis of polymers with a minimum organic fuel component. Such polymers give off little or no fuel gases during decomposition although some of them are able to decompose almost completely without formation of a residue.
2. Synthesis of polymers with aromatic and heterocyclic structures. Such polymers do not decompose completely because they form a larger quantity of carbonised residue.

The first class can be divided into two categories:

(a) Synthesis of inorganic polymers and heteroelement polymers [1, 2]. There is a group of inorganic polymers that contains phosphorus-nitrogen bonds (polyphosphazines), silicon-nitrogen bonds (polysiloxane) and boron-containing polymers (the best known are the siloxane polymers with carborane groups). One of the disadvantages of inorganic polymers is their hydrolytic instability, especially for polymers with P-N-P, P-O-P and Si-O-Si bonds.

(b) Synthesis of organic polymers that liberate primarily incombustible gases during decomposition. Practical applications have been found for groups containing the polymers polytetrafluoroethylene, polyperfluoroalkylenetriazines and nitrosofluorocarbon elastomers. The latter polymers decompose at relatively low temperatures (~200°C) but the volatile products of decomposition are very difficult to ignite. Materials based on these rubbers do not burn even in an atmosphere of pure oxygen at a pressure of 82 atm, a situation which surpasses that of teflon [3].

Synthesis of polymers with aromatic and heterocyclic monomers in the polymer chain has been successfully developed since the 1960s. Nowadays, a practical application has been found for polymers with a linear, ladder and three-dimensional network structure. These are used as fibres, films, coatings, plastics, glues, composites, foams and other materials and include such well-known thermostable materials as polyphenylenes, polyimides, polybenzimidazoles, polybenzoxazoles, polyquinoxalines, polybenzimidazopyrrolones and polymers that contain various links and bonds such as amidoimides, quinoxalineimides and others. A survey of the methods of determination and description of the properties of such polymers would be beyond the scope of this monograph. Publications on thermostable polymers are numerous [4-6]. Such polymers are characterised by high ignition and self-ignition temperatures and by resistance to flame action. In the case of short-term action of a gas-jet flame with a temperature of 900-1200°C, the polymers do not melt but become carbonised, retaining their initial physical state.

There is a simple relationship between the yield of carbonised residue (during decomposition) and the rate of polymer ignitability (oxygen index). If one considers the contribution of each individual carbon group in the polymer's chain structure to the formation of carbon residue, the rate of ignitability may be predicted for a thermostable, synthesised polymer [7] (see Chap. 3).

The nitrogen-containing macroheterocycles within the polymer group called polyhexazocyclanes are an interesting group of polymer materials. Methods have been developed for the synthesis of these polymers based on aromatic acid tetranitriles and diamines [8, 9]. The synthesis of these has also been developed for polymers based on more readily available raw materials and using dichloroderivatives of macrocycles during polycondensation [10].

The latter method permits the production of colourless or slightly coloured polymers, soluble in organic solvents. Polyhexazocyclanes can undergo chelation by different metals, a process which permits regulation of the properties of polymer substances and their resistance to the effect of a flame. Chelation of the polymer based on oxamidrazone and terephthalic acid dichloroanhydride is an example of the variation in the ignitability of thermostable fibres as a result of a change in the nature of the metal:

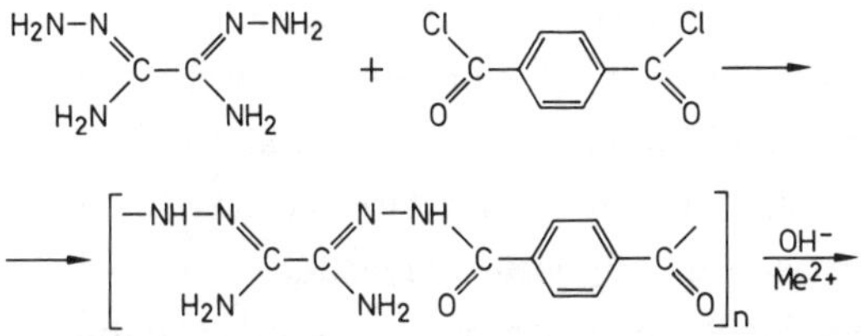

Nitrogen atoms and amino groups [11] may participate in the chelation of oxygen. When chelated compounds decompose they liberate nitrogen, carbon dioxide and water; metal oxides and carbon residues are also formed. The metals Ca, Zn and Sr reduce the flammability of polymer fibres for flames with a temperature of 1500°C and increase the OI to 40%. The metals Cu and Fe cause luminosity in the carbon residue and catalyse complete combustion. By varying the concentration of metal it is possible to change the OI value. This example illustrates the importance of the heterogeneous reaction of carbon residue oxidation during the combustion of thermostable polymers.

Chemical Modification of Polymers

Chemical modification of polymer materials is widely applied for reducing flammability. As has already been mentioned, modification can be in bulk (i.e. through the whole bulk of a substance) or superficial. Surface modification is applied in order to change the flammability of materials; such materials are characterised by a high ratio of material surface area to volume (fibres, fabrics, films, foam materials). It has thus been proposed to treat the surface of films, fabrics and cellular materials made from aromatic polyamides with chlorine or bromine in the absence of oxygen at a temperature below the annealing temperature of the polymers. The materials produced are more resistant to the effect of a flame and retain their elasticity. After being treated with chlorine, for instance, a fabric made from polyamide (Nomex) retains its dimensions and does not ignite but only blackens when in a flame at a temperature of 1100-1200°C [12].

When Nomex is treated with sulfuryl chloride its oxygen index increases from 20-30% to 35-37%. Even lower flammability is attained when the fabric is treated with a mixture of phosphorus oxychloride and bromine oxychloride followed by the last stage of the treatment, i.e. treatment by gaseous bromine. Surface modification reduces the shrinkage of the fabric during the action of the flame. With the help of electron microscopy, the presence of a film surrounding each individual fibre and containing concentrated halogen was discovered [13]. We note that surface halogenation is ineffective for fibrous materials made of aliphatic polyamides.

The potential of surface modification of polymer materials in reducing flammability is also illustrated by the example given above for the surface treatment of fibres based on poly(terephthaloyloxamidrazone) with metal salts.

Surface chemical modification noticeably reduces the flammability of materials when this treatment is performed with the aid of agents containing flame retardant elements. Otherwise, surface modification is less effective. It should be mentioned that the modification of polyether fibres through grafting with reactive flame retardants was found to be most effective when the grafted polymer was localised within the fibre and did not form a surface coating [14].

Bulk chemical modification of a polymer can be carried out at different stages of the synthesis. The general trend in developmental research in the field of polymer modification with a view to reducing flammability is the introduction of fragments with more stable bonds, aromatic and heterocyclic links, in order to change the thermal stability and pyrolysis of polymers in the molecular structure.

Extensive research has been undertaken in the area of polyurethane modification. Polyurethanes, derived from aromatic polyols (phenol-formaldehyde oligomers of resol and novolac types), possess a higher resistance to the action of flames than polyurethanes based on common hydroxyl compounds. It has been recommended that use be made of polyurethanes containing benzimid and benzimidazole rings in the chain [15] and polyurethanes with isocyanurate rings introduced both into the diols [16] and into the diisocyanate component [17].

The Japanese firm Kawasaki patented a process for the manufacture of noncombustible rigid foam polyurethanes based on the use of polyols with isocyanurate rings and on catalysts for trimerisation of isocyanates [18]. Foam plastics with urethane urea, and carbodiimide groups [19] and also polyurethanes with an increased content of allophanate groups [20] and polymers containing up to 45% of carbodiimide groups [21] have been developed.

In addition, polyurethanes with oxazolidine groups have been derived from the reactions between epoxy compounds and polyisocyanates [22]. The introduction of isocyanurate groups into phenol-formaldehyde oligomers and unsaturated maleate-type oligoethers enables one to obtain materials with an increased thermal stability and resistance to heat and with reduced flammability [23]. The modification of polyamide using compounds with heterocycles during polymerisation is described in review [24].

Yet another method of modification of polymer materials to reduce their flammability by the application of thermal treatment should be mentioned. This method permits the manufacture of materials with various properties by regulating the degrees of temperature treatment. It was noticed, for example, that poly(*m*-phenylene-bis-(*m*-benzamide) terephthalamide) becomes incombustible an atmosphere of air when ignited after preliminary thermal treatment in air at 425°C over a period of 150 minutes [25]. A similar effect occurs with thermal treatment of cellulose fibres, polyacrylonitrile, and other polymers. The gradual thermal treatment of polymer materials at high temperatures leads to the formation of carbonised and graphitised materials. This method requires special thermal treatment technology and is not applicable to all types of polymer materials.

Use of Flame Retardants

An analysis of scientific and patent [26, 27] information on the use of different substances for the reduction of flammability in polymer materials indicates that such inorganic elements as halogens, phosphorus, nitrogen, boron, metals and groups with a combination of these elements are included in the group of flame retardants.

The separation of potential flame retardants from the great number of existing chemical compounds is based on a purely empirical foundation. Nevertheless, it enables one to consider the use of flame retardants as an independent approach to the reduction of the combustibility of polymer materials. At the same time it makes possible the determination of common features in the mechanism of flame retardant action, which is due to the presence of the element mentioned above. Quantitative criteria of the effect of flame retardant action and the relative evaluation of the different flame retardant efficiencies become very important. We shall return to this problem later.

The subdivision of flame retardants into two groups, inert and reactive, is a conventional approach and is only used in relation to a specific polymer substrate and specific conditions of polymer material production. The overwhelming majority of organic or heteroatom-containing organic flame retardants belong to the group of reactive flame retardants. Inorganic flame retardants are rarely introduced into the macromolecular structure of the polymer, and only when groups that provide a chemical reaction between components (for instance, in the case of chelate formation) are present in it. Inert flame retardants (additive-type) are the leading retardants in terms of production volume of polymer materials. Among them, first of all, are the inorganic compounds: oxides, hydroxides and metal salts, followed by derivatives of phosphoric acids and halogen-containing compounds. Important flame retardants used in the production of plastics are given in Table 5.1 [28].

Additive-type flame retardants attract manufacturers of materials with reduced flammability because their application is not necessarily directly linked to the manufacture of the polymers. The introduction of flame retardants into the composition can be achieved at the stage when the polymeric materials are processed into articles without changing the technology of material manufacturing. This greatly widens the production possibilities for new materials.

At the same time, additive-type flame retardants have disadvantages which have already been mentioned. In addition they have a greater effect on physical and chemical properties and the thermal stability of polymer materials than the reactive retardants. Inorganic flame retardants are more available and cheaper by comparison with organic ones: most of them are nonvolatile and form low-toxic gases during decomposition.

It may be said that there is no such substance as a universal flame retardant that would be good for the reduction of flammability in all polymers. Over the last few years an increasing interest has become apparent in organic flame retardants and, in fact, in such compounds as borates, fluoroborates of ammonium, alkali and alkali earth metals [29, 30]. For example, a method has been developed for producing a crystalline zinc borate with a low degree of hydration ($2ZnO \cdot 3B_2O_3 \cdot (3.3 \text{ to } 3.7)H_2O$), which has been recommended for use in the reduction of the flammability of halogen-containing polymers [31]. Borates of alkali earth metals and zinc are usually substitutes for antimony trioxide which is a more expensive

Table 5.1 Important flame retardants in the production of plastics

Flame Retardant
Inert:
Aluminium hydroxide
Antimony trioxide
Nonhalogenated phosphates
Halogenated phosphates
Chlorinated hydrocarbons
Bromine-containing substances
Boron-containing substances
Others

Flame Retardant
Reactive:
for: Foam polyurethanes
- flexible
- rigid
Polyethers
Epoxy resins
Polystyrene
Polycarbonates
Other resins

product. The efficiency of zinc and calcium borates increases according to references [32, 33] when they are combined with aluminium hydroxide.

There is a noticeable tendency among manufacturers to attain a higher reduction of flammability by the use of mixtures of various flame retardants and their synergists. In order to make it easier to introduce flame retardants into the compositional materials, concentrates are used. These also contain additives of a different type such as stabilisers, plasticisers etc. Manufacturers are looking for flame retardants with multifunctional effects, which in addition to their chief function must play a role as surface-active agents, plasticisers, frothing agents and solidifiers or structure formers. Insoluble and infusible (at the material reprocessing temperatures) flame retardants are often used as fillers. Such a situation developed, for example, in the case of aluminium hydroxide, carbonates of alkaline earth metals and aluminium, ammonium phosphates and others. Although potential flame retardants are also responsible for other functions inside the polymer materials, their major function in the change of the material's physical and chemical properties is the function that is considered when they are classified. It is then possible to separate out from the total number of flame retardants the flame retardant plasticisers, the flame retardant fillers and the flame retardant structure formers etc.

The tendency towards a more efficient reduction in the combustibility of polymer materials leads either to the use of mixtures of substances with different flame retardant elements or to the use of substances in which such elements are present simultaneously. There is also a clear tendency towards complex combinations of these elements (P + Cl + Br, P + N + Hal, P + MeX, and others). This tendency is not, however, always justified. The problems of additivity, synergism and antagonism of certain flame retardant combinations are discussed in great detail in Weil's review [34].

Among the organic derivatives of phosphoric acids, which belong to the inert flame retardants, the most well-known are alkyl- and arylphosphates and their halogen derivatives. Many of them display plasticising effects. They are recommended for the production of elastic and plasticised materials (PVC, polyolefins, flexible foam polyurethanes). The plasticising effect of such flame retardants depends on the structure of the hydrocarbon substituent and the nature of the halogen and of the polymer substrate itself. Cyclic and branched groups strengthen the compatibility and plasticising effect [35].

Along with ethers of phosphoric acid, ethers of phosphonic and phosphorus acids are being more widely used. It should be mentioned that some compounds from this group, in particular halogen-containing alkylphosphates [36], show strong toxic properties, which is why it is necessary to carefully examine the effect of the flame retardants themselves on humans and not only the effect of the decomposition products of combustion.

Of the halogen-containing compounds that belong to the inert flame retardants, the most widely used are the inexpensive chlorine-containing low-molecular-weight aliphatic hydrocarbons, used in combination with the synergists (Sb_2O_3 and other antimony compounds). However, there is a tendency towards the substitution of these hydrocarbons by the more stable cycloaliphatic and aromatic halohydrocarbons and by more efficient flame retardants with a high concentration of bromine (for instance, decabromodiphenyloxide and octabromodiphenyl for thermoplastics).

The polymeric halogen-containing flame retardants are preferred in comparison to the low-molecular-weight substances, because they do not tend to migrate and they also reinforce many of the properties of the polymer materials. Polyvinyl chloride, chlorinated polyethylene and halogenated polyethers are used as polymeric halogenated flame retardants. However, the use of high-molecular-weight flame retardants produces additional problems during the reprocessing of the compounded materials into articles.

Reactive flame retardants contain functional groups or atoms that participate in different polyreactions (polymerisation, polycondensation, polyaddition). In order to obtain polymers with reduced flammability (polyolefins, acrylic polymers, polystyrene), flame retardants with hydroxyl, carboxyl, anhydride, isocyanate and other groups are used. The same is true for the inert flame retardants for which halogen compounds and phosphoric acid derivatives are used to reduce polymer flammability. It should be mentioned that there is a tendency to use substances with a maximum relative content of flame retardant. For example, tetrabromophthalic anhydride is used in order to produce polyetherpolyols, designated for self-extinguishing rigid foam polyurethanes [37], hexachloronapthalenedicarboxylic acid and its anhydrides [38], tetrabromohexanediol [39] and tetrabromobisphenol A. The partial or complete substitution of chlorine by bromine in the halogen-containing compounds is dictated not only by the requirements for increased efficiency of flame retardants but also by requirements for a reduction in the cost of manufactured materials.

Reactive flame retardants - halogenated ethers of phosphoric acids - are presumed to be more effective than their analogues, which contain only a halogen or phosphorus atom. Reactive flame retardants with phosphate, phosphonate and phosphite groups are finding their way into the production of reduced-flammability materials. Nowadays, there is a larger assortment of flame retardants which include not only phosphorus and halide but also atoms of nitrogen, boron and other elements.

We believe that one of the promising new trends in the reduction of polymer material flammability is the use of metallic complexes with organic ligands, containing phosphorus, nitrogen, oxygen and other atoms with unshared electrons as flame retardants. Use of complex platinum compounds has been suggested for reducing the flammability of polyorganosiloxane elastomers. These flame retardants are effective in very small concentrations (1-150 parts per million). Chloroplatanic acid, platinum chloride complexes with nitriles, phosphines and aminophosphines have been patented [40]. Information is available on the use of a zinc chloride complex with dimethylformamide for the reduction of the flammability of polyvinyl chloride and halogenated polyethers [41] and complexes of the metals with a variable valence in combination with halogen-containing flame retardants for polystyrene [42]. The mechanism of their action has not so far been determined.

Use of Fillers

Three groups of fillers are used in order to reduce the flammability of polymer materials:

1. Inorganic-type fillers
2. Noncombustible thermostable organic fillers
3. Modified organic fillers.

In the majority of cases the inorganic type of filler is used. Since fillers are introduced into polymers in vast quantities (>20%), their presence reduces the relative fuel component content of the material. At the same time, a change occurs in the thermophysical properties and the conditions of heat exchange and mass exchange during combustion. The most interesting are the active types of fillers. In this case a specific reaction occurs between the polymer substrate and the surface of the filler, and this reaction affects the nature of the polymer's pyrolysis.

Fillers can be dispersed with particles having granular shapes (sand, chalk, kaolin) and flaked shapes (graphite, mica, talc) and can be fibrous (glass fibre, asbestos) or porous (glass microspheres, vermiculite, perlite).

Nowadays there is a tendency for multifunctional fillers to be used. An effort is underway to develop fillers that could more effectively reduce the flammability of polymer materials, the formation of smoke and the concentration of the toxic pyrolysis and combustion products formed. At the same time these fillers should improve the technological properties of the plastics composites during their reprocessing, the physical and chemical properties of the polymer materials and their resistance to external effects (atmospheric or thermal stability).

Even more important are the fillers that not only reduce the fuel component of the material but also exhibit flame retardant properties. Aluminium hydroxide is the most widespread among the fillers that possess flame retardant properties. This filler not only reduces the flammability of polymer materials but also cuts down the formation of smoke during combustion. Aluminium hydroxide is nontoxic and liberates only water vapour during decomposition at temperatures above 220°C. The powdered filler is not hydroscopic and does not clot, which is very important from the viewpoint of processing the material. Aluminium hydroxide is used for the development of materials with reduced flammability based on thermoplastic and thermoreactive polymers. It was discovered that treating materials based on epoxy polymer with aluminium hydroxide reduces their flammability when nitrous oxide or oxygen are used as an oxidising agent [43]. In other words, the filler does not affect the free-radical gas phase reactions within the flame. A decrease in the material's flammability is caused by the liberation of water vapour during decomposition of the filler, by cooling the combustion wave zones [44] and possibly by the formation of an oxide film on the burning surface.

By way of flame retardant fillers, use is made for example of ammonium pyrophosphate and hydrated metal carbonates, which decompose under the action of a flame on polymer materials and form carbon dioxide and water vapour.

The effectiveness of a filler during combustion generally increases as the area of the contact surface between the filler and the polymer substrate increases. This is why there is a tendency to use either highly dispersed fillers or ones that undergo special activation in order to enlarge their specific surface. It has been shown that highly dispersed silica-type fillers increase the yield of carbon during the decomposition of polystyrene and significantly reduce the formation of smoke during its combustion. This effect increases with an increase in the concentration and the specific surface of the fillers. However, a concentration of inorganic fillers of a different nature ($Al(OH)_3$, TiO_2, $CaCO_3$, soot, carbosil and aerosil) led the authors of reference [45] to conclude that the physical and chemical properties of fillers are the more important ones. According to these authors, adsorption of the aromatic compounds occurs on the surface of the filler; these compounds are the products of polystyrene decomposition and they are the predecessors of the soot particles.

The modification of fillers is of special importance because it significantly widens their functions in composite materials. Kisselev and Lyghin studied the properties of the surface of inorganic substances [46]. It has been discovered that the surface hydroxyl groups of different metal oxides, silica among them, serve as centres of adsorption of sorbate molecules. At the same time, these groups are reactive. They react with halides, ammonia, alcohols, diazomethane, halogenated silanes and others [47]. The condition of the fillers' surface not only affects their reinforcing properties but also the thermal stability and thermo-oxidising stability of the polymers [48]. Use of modifiers or surface treatments of fillers is an effective means of regulating the technological and functional properties of composite materials. Such substances can be introduced directly into the composite or they can be used for preliminary treatment of the filler.

Union Carbide is manufacturing an oganosilicon modifier for aluminium hydroxide and also a ready-modified filler. The use of aluminium hydroxide modified in such a way reduces the viscosity of the polyether-resin-containing composite and also reduces the concentration of an added bromine-containing flame retardant by as much as 75% while maintaining the ignitability and combustibility characteristics at the same level [49].

A new class of modifiers, organotitanates, has been developed. They are used for fillers of a silica type, for calcium carbonate, clay, mica, talcum powder and others. Modified by organotitanates, the fillers are recommended for the manufacture of polymer materials with reduced flammability based on polyvinyl chloride, polyolefins, polystyrene and various polymers of the thermoreactive type. Modified calcium carbonate used, for instance, in the manufacture of polyvinyl chloride pipes and finishing materials, reduces the viscosity of this composition, accelerates the extrusion process and increases the material's resistance to impact [50].

Depending on the type of titanate, modified fillers, in addition to their flame-retarding properties are used as hardeners for epoxy resins which contain amino groups; as inhibitors for the reaction of transesterification of epoxy, alkyd polyether and polyurethane systems; and as accelerators for the vulcanisation of elastomers.

Organotitanates provide formation of chemical bonds between the filler and the polymer [51]. There is the possibility of modification of highly dispersed silica-type fillers through the polymerisation of tetracyano-ethylene and tetracyanobenzene on their surface [52]. The application of flame retardants to the surface of fillers is a very promising method [53] although the cost of the modification of fillers increases, the improvement in the properties of the composites and of the materials justifies the expenditure.

Carbon fibre fillers are more and more widely used today [54]. The development of the technology for the manufacture of carbon fibres from petroleum, coal and synthetic and mesophase pitch is the reason for the very significant reduction of the fibre's cost, which is responsible for the growth in this field.

Common organic fibrous fillers (sawdust, cellulose and cotton threads) increase the flammability of polymer materials. It is interesting in this regard to look at incombustible thermostable fibres and organic fillers, which are modified by various flame retardants. It was discovered, for instance, that cellulose fibres, modified with phosphorus-containing flame retardants, reduce the flammability of polymer materials more efficiently than flame retardants used independently [55, 56]. The flammability of epoxy plastics decreases with an increase in the content of phosphorus in the fibrous filler (Figure 5.1).

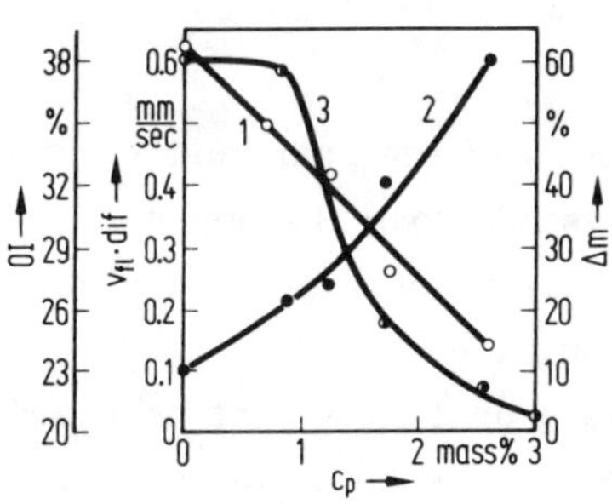

Fig. 5.1 Relationship between the flammability rate of an epoxy plastic and the percentage of phosphorus in the filler. (1) Rate of flame diffusion $\nu_{fl.diff.}$, (2) oxygen index, OI, (3) loss of weight during combustion (Δm).

Among the thermostable fibrous fillers, fabrics and fibres made from polymers with aromatic and heterocyclic links inside macromolecules of the linear, semiladder, ladder, and three-dimensionally crosslinked structure are of practical value. These fibres are made of aromatic polyamides, polyimides and polyimidoamides, polyoxadiazoles and pyrrones, phenol-formaldehyde polymers and others.

Such fibres are either combustible with difficulty or incombustible, according to the accepted classification [57]. By comparison with inorganic fibres (glass fibre) they have a lower density. This circumstance, in conjunction with the high rates of the modulus of elasticity and safety factor, which are unchanged to a large degree at high temperatures, provides the advantages of the use of thermostable organic fibres for composites.

Unfortunately, there is practically no quantitative data in the literature on the effectiveness of such fillers as retardants of combustion for polymer materials. This area has not been thoroughly explored, although it deserves attention. Data exists on the use of such fibres in order to obtain technical rubber products and plastics, designed for applications in aircraft, cars and the ship-building industry [58, 59].

Application of Fireproof Coatings

Coatings that are applied to the surface of the objects to be protected and that either prevent or retard ignition and destruction under the effect of fire may be subdivided into two groups:

1. Coatings ignitable with difficulty and incombustible coatings
2. Heat-insulating incombustible coatings

The first group includes various polymer coatings of the paint and varnish type, and sheets and films that are less combustible than the objects protected. Use of these coatings reduces the hazard of ignition and, at times, the flammability of the whole construction.

All the means of reducing polymer material combustibility that have been discussed above may be used for the development of such coatings. Paint and varnish coatings are widely used and there is no need to discuss their importance.

We think that it is important to mention here the tendency towards using polymer coatings in order to reduce the flammability of fibrous, film and cellular polymer materials. The application of fluoroelastomer-based or neoprene-based coatings to the surface of polyamide fabrics with a subsequent extraction of solvent may be used as an example. Fluoroelastomer coatings increase the oxygen index of foam polyurethane materials by up to 30% with a moderate increase in density.

The use of flame retardants alone is less effective for foam polyurethane. The materials developed are designed for use under rigorous conditions [60].

The second group includes coatings that reduce the rate and the depth of heating through the protected surface under the effect of a flame. These coatings are designed to protect not only combustible but also incombustible materials from destruction during a fire. Coatings of the group are subdivided into swelling and nonswelling coatings.

The heat insulating effect of the nonswelling coatings is due to the use of materials with a high reflectivity and low thermal conductivity, materials that undergo endothermic degradation with liberation of incombustible gases (water and others). Such coatings are essentially derived from a base of inorganic compounds through the use of fillers of mineral origin (asbestos, vermiculite). The application of coatings based on mineral binding materials and liquid glass is well known. The temperature of the structures protected by such coatings does not exceed 100°C over a very long period. Coatings based on magnesium oxychloride protect steel structures against fire for one hour [61] (for a coating thickness of 18 mm).

The heat insulating capacity of nonswelling coatings increases as the thickness of the coatings rises. The effectiveness of such coatings is insufficient if thick objects are considered, a situation that limits their applications.

The most promising are fireproof coatings of the swelling type. Under normal working conditions these coatings differ little from common paint and varnish coatings. In critical situations, however, such as in the case of high temperatures, they bulge and form carbonised foam layers. The low heat conductivity of the layer provides protection of the object against the heat of the flame.

Fireproof coatings of the swelling type include several components with different purposes:

1. Substances that are the source of the carbon skeleton
2. Substances that catalyse the reaction of carbon skeleton formation
3. Swelling agents

Various compositions have been proposed for swelling, fireproof coatings. A number of carbohydrates have been suggested as a source of carbon skeleton formation, such as polysaccharides, polyols (pentaerythritol), nitroamino derivatives of aromatic compounds and thermoplastic polymers (polyvinyl acetate, polyvinylidenechloride, chlorinated rubber, latexes of vinylchlorides and vinylidenechloride copolymers and polyurethanes). Among the thermoreactive polymers the epoxy and phenol-formaldehyde oligomers, which form a small number of links, are the preferable ones. Rapid solidification of thermoreactive binders inhibits foam formation during combustion. The catalysis of carbon formation reactions occurs in the presence of substances that cause formation of strong acidic agents of dehydration, cyclisation and crosslinking (phosphates, sulphates) in the system.

Substances such as melamine, dicyandiamide and ammonium polyphosphate, which liberate incombustible gases, such as H_2O, CO_2 HCl and NH_3, are used as high-temperature pore formers. Polyols and chlorinated hydrocarbons at the same time serve as sources of carbon and pore-forming agents.

Substances used in fireproof coatings generally display a multifunctional effect. Exhaustive studies on fireproof coatings involving the swelling type have been conducted by Vandersall [62].

The methods discussed for the reduction of flammability of polymer materials are all equally promising. However, it is flame retardants that are used in order to reduce the flammability of materials based on traditional polymers manufactured on a multi-ton scale. In this case, manufacturers of such materials are very interested in the criteria applied in the selection of flame retardants and also in which ones are most effective.

Leaving aside such important problems as the influence of flame retardants on the technological and working properties of materials and their change during the ageing process, we believe major attention should be focussed on a proper evaluation of the effectiveness of flame retardants.

Criteria for Evaluating Flame Retardant Effectiveness

The determination of an efficiency criterion for flame retardants is important for a quantitative evaluation of their effect, for their comparison and for the determination of their mutual influence. This task is not as simple as it may seem at first sight. It is true that the mechanism of action is different for different flame retardants, which is why macroscopic characteristics of polymer combustion can change unpredictably after the introduction of flame retardants.

An analysis of the published data leads to the conclusion that the effect of flame retardants on polymeric material flammability is generally determined according to two major characteristics: the stability of the flame (ignition and extinguishing conditions) and the rate of stationary combustion. In practice, various efficiency criteria are used. The concentration of a flame retardant at which the flame is extinguished under standard conditions can serve as a quantitative measure of the flame retardant: $[An]_B$ (An = antipyrene = flame retardant) [63].

The efficiency criterion can also be defined as the ratio of the concentration of the flame retardant selected as a standard to the concentration of flame retardant under study which will give the same reduction in material flammability: $[An]_{stand.}/[An]_{stud.}$ at P = constant (where P is a flammability characteristic such as OI). Another measure of flame retardant effectiveness is the difference in the effect of reducing material flammability (ΔP) in the case of equal concentrations of the flame retardant under study and the standard reagent ($[An]_{stand.} = [An]_{stud.}$ = const).

The shape of the curves which show the relationship between the characteristics of polymer material flammability and the concentration of flame retardant may be very varied. The most typical shapes of curves showing a change in the OI of polymer materials due to the concentration of flame retardant are shown in Fig. 5.2.

The concentration of flame retardants is expressed in the form of the relative content of flame retardant substance inside the polymer material, although the concentration is usually converted into the relative content of flame retardant element that reduces the flammability of the material.

Studying the effect of different flame retardant concentrations (125 items) on the change in oxygen index ΔOI for a number of polymers, VanKrevelen and coworkers [64] discovered that ΔOI increases linearly

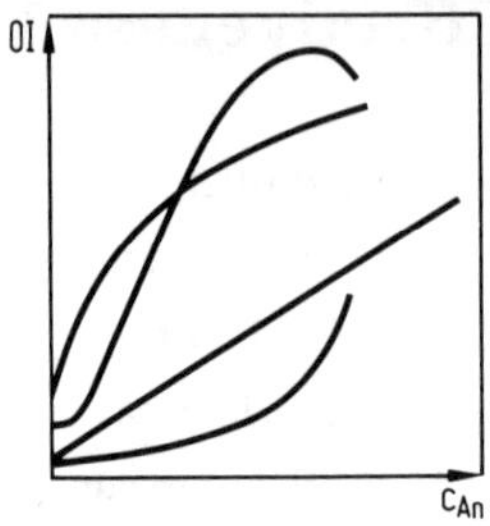

Fig. 5.2 Typical curves showing the dependence of the polymer oxygen index, OI, on the concentration of flame retardant (antipyrene) c_{An}.

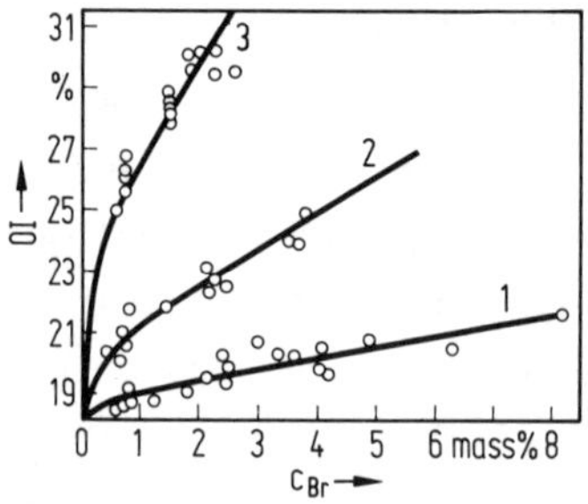

Fig. 5.3 Relationship between the oxygen index OI of polystyrene and the concentration of flame retardant (in relation to the element Br) c_{Br}. (1) Bromoaromatic flame retardants, (2) bromoaliphatic and bromoalicyclic flame retardants (3) bromoaliphatic flame retardant with addition of diisopropylbenzene.

with an increase in flame retardant concentration within the limits of 1-10 weight % for bromine and chlorine and of some 0.5-2.5 weight % for phosphorus. In general, this relationship is expressed in the following form:

$$\Delta OI = a(c + b)$$

where

- a efficiency coefficient of the flame retardant - this also depends on the polymer
- b constant characteristic of a given flame retardant ($b = 0.02A$, where A is the atomic mass of the flame retardant element)
- c concentration of the flame retardant element in weight %

Bromoaliphatic and bromoalicyclic flame retardants reduce the flammability of polystyrene more effectively than bromoaromatic hydrocarbons (Fig. 5.3). The efficiency coefficient shows a sharp increase after introduction into the system of antimony oxide or of a compound that forms free

Table 5.2 Efficiency coefficient *a* of flame retardants in various polymers.

Polymer	Flame retardant containing: Phosphorus (b = 0.6)		Bromine (b = 1.6)		Chlorine (b= 0.7)	
	a	compound type	*a*	compound type	*a*	compound type
Polyethylene-terephthalate	1.5	Phosphates	0.5	Aromatic	0.8	Aromatic
	2.6	Amido phosphates	0.7	Aliphatic		Aliphatic
Polyamide	1.1	Phosphates Amido-phosphates	0.1	Aromatic	0.1	Aromatic Aliphatic
Polystyrene	1.5	Phosphines Phosphinoxides	3.0	Aromatic	0.1	Aromatic
			1.1	Aliphatic	0.5	Aliphatic
			0.3	Aliphatic, activated	1.1	Aliphatic + Sb_2O_3
Polypropylene	1.3	Phosphates	0.45	Aromatic	0.2	Aromatic, aliphatic
			0.6	Aliphatic		
			1.0	Aromatic + Sb_2O_3		
Polyacrylonitrile	1.5	Phosphates	0.35	Aliphatic	0.1	Aliphatic
Cellulose	1.5	Phosphates				
	2.4	Phosphates				

radicals during decomposition. The efficiency coefficient represents the degree of relative change in the value of the oxygen index for a change in the concentration of flame retardant: $a = \Delta OI/\Delta c$. The efficiency coefficients of different classes of flame retardants for a number of polymers are given in Table 5.2. The table shows that the least efficient flame retardants are the ones used in polyamides. The efficiency coefficient of phosphorus-containing flame retardants exceeds the efficiency coefficient of halogenated flame retardants several times over. This efficiency of phosphorus-containing compounds is not intrinsic but a function of the polymer.

A study of the effect of halogenated flame retardants, which combine with antimony oxide or with other compounds to form free radicals during decomposition, shows that the rate of polymer decomposition increases in the presence of such substances.

The increase in the oxygen index of thermoplastic polymers is caused to a large degree by an increase in the elimination of heat from the combustion zone by the flowing melt. According to reference [65], the ratio of the weight of polymer melt formed during the test to the weight of the burned sample increased from 1.8 for initial polypropylene to 3.0 and 6.3 with the introduction of 2% Br + 1% Sb_2O_3 and 2% Br + 1% Sb_2O_3 + 1% dicumyl peroxide, respectively.

As a result, the oxygen index increases from 18 to 24.5 and 28.5. The effect of different factors on the reduction of polymer flammability as related to the removal of heat by means of the melt is discussed in reference [66].

In order to reduce or eliminate the effect of drop-falling on the oxygen index and to increase the value of the efficiency coefficient of the flame retardants, the tested samples were wrapped in glass fabric. Ravey and Fishler [67], using a similar method, discovered that the effectiveness of bromo-organic flame retardants in the reduction of the ignitability of polystyrene and polymethylmethacrylate does not depend on the chemical nature of the organic radical.

The general effect is determined only by the concentration of bromine in the composition (Fig. 5.4). The efficiency coefficient of bromine depends on the nature of the polymer substrate. In the case of polymethylmethacrylate and polystyrene, the values are near 0.09-0.12. According to other data, neither the nature of an organic radical (the type of carbon-bromine bond) nor the type of flame retardant (reactive or additive) influences the effectiveness of the flame retardant element in the process of changing the rate of ignitability [68].

In the overwhelming majority of cases, the dependence of the studied rate of a polymer's flammability on the concentration of flame retardant is not linear, and there are limitations on the effectiveness of the substances introduced. It is obvious that the slope of the tangent to the curve, which expresses the efficiency of the flame retardant, can change in this case with a change in the concentration of the flame retardant over a broad range (see Fig. 5.2).

Formally, the nonlinear nature of the change in effect on polymer

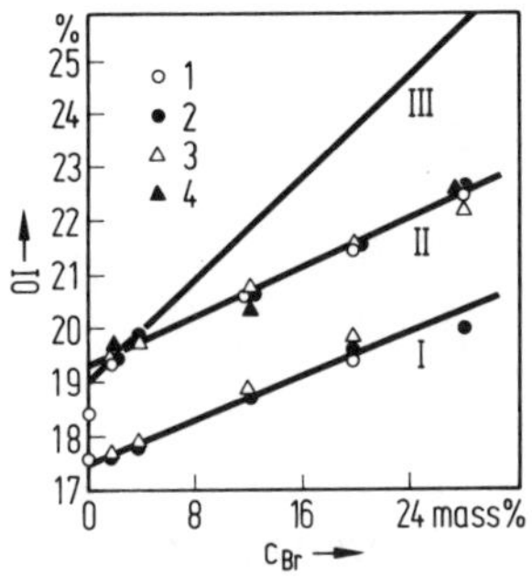

Fig. 5.4 Relationship between the oxygen index OI of polymethylmethacrylate (I), impact-resistant polystyrene (II), and a polymer based on nonsaturated polyether (III), and concentration of flame retardant based on the element Br), c_{Br}. (1) Tetrabromoethylene, (2) tetrabromoethane, (3) pentaerythritoltetrabromide, (4) tetrabromoxylene.

flammability with the concentration of a flame retardant may be linked to two factors:

1. A change in the efficiency coefficient of the flame retardant, due to an increase in its content in the polymer material
2. A deviation of the kinetic law of flame retardant consumption (for the reduction of flammability) from a first-order law.

Assuming that VanKrevelen's equation is correct and accepting that b = 1.6, the authors of reference [69] calculated the efficiency coefficient of bromine for the reduction of polypropylene flammability for a series of bromophosphorus-containing compounds. It was observed that the efficiency coefficients of bromine decrease with an increase in its concentration in the composition. For example, with the introduction of tris-(2,4,6-tribromophenyl) phosphate into polypropylene, the coefficient fell from 2.0 to 1.52 with a change in the concentration of bromine from 2.9 to 5.6 mol.%. The results derived, however, only indicate the nonlinear nature of the relationship between the oxygen index and the concentration of flame retardants; the causes are still unknown.

The nonlinear nature of the relationship between flammability rate and concentration of flame retardant may result in some cases from insufficient sensitivity of the testing method employed. For example, when testing by the "flame tube" method, the relationship between the loss in sample weight and the concentration of flame retardant in the polymer is non-

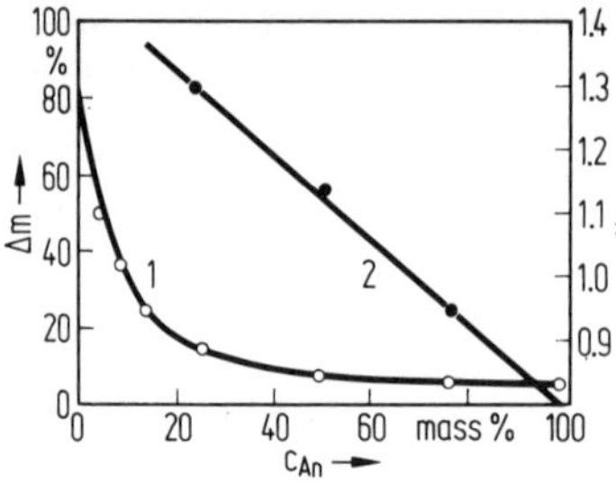

Fig. 5.5 Relationship between the characteristics of flammability for polyoligoether PN-1 and concentration of phosphoacrylate flame retardant, C_{An}. (1) Loss of weight during combustion (Δm), (2) rate of ignitability (K).

linear. When the percentage of flame retardant is above 25 weight % the loss of weight during combustion undergoes practically no change. At the same time, the rate of ignitability K, determined according to the calorimetric method at $C_{An} > 2.5$ weight %, decreases proportionally to the concentration level of the flame retardant (Fig. 5.5), which is why the selection of a flammability characteristic as a criteria of the effectiveness of the flame retardant is so important. Holve and Sawyer [70] attempted to quantitatively determine the effectiveness of flame retardants on the basis of an analysis of the effect of the flame retardants on the rate of stationary combustion, during convective heat transfer and under conditions of flame quenching.

As was mentioned in Chap. 2, the rate of stationary combustion is proportional to $\ln(1 + B)$, i.e. it is determined by the mass-transfer number B, which is the most important characteristic of the material. The effectiveness of a flame retardant in reducing polymer material flammability can therefore be determined from a change in mass-transfer number B.

On the other hand, the effect of flame retardant extinction of a flame has been studied under the critical conditions of the gas-phase combustion reaction. This type of condition is denoted by the Damkeler number, De, which characterises the ratio of time that the reagent stays in the reaction zone to the time of its chemical reaction. This number must be smaller than the critical number. Small changes in flame temperature can greatly affect the rate of the gas phase combustion reaction and may lead to extinction of the flame.

The theoretical analysis shows that changes in flame temperature may be caused by changes in the concentration of oxygen in the oxidising gas flow or by preliminary heating of the latter [70]. In the case of the diffusion combustion of polymers with an oxidiser counterflow, a functional link has been discovered between the kinetic characteristic of the gas phase reaction of a polymer's combustion, the change in concentration of oxygen, the temperature of the oxidiser flow and the change in the rate of extinction.

According to the authors of reference [70] the activation energy of the gas-phase combustion reaction may serve as a quantitative measure of flame retardant effectiveness. E values are computed from the slope of the straight lines obtained in the coordinates: relative variation in rate of extinction $\Delta v_e / v_e$ against the relative change of oxygen concentration in the flow of the oxidiser ($\Delta Y_{ox,0} / Y_{ox,0}$) or against the relative change in its temperature ($\Delta T_0 / T_0$). The authors note a direct analogy between the method discussed for the evaluation of flame retardant efficiency and the method proposed earlier for measuring the effectiveness of combustion inhibitors in gas phase systems [71]:

$$F_v = [O_2]\Delta v / [I] v$$

where

$[O_2]$ oxygen concentration

Δv change in the rate of flame distribution for the introduction of specific amounts of inhibitor [I]

v rate of flame diffusion in the absence of inhibitor

It should be mentioned that Holve and Sawyer found two principally different types of flame retardant. The first type affects the stability of a flame participating in gas-phase flame reactions. With excess oxidiser, such a flame retardant further increases the rate of thorough stationary combustion of the polymer (mass transfer number B increases). The other type of flame retardant does not noticeably affect the stability of the flame but reduces the rate of thorough combustion (mass transfer number B decreases). The latter situation indicates the necessity for consideration of the flame retardant's action mechanism in addition to determining its effectiveness.

The effectiveness criterion must include a certain complex quantity for the macroscopic characteristics of polymer material flammability to reflect the effect of the flame retardant on the change in the condensed and gaseous phases.

Attempts to account for the effect of the flame retardant on the processes in the C phase during the combustion of polymers are presented in the discussion of the thermal and thermochemical properties of polymer compositions. The efficiency of a flame retardant can be determined, in particular, by establishing the potential heat release of a polymer sample without a flame retardant (Δ) and that of a sample with a flame retardant (Δ_{An}):

$$\mathbf{F} = \frac{\Delta}{\Delta_{An}} = \frac{\Delta H_{com.} - r'\Delta H'_{com.}}{\Delta H_{com.An} - r'_{An}\Delta H'_{com.,An}},$$

where

$\Delta H_{com.}$ and $\Delta H'_{com.}$ heat of combustion of the sample in an atmosphere of oxygen with and without addition of flame retardant

γ' and γ'_{An} amounts of residue (char) after either combustion of the samples in an atmosphere of air [72], or under special conditions (e.g. two hours in an air atmosphere inside a muffle furnace at a temperature of 750°C [73].

$\Delta H'_{com.}$ and $\Delta H'_{com.An}$ heat of combustion in an oxygen atmosphere for the residues (char).

If one of the flame retardants is selected as the standard, then the effectiveness of the flame retardants in the same polymer substrate will correspond to the ratio F = $\Delta_{An,stud.}/\Delta_{An,stand.}$ where $\Delta_{An,stand.}$ is the potential heat release from the composition with a standard flame retardant and $\Delta_{An,stud.}$ is the heat release with the same amount of the flame retardant being tested. According to this efficiency ratio, the effectiveness of tris-(2,3-dibromopropyl)phosphate is twice that of triphenylphosphate [74]. However, in this case too, the concentration dependence of the combustion properties is of a non-linear nature, which complicates the situation.

Flame retardant mixtures, containing two or more flame retardant elements are normally used to reduce the flammability of polymer materials. In addition, different admixtures are introduced into the compositions. The evaluation of their effect and counter-effect on flammability reduction constitutes a major problem for the manufacturers of polymer materials.

In this case, the statistical method of regressive analysis is the method most frequently used. The overall effect is expressed as the sum of the effects of each element (substance) regardless of the contributions from their interaction. For example, if the composition contains P and Br, then a change in the oxygen index may be represented as the sum:

$$\Delta OI = k_{Br} c_{Br}^{a} + k_{P} c_{P}^{b} + k_{P,Br} c_{P}^{c} c_{Br}^{d}$$

where

$k_{Br}, k_{P}, k_{P,Br}$ constants characterising the effectiveness of each flame retardant element separately and in conjunction with the others.

c_{Br} and c_{P} concentration of elements

a, b, c, d numerical constants that determine the shape of the curve expressing the relationship between the degree of flammability and the concentration of the flame retardants

The use of computers makes it possible to interpret the data on a statistical basis. Weil [34] analysed the relationship between the oxygen index of polyesters and the concentration of halogenated phosphates (25 experiments) in this way. He discovered that

$$\Delta OI = k_{P} c_{P} + k_{Br} c_{Br} + k_{Cl} c_{Cl} + k_{P,Br} c_{P} c_{Br}$$

where the coefficient of interaction $k_{P,Br}$ is either small or negative. A similar approach was adopted to determination of the synergistic behaviour of Br and P in the bromophosphorus-containing compounds introduced into polypropylene [69].

Such an analysis is not an exact one, however, since the concentrations of flame retardant elements do not represent independent variables. The efficiency coefficients for the elements at smaller concentrations are too high. The above data indicates that, at present, the question of determining the effectiveness of flame retardants on the basis of the macroscopic characteristics of polymer material flammability is not yet sufficiently clear, and this problem calls for a sounder scientific approach.

Polymer Materials with Reduced Flammability

World consumption and production of polymer materials is steadily increasing, and production of thermoplastic polymers is currently heading the list. Thermoplastic polymers account for 60-85% of total world production of synthetic resins and plastics. The growth of thermoplastics depends on the availability and application of raw materials that stem from the switch from coal-based chemicals to chemicals obtained directly from petroleum and gas, highly progressive and economical methods of synthesis and reprocessing, and the possibility of by-product regeneration. A leading position among polymers is held by the polyolefins, polyvinyl chloride and styrene polymers and copolymers [75-77].

The areas of polymer material application are many and varied and they are responsible for the production of materials with complex and diverse properties for different purposes. They cover construction materials, organic glasses, films, fibres, foam materials and coatings, resin industrial components, adhesives and many others. Until 1975, an especially high rate of production was attributed to the manufacture of chemical fibres, including a steady growth in the production of synthetic fibres. In the late 1970s, some 88-90% of all types of synthetic fibre came under the polyester, polyamide and polyacrylonitrile categories [78].

Synthetic fibres are used in the manufacture of everyday articles (clothing, fabrics, rugs) and in industrial components (tyre cord, resinotechnical components, composite construction materials). Demand for chemical fibres will reach 30 million tons by the end of the century [79]. Development of the most progressive industrial areas (automobiles, aviation) is closely linked to progress in the production of synthetic elastomers, which not only substitute natural products but surpass them in a large number of ways. Butadiene-styrene, butadiene resins and butyl rubber are still in the lead.

Alkyd resins (some 60% of all the polycondensation resin types) are those used predominantly in the manufacture of paint and varnish coatings. The availability and relatively low cost of the initial raw materials, good technological properties and the potential for material modification explain this situation. The paint and varnish industry consumes up to 40% of the epoxy resins produced and nearly 30% of the polyvinylacetate. Polyurethane and organosilicon film-forming substances are even more widely applied. Many coatings with valuable properties are derived from these substances.

Various gas-filled polymer materials (foams) are highly important. On account of their low density and high thermal, electrical and noise insulation characteristics, the gas-filled materials are used in various industrial branches (furniture, refrigeration engineering, radio-electronics, electrical engineering, the automobile industry, aviation, shipbuilding, the building industry and others).

The share of foam plastics in total international production of plastics is nearly 10%. Foam polyurethanes rank top of the list, while foam polystyrene and foam polyvinyl chloride take second place. In the USSR, the production of foam phenoplasts is significant [80]. Aminoaldehydes, phenol-formaldehydes and epoxy-reactive oligomers are used for most of the production of various composite materials for construction purposes. An increase in the manufacture of epoxy resins [81] is particularly apparent.

A brief survey of the development of production and consumption of polymers indicates that in the USSR, it is most frequently materials with a combustible base that are used. The problem of reducing flammability in traditional polymer materials produced on a large scale is of paramount importance.

The essential growth in consumption of materials with reduced flammability is expected to come about in the fields of transport, electrical engineering and electronics, and in the production of furniture and construction materials. According to one forecast, all clothing in the US will be manufactured from low-flammability fibres by the end of the 1980s [82]. We believe it is appropriate in this section to give some concrete examples of the ways in which flammability is reduced in the most widely used polymer materials.

Polyolefin-Based Materials

The leading representatives of this class are homopolymers and copolymers of ethylene and propylene. Such materials possess high mechanical strength, flexibility at low temperatures, high resistance to impact, resistance to the effect of severe media and moisture, and excellent electrical insulation properties. These polymers burn without pronounced liberation of smoke and form a melt that can spatter in the form of burning drops.

The methods used most frequently to reduce flammability in polyolefin materials involve the application of additive-type flame retardants. Surface chemical modification is used considerably less. Bulk modification of the polymer is achieved through the reaction of olefins with other compounds, including reactive flame retardants, and requires intervention in the synthesis process and in the technology of polymer production. As a rule, this makes the cost of materials significantly higher.

Modification brings about essential changes in the physical, chemical and mechanical properties of polymers. Unlike common polyethylene, the chlorinated or brominated product with halogen percentages of 25-40% and 55-65% respectively, possesses resin-like properties. Fluorinated polyethylene (76% F) has properties similar to those of polytetrafluoroethylene. Chlorosulfonation of polyethylene also leads to the formation of elastomers. Resins based on such modified polymers have flammability values close to those of chloroprene vulcanised rubbers but are characterised by low gas penetrability, good chemical and atmospheric stability and better electric insulation properties.

The conditions and degree of chlorination reflect the flammability characteristics of polymer materials in a very complex way. Chlorinated in solution and in suspension, polyethylene with a chlorine content of 40% and 25% respectively becomes self-extinguishing in an atmosphere of air [83].

Polypropylene containing 7.3% and 17.7% of chlorine is characterised by OI values of 26.2% and 20.5% respectively [84]. This is determined by the special properties of the halogenation and dehydrohalogenation reactions of the product. Polymers with a low percentage of chlorine eliminate HCl more readily. Random copolymers of ethylene with vinyl chloride, vinylidenechloride, chlorotrifluoroethylene, tetrafluoroethylene and other halogen-containing olefins occupy an intermediate position between the homopolymers in terms of their degree of flammability [85].

Copolymers with reduced flammability have been derived from ethylene and esters of phosphoric acid, such as from vinyldichloroethylphosphonate [86]. Halide-phenylacrylate has been used as a modifier for polyethylene [87]. The possibilities for chemical modification of polyolefins are virtually unlimited. In fact, this type of modification leads to the creation of new polymer materials with new properties.

Graft copolymerisation of polyolefins with reactive flame retardant monomers is of special importance. Through gas phase graft copolymerisation, for instance, a graft is made to polyolefin fibres and films of various monomer pairs: acrylonitrile-vinylidenechloride; vinyl chloride-vinylidenechloride, and others have been produced. Such materials, which have a three-layer structure, are characterised by a high orientation of polymer molecules [88]. Materials with reduced flammability have been derived by grafting gaseous vinyl chloride and vinylidenechloride onto polypropylene fibres; the graft was initiated by radiation [89].

The graft of vinyl chloride to polyvinylacetate is important in industry. Impact-resistant materials of this type were created for the first time in West Germany [90]. Both suspension [91] and emulsion [92] methods of polymerisation are used for this purpose. Methods have been developed for grafting vinyl chloride to mixtures of copolymers of ethylene and vinylacetate and atactic polypropylene [93] or chlorinated polyethylene [94]. Grafted copolymers are characterised by high physical and mechanical characteristics, in particular by high impact viscosity, and they also blend well with polyolefins.

As a rule, the modified olefins also include flame retardants and synergists. For example, in a composition based on copolymers of ethylene and chlorotrifluoroethylene, the introduction of tin dioxide, phosphate or oxalate of tin is recommended [95]. Antimony trioxide is introduced into compositions based on chlorinated polyethylene or polypropylene [96]. The antimony trioxide will affect the degree of flammability of the composition in different ways, depending on the percentage of chlorine in the initial product (Fig. 5.6). Materials with reduced flammability based on copoly-

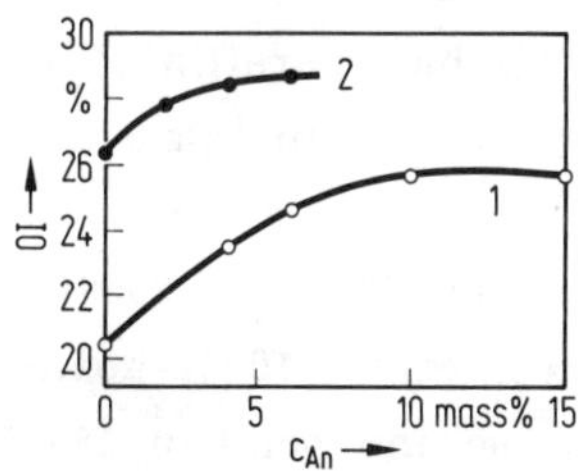

Fig. 5.6 Relationship between the oxygen index OI of chlorinated polypropylene (CPP) and concentration of Sb_2O_3 (c_{An}). Concentration of chlorine in CPP: (1) 16.7% (thermostable), (2) 7.3% (easily dehydrochlorinated).

mers of ethylene and vinyl chloride include hydrogenated aluminium oxide [97].

When flame retardants are used in order to reduce the flammability of polyolefins, one problem is the blending of flame retardants with nonpolar polymer substrates. Polyolefins are reprocessed into articles through die casting, extrusion and compression moulding at temperatures that exceed by 10 to 15 degrees the temperature at which the polymer is converted into a viscous-flowing state. The temperatures applied in the reprocessing of polyolefins are very high. They fluctuate in the range of 180-300°C. Polypropylene is reprocessed at even higher temperatures. Flame retardants must not decompose under these reprocessing conditions. It is essential that migration - the separation of the flame retardant from the material - be suppressed or retarded.

The introduction of flame retardants into the composition of polyolefins is usually performed at the granulation stage, although a method of "dusting" the granules, with subsequent blending of melt and flame retardant directly in the chamber of the extruder or casting machine, is also applied.

In addition to the introduction of flame retardants into the polyolefin melt, surface treatment is also applied. In general, such a method is used in the production of fibres for rugs. Polyethylene, especially the highly crystallised form, is less miscible with flame retardants than polypropylene.

Stable flame retardants that have softening and melting points close to the temperature at which the polymer is converted into a viscous-flowing state, and that will also not diffuse out of the material during service, are the preferred flame retardants. This explains the tendency to use the halogenated acyclic and aromatic compounds, and also high molecular weight halogen-containing flame retardants, instead of chloroparaffins. It has, for instance, been suggested that the following be used in this application: tetrabromobisphenol A and its bromoalkylated ethers [98], tetrabromodiphenyl-sulfone [99], decabromodiphenyloxide ($T_{melt.}$ = 295°C) in combination with antimony trioxide and methyltetrasiloxane [100], hexabromodiphenyl [101], hexabromocyclododecane with a melting point of 185°C [84] and adducts of hexachlorocyclopentadiene [92]. When 4% hexabromocyclododecane and 2% antimony trioxide is introduced into polypropylene, the oxygen index of the material increases from 17.8% to 30.9% [84].

Physical and mechanical properties of materials change slightly during the process [91].

Chlorinated polyethylene and polypropylene, chlorinated polyisobutylene, polyvinyl chloride and various copolymers of vinyl chloride and vinyl bromide are used as high molecular weight halogenated flame retardants for polyolefins. For instance, a self-extinguishing material in terms of standard UL 94 (group SE-0) is obtained from grafted copolymers of vinyl chloride with low density polyethylene containing 26-27.5% chlorine, compounded with polyethylene (2:3) and antimony trioxide (15 parts by weight).

According to reference [84], the efficiency coefficients of bromine and chlorine atoms in the reduction of polyolefin flammability are equal. They are 1.6 for polypropylene and 0.2 for polyethylene for the particular case of using chlorinated polyisobutylene or poly(β-tribromoethylmethacrylate). The introduction of antimony trioxide increases the efficiency coefficients of the halogen to 2.2-2.4 and 1.25-1.3 respectively for these polyolefins. Thus the effectiveness of flame retardant elements in the polyethylene is essentially lower than in the polypropylene.

Lately, there has been a tendency to use substances containing halogen and phosphorus simultaneously in polypropylene materials. Such compounds make it possible to create materials that do not form drops during combustion. Various esters of phosphorus acids (phosphites, phosphonates, phosphates) are introduced in quantities of 5.0-20 weight % [69, 86]. The compatibility of organophosphorus compounds and polyolefins is also important.

Compounds of polyolefins with polymeric phosphorus substances and phosphorated phenol-formaldehyde novolac-type oligomers [102] have been proposed. The latter are recommended for the insulation of electric cables. In reference [103] it was shown that the reduction in flammability of polypropylene fibres may be obtained by using a mixture of organic compounds and bromoaromatic compounds and also by using either of the phosphoric acids. Infusible flame retardants migrate from the polymer substrate to a lesser degree. A method of casting thermoplastic materials based on polyolefins containing red phosphorus and nitrogen-containing melamine or polyacrylonitrile has been developed [104].

It should be mentioned that the use of common inorganic fillers for polyolefin materials is of little value in the reduction of combustibility,

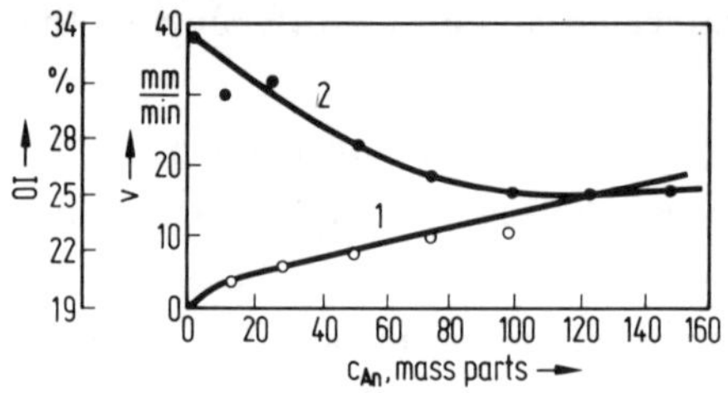

Fig. 5.7 Relationship between the oxygen index OI (1), rate of combustion in an atmosphere of air (2), of polyethylene, and the concentration of aluminium hydroxide c_{An}.

which is why the use of flame-retardant fillers, such as aluminium hydroxide and calcium carbonate has been advised. For materials with reduced flammability there are known compositions based on either polyolefin and aluminium silicate [105] or on mixtures of organic and inorganic fillers [106].

When up to 150 parts by weight aluminium hydroxide are introduced into low-density polyethylene, the material becomes very rigid and brittle. Even for such large filler to polymer substrate ratios, however, the rate of combustion, measured according to the ASTM 635 method, decreased by a factor of two when the samples were burned in air (Fig. 5.7). Derivatographic research into compounds of polyethylene and $Al_2O_3 \cdot 3H_2O$ showed that the formation of water during decomposition of the filler was completed before the polymer decomposed, which is why the flame retardant filler is not effective [107].

Filled materials also contain other flame retardants. Polypropylene filled, as a rule, with asbestos and admixtures of inorganic flame retardant substances, is used as a reduced-flammability material for the production of engine bodies, sealing rings and rear panels for television sets.

Polymer materials contain various ingredients. The interaction of certain flame retardants and stabilisers, their effect on flammability and on the thermal and thermo-oxidative destruction of polyethylene are discussed in reference [108].

Polyvinyl Chloride-Based Materials

Materials based on polyvinyl chloride (PVC) are subdivided into rigid, nonplasticised and plasticised groups. Rigid PVC is used for the production of construction materials and pipes, plasticised PVC is used for film, sheet materials and coatings. Nearly 50% of PVC is used in the plasticised form at present.

Nonplasticised PVC self-ignites at 555-560°C in an atmosphere of air at atmospheric pressure [109]. It has an OI of 42 to 49%. On exposure to a flame it burns with the formation of large amounts of smoke. However, after removal of the flame, it will self-extinguish in an atmosphere of air. Stabilisers, pigments, fillers and other ingredients affect the flammability characteristics of the material. In particular, stabilisers of PVC degradation (salts of lead) increase the OI from 42.2% to 47.7% [110].

Plasticised PVC containing common fuel organic plasticisers surpasses rigid PVC in terms of flammability. Plasticised by diisooctylphthalate, PVC has an OI of 20 to 24%. The use of flame retardant plasticisers is a widely applied means of reducing the flammability of such materials. Chemical modification of PVC permits the creation of materials with new properties. The modification of PVC through chlorination has long been familiar.

Chlorinated PVC (CPVC) is used for manufacturing films, fibres, paints and varnish coatings because it dissolves readily in organic solvents. The OI of chlorinated PVC increases to 60% with a chlorine percentage of approximately 65%. This CPVC liberates less smoke during combustion than PVC. As in the case of PVC, CPVC burns in a flame in an atmosphere of air but when the flame is removed the combustion ceases.

Copolymers of vinyl chloride with vinylidenechloride, vinyl bromide, tetrafluoroethylene and trifluorochloroethylene have higher OI values than PVC.

In terms of their flammability characteristics, copolymers occupy an intermediate position between homopolymers. The OI values for copolymers of vinyl chloride with vinylidenechloride, for instance, fluctuate in the 50-60% range. The copolymers of vinyl chloride in latex form are often used as high molecular weight flame retardants in order to reduce the flammability of various polymer materials. The many valuable properties of copolymers are responsible for their wide application in various branches of industry.

Latexes of copolymers of vinyl chloride with phosphorus-containing monomers, for instance, bis-(β-chloroethyl)vinylphosphate [111] and other derivatives of vinylphosphoric acid [112], are very important. Latexes of copolymers of vinyl chloride are used for the manufacture of film materials and coatings. Nonplasticised PVC has a low fluidity at temperatures close to the decomposition temperature of the polymers. This makes the reprocessing of such polymers into articles difficult.

The polymer stability, the deformation temperature during heating and the specific impact viscosity of materials based on nonplasticised PVC are relatively low. In order to improve casting and increase thermal stability and to obtain materials with a high impact strength, copolymers of acrylonitrile, butadiene and styrene are introduced into the PVC composition. However, such polymers increase the flammability of polyvinyl chloride materials and special measures are required [113]. In view of this problem, the use of high molecular weight plasticisers with reduced flammability is preferred. In order to obtain high-impact PVC, for instance, a grafted block copolymer of vinyl chloride with vinyl acetate [114], polychloroprene and other chlorinated resins and chlorinated polyethylene [115] has been recommended.

A highly original principle for improving PVC castings has been used in the manufacture of polyvinyl chloride construction materials with increased thermal stability and heat resistance. The principle of intermittent plasticisation of the polymer by means of oligomer systems allows the polymer to be reprocessed into articles by extrusion at a sufficiently low temperature (70-80°C).

Subsequent hardening of the oligomer, which is capable of polymerisation as a result of additional thermal treatment (~100°C) leads to the creation of materials with improved properties [116, 117]. It ought to be expected that the use of low-flammability polymerisable oligomers will increase application of the plastification method based on this principle in the production of difficultly combustible PVC materials.

Reduction of the flammability of plasticised PVC is achieved through partial or complete substitution of fuel plasticisers by less combustible halogens or phosphorus-containing substances [109] or through the use of flame retardant fillers. For systems including only halogens, the use of Sb_2O_3 is common practice. The use of zinc borates and aluminium hydroxide has also been proposed. Thus the use of substances that react and eliminate hydrogen chloride during decomposition is less expedient from the viewpoint of reducing material flammability. However, on account of the harmful effect of HCl, not only on the human body but also on metal components in fittings, electric wires and other armatures, the

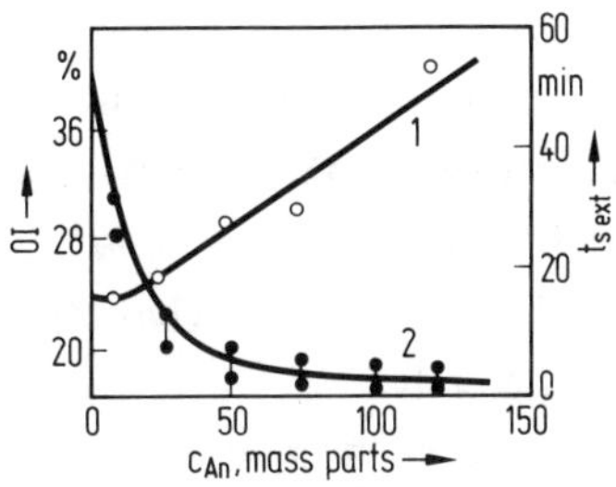

Fig. 5.8 Relationship between the oxygen index OI (1) and self-extinction time at 120°C in an air atmosphere after removal of the flame $t_{s.ext.}$, (2) of polyvinyl chloride and the concentration of aluminium hydroxide c_{An}.

developers of polyvinyl chloride materials use various substances which absorb HCl. For electric insulation, plasticised compounds ("chlorstone") have been developed that contain calcium carbonate as a filler [114]. When heated for 30 minutes at 650°C, HCl is converted almost completely into residue.

Carbonates of other metals and calcium cyanamide have been used for similar purposes [118]. The cyanamide that is formed reacts with the product of PVC degradation. In this case, the degree of polymer unsaturation increases, the dehydrochlorination process slows down and the yield of carbonised residue increases. Contrary to the case with polyethylene, flame retardant fillers, such as aluminium hydroxide, reduce the ignitability and combustibility of polyvinyl chloride more effectively (Fig. 5.8). PVC plasticised by means of diisooctylphthalate (50%) decomposes highly endothermally in the same temperature range as aluminium hydroxide [107]. The efficiency coefficient of the flame retardant filler in PVC is 0.13.

Materials Based on Styrene Polymers and Copolymers

Such materials are characterised by a large variety of properties. Apart from homopolymers, the best-known materials are those based on copolymers of styrene with acrylonitrile and so-called ABS plastics. The latter are impact-resistant materials. ABS plastics have a microheterogeneous structure and, in fact, are two-phase systems in a continuous thermoplastic matrix of the styrene and acrylonitrile copolymer in which the dispersed elastomer particles are distributed. This type of system can be created either by the mechanical mixing of polybutadiene, butadiene-styrene or butadiene nitrile rubbers with the copolymer of styrene and acrylonitrile, or by blending a mixture of latexes of rubber and thermoplastic. The most frequent practice in the manufacture of ABS plastics is to use styrene and other monomers copolymerised in bulk or in emulsion in the presence of the rubber. Copolymerisation of styrene and acrylonitrile and bonding to the rubber particles occurs simultaneously.

Direct copolymerisation of styrene, acrylonitrile and butadiene does not lead to the creation of materials with the required properties. To form a microheterogeneous system, at least one of the monomers must polymerise in the presence of the copolymer or homopolymer that has already formed.

There are three types of ABS plastics: those with average, high and ultra-high impact strength (the impact strengths are up to 15, 15-25 and 25-35 $kg \cdot cm/cm^2$, respectively). A change in the ratio of the monomer to the elastomer and the conditions of their copolymerisation can modify the technological and mechanical properties of the materials.

The ABS plastics differ favourably from binary copolymers of styrene with acrylonitrile (SAN) and from polystyrene not only in terms of their increased impact strength but also on account of their chemical stability, and surface hardness and their suitability for metallisation. ABS plastics are used as construction materials and in the manufacture of materials for the transport industry, instrument engineering and other sectors of industry. The properties of ABS plastics and styrene polymers manufactured in the USSR are given in [119].

Homopolymers and copolymers of styrene burn with pronounced liberation of sooty smoke. Oxygen indexes fluctuate in the range of 17.7-18.3% (PS and SAN) and 18-20% (ABS). Those types of ABS plastics with the highest impact strength usually have the lowest OI values [120].

The method for manufacturing materials with reduced flammability based on styrene polymers and copolymers is selected according to the nature of product application and the special processing requirements of the material. Chemical modification of the polymers is one efficient means of changing the level of polymer flammability. Reactive flame retardants are used for the chemical modification.

Flame retardants capable of polymerisation are introduced at the stage of styrene polymer synthesis and form random or grafted copolymers. The use of halogenated derivatives of styrene, pentabromophenylmethacrylate, methyl-α-bromoacrylate, halogen derivatives of unsaturated acids (for instance, fumaric acid), vinyl bromide, vinyl chloride or vinylidene-chloride, and also unsaturated phosphorus-containing compounds, is well established [26, 121].

For the synthesis of fire-proof type ABS plastics, reactive flame retardants may be introduced into the bulk or emulsion of styrene and acrylonitrile as comonomers. The microstructures of plastics derived in bulk and by the emulsion method are different. In the latter case, the dispersion of the particles of the rubber gel also includes particles of thermoplastic, thereby strengthening the properties of the material.

Reactive flame retardants are used in large amounts of up to 20 weight % or, converted into the percentage of phosphorus, of up to 4 weight % [122]. In many cases, the characteristics of melt fluidity change and the thermal stability drops at the same time. This is especially characteristic of halogen-containing reactive flame retardants. In the latter case, synergists such as Sb_2O_3 are used as additives in order to increase the effectiveness of flame quenching. As in the case of other systems, bromine-containing flame retardants are more effective than chlorine-containing flame retardants.

It should be mentioned that the introduction of reactive flame retardants into the polymeric structure affects the nature of decomposition and the flammability characteristics of a polymer material in different ways.

Hence, in the case of copolymerisation of vinyl bromide with methyl-methacrylate, when the copolymer decomposes, vinyl bromide reforms. Although the OI values increase, the linear rate of combustion also increases when there is an excess of oxygen. In the copolymerisation of styrene with vinyl bromide, HBr is liberated during decomposition. The oxygen index of a sample with 6.65 weight % vinyl bromide increases from 18.3 to 41%, and the rate of combustion in the atmosphere with an increased content of oxygen is diminished by almost a factor of two [123].

A few reactive phosphorus flame retardants should be mentioned among those that are recommended for the reduction of styrene polymer flammability: bis-(2,3-dibromopropyl-)allylphosphonate, amides and esters of α-halogenvinylphosphonic acid and phosphorylated styrene [26, 124]. It was discovered, for instance, that bis(β-chloroethylene)-α-β-dibromoethylphosphonate slightly reduces the relative viscosity of the copolymer melt but does not impair the physico-mechanical properties of the foam plastic that forms during copolymerisation, and greatly reduces flammability [125].

A further approach is also adopted in order to obtain impact-resistant materials with reduced flammability on the basis of styrene polymers: grafted block polymerisation of styrene with a high molecular weight chlorine-containing polymer. A method has, for instance, been developed for producing materials with a reduced flammability, high gas impenetrability and high impact strength through the use of grafted block polymerisation of styrene with polyvinyl chloride containing aldehyde, carboxyl and sulfo groups.

When peroxides react with PVC, oxidation of the above-mentioned groups occurs, and active centres are formed that participate in the grafted block copolymerisation of styrene. The derived product is a two-phase system. It includes the homopolymer and block polymer forms of styrene and vinyl chloride and has properties superior to a physical mixture of PVC and polystyrene [126].

Impact-proof styrene plastics with reduced flammability have been obtained by partial or complete substitution of butadiene rubber for polychloroprene and chlorinated polyethylene. The synthesis is carried out in such a way that grafted copolymers of styrene and acrylonitrile with a chlorinated elastomer are formed [127, 124]. In addition, oxides of antimony, bismuth, molybdenum, lead and tungsten or their mixtures are introduced into the composite material [127].

Along with polymerisable flame retardants, the chemical modification of polystyrene materials by direct chlorination or bromination is also applied. This method is preferred for the production of latex materials and is designed for treatment of carpet and textile materials in order to reduce their flammability. A method was proposed, for example, for the manufacture of self-extinguishing latexes from copolymers of styrene and butadiene by treating them with 20-30% of aqueous bromine emulsion [128].

The additive-type flame retardants allow production of materials with reduced flammability on the basis of styrene polymers without substantially changing the method of production or the method of reprocessing the materials into products. This then accounts for the recent interest in this field.

Flame retardants of this type are inorganic and organic compounds with halogen, nitrogen and phosphorus atoms in their molecular structure. Low and high molecular weight organic compounds are used. The list of flame retardants used is given in Lindemann's study [121].

There is a tendency to select flame retardants with an elevated percentage of halogen and phosphorus. Use is also made of substances such as adducts of hexachlorocyclopentadiene with unsaturated halogenated compounds (dechloranes), hexabromodiphenyloxide, hexabromocyclododecane and others. In combination with synergists (Sb_2O_3), these permit the creation of self-extinguishing polystyrene materials.

Low molecular weight flame retardants generally have a marked effect on the physical and mechanical properties of materials and reduce the specific impact viscosity. When tetrabromobisphenol A (20 parts) and Sb_2O_3 (7 parts per 100 parts ABS) are introduced into ABS, the oxygen index of the material increases to 27.7% but, in this case, the specific impact viscosity decreases from 22.6 to 8.1 $kg \cdot cm/cm^2$ [120].

The introduction of flame retardants changes the morphological structure of a two-phase material. For flame retardants mixed with a polymer melt and interacting with individual polymer phases, the flame retardant distribution within these phases is an important factor.

It was discovered that localisation of admixtures within the continuous rigid polystyrene matrix makes it easier to maintain the impact viscosity properties of impact-resistant polystyrene [129]. This is why flame retardants should be more compatible with the continuous phase of the styrene thermoplastic and not with the elastomer. Recently, much attention has been focussed on flame retardants containing both posphorus and halogen in their molecules. Benbow and Cullis [130] showed that the effectiveness of such flame retardants in reducing polystyrene flammability depends on their volatility and the nature of their thermal transformation. Bromine-containing phosphorus compounds are also more effective than chlorine-containing ones in this case (Fig. 5.9).

Substances that form radicals (peroxides, nitrocompounds) during decomposition increase the OI. This condition is, however, the result of an acceleration of polymer decomposition and a decrease in the stability of the flame as a result of the removal of heat from the combustion zone by the polymer melt. An apparent reduction in the combustibility of the polymer is observed. This method does not, however, reduce the materials' flammability.

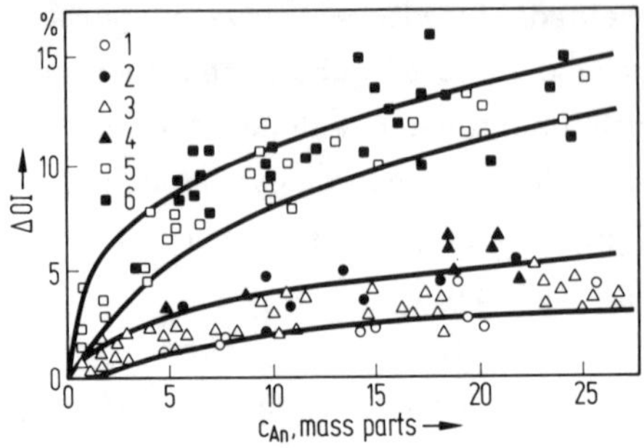

Fig. 5.9 Relationship between the change in oxygen index ΔOI of polystyrene (PS) and the concentration of flame retardant c_{An}. (1) P, (2) P + free radicals, (3) P + Cl, (4) P + Cl + free radicals, (5) P + Br, (6) P + Br + free radicals, (5) P + Br, (6) P + Br + free radicals, OI_{PS} = 24%.

Materials Based on High Molecular Weight Linear Polyesters

In this groups of polymers, the fibre and film-forming polyesters based on phthalic acid and, in particular, polyethyleneterephthalate, are particularly important in terms of the tonnage produced and consumed.

An extensive source of raw materials, the low cost and many valuable properties are responsible for the high rate of polyethyleneterephthalate (PETP) fibre production. The substitution of ethylene glycol by higher homologues of normal glycols during the polyester synthesis process leads to the formation of crystallising polymers with a melting point and a hardness that decrease when the molecular weight of the glycol increases. Polybutyleneglycolphthalate has a melting point of close to 210°C and polyhexamethyleneglycolphthalate of close to 150°C. By comparison with PETP, therefore, it is possible to reduce the temperature and increase the process rate for processing the polymers into products.

Polyalkyleneglycolterephthalate fibres melt during combustion. The oxygen index for PETP fibres varies within the 20-22% range. A relationship has been discovered between the concentration of terminal carboxyl groups in polyethyleneterephthalate (or the molecular weight of the polymer) and the sample's oxygen index (Fig. 5.10) [131]. According to these authors, the increase in the concentration of carboxyl groups accelerates the acidic catalysis of the hydrolytic reactions of polymer degradation.

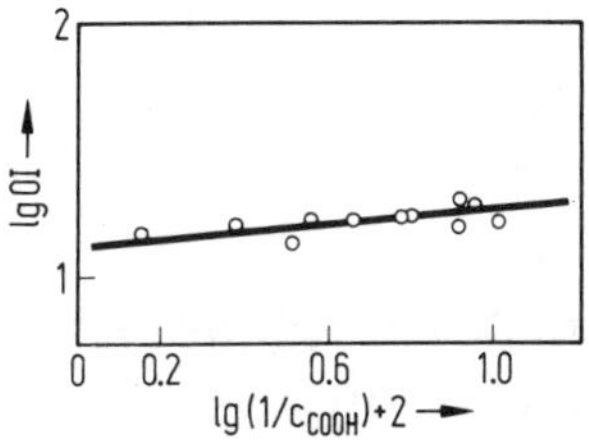

Fig. 5.10 Logarithmic relationship between the oxygen index (lg OI) of polyethylene-terephthalate and inverse concentration of terminal carboxyl groups [lg $(1/c_{COOH})$ + 2] inside the polymer.

The reduction in combustibility of polyester fibres is achieved by using reactive flame retardants during synthesis, by including additive-type admixtures in the polymer melt before processing, or by special treatment of the products (fibres and fabrics).

Direct bromination or chlorination of polyesters (both surface and in bulk) causes an increase in the solubility of the polymer in organic solvents, which is why this method of reducing the flammability of PETP fibrous materials has not been implemented.

Polymers with reduced flammability (copolyesters) have been derived through the use of halogen-containing or phosphorus-containing diols or by using dibasic acids and their esters [26]. Since the synthesis of polyesters is carried out in the melt under the conditions of a vacuum and in a temperature range of 150-290°C, the flame retardants must have low volatility, be stable and not affect the hydrolytic stability of the copolyesters. For copolycondensation, for instance, the use of bromine-substituted terephthalic acid in combination with additions of Sb_2O_3 [132] is advised. The use of 10 mol.% of 2,5 dibromoterephthalic acid during the copolycondensation of PETP allows the OI to be raised to 25%. When the latter acid is combined with 4 weight % of Sb_2O_3 the OI increases to 28%. Random copolyesters have lower melting points in contrast to PETP homopolymers.

In the work of Lawton and Setzer [133] a few reactive flame retardants are given that are suitable for industrial products. Flame retardant diol, which is the product of the reaction between ethylene oxide and tetrabromobisphenol A, when it is present in PETP at a level of 4 mol.% (6 weight % Br), shows the same effect on the reduction of copolymer flammability as 9 mol.% of 2,5-dibromoterephthalic acid. In terms of physical and chemical properties, the fibre made from this type of polymer is similar to the PETP fibres, but the former is less resistant to deterioration [134].

Methods have been developed for the production of copolymers with reduced flammability using arylene and alkylene-di(methylphosphonic acid). These compounds are introduced at the point of cross-esterification of PETP [135]. Various phosphorus-containing dibasic alcohols and acids and their esters have been employed for the production of polyesters with reduced flammability by the West German firm Hoechst [26].

It should be mentioned that copolyesters prepared by the polycondensation of bis(2-oxyethyl)terephthalate and phenylphosphonyl-di(*n*-methoxybenzoate) in a molecular ratio of 1:1 (the phosphorus content in the polyester is ~5 mol.%) were similar in terms of ignition to a sample prepared by a simple mixture of PETP melt and the above-mentioned phosphorus monomer (OI = 29 to 30.8%). It is assumed that this result is due to a fast esterification reaction inside the polymer melt [131]. Halogenated compounds have been proposed as additive-type flame retardants. Basically, those halogenated compounds are aromatic substitutes that do not decompose at the temperature at which fibres form from the melt; examples include hexabromobenzene, decabromodiphenyl, esters of tetrabromophthalic acid, bromo-substituted bisphenoxy compounds, adducts of hexachlorocyclopentadiene and others. These compounds are used in combination with metal oxides which play the role of synergists. Self-extinguishing samples that do not form drops during combustion have, for example, been prepared according to ASTM D-635. These samples are prepared on a base of polybutyleneterephthalate through a Diels-Alder reaction between hexachlorocyclopentadiene and 1,5-cyclooctene. Flame retardant and Sb_2O_3 were mixed with a polymer melt in the extruder during the granulation process [136].

The thermal properties of polymers and admixtures determine the effect of flammability reduction in polyester materials to a large degree. This is why PETP fibres with the addition of 10 mol.% hexabromobenzene, 0.5 weight % Sb_2O_3, and 0.5 weight % TiO_2 have a higher oxygen index (28.1%) than when the hexabromobenzene is substituted by an equal amount of decabromodiphenyl (OI = 22.6%) [137].

As in the case of halogen-containing flame retardants, phosphorus and its compounds are used in reducing the combustibility of linear polyesters.

Recommendations have been made, in particular, for the use of finely pulverised red phosphorus and an inert filler (asbestos) in order to obtain materials that are self-extinguishing in an atmosphere of air when the flame is removed. Such materials do not form drops during combustion [138]. It has been discovered that the effectiveness of compounds that oxidise phosphorus to different degrees (organophosphates, phosphonates, phosphineoxides) in reducing polyester flammability is practically identical [133].

Alkali or alkali-earth metal salts of either phosphoric or diphosphoric acids [139], metallic salts of sulfoacid, i.e. derivatives of alkylphosphineoxides [140], arylpolyphosphonates containing diphenyloxide, diphenylsulfide, diphenylsulfonic fragments or their halogen derivatives [141], are compounds which may be cited as flame retardants that can be introduced into polyesters at the last stage of synthesis, i.e. during cross-esterification, or into the melt prior to granulation.

Many recommendations have been made for the use of flame retardant mixtures that contain halogen, phosphorus, ammonium or sulphur. This approach is not always justified, however, because the increased complexity of the compound does not always lead to a reinforcement of the effect. The way in which a mixture of triphenylphosphineoxide and hexabromobenzene affects the values of the oxygen index of PETP fibres is shown in Fig. 5.11. The efficiency coefficient of phosphorus hardly changes in the presence of bromine. In other cases, an antagonistic effect has been discovered when bromine is combined with phosphorus [137].

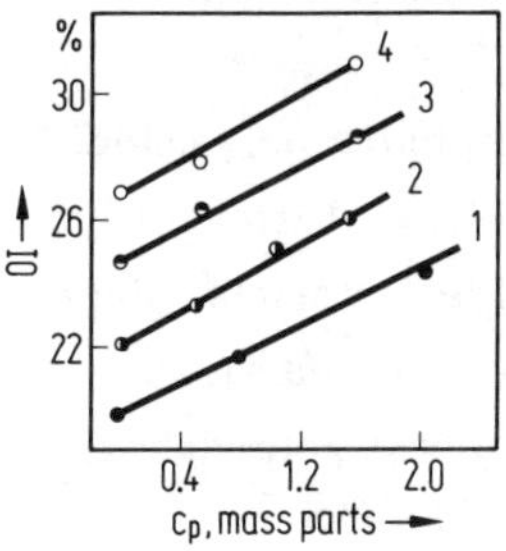

Fig. 5.11 Relationship between the oxygen index OI of PETP fibres and concentration of phosphorus c_P introduced together with flame retardants. Concentration of bromine, weight %: (1) 0, (2) 3.7, (3) 7.4, (4) 11.1.

The introduction of flame retardants into polyester melts makes it possible to prepare materials that are more resistant to different treatments (laundering, dry cleaning) than materials that have been coated by the same flame retardant. This is why polyester fibres, fabrics and films should be surface-treated with reactive flame retardants which become insoluble on the surface. After materials have been soaked in these flame retardants they undergo thermal treatment. As a result, the flame retardants undergo either polycondensation or polymerisation reactions. Their solubility in water and organic solvents falls sharply.

It has been proposed, in particular, that for fireproofing polyester materials use should be made of such compounds as tetra(oxymethyl)-phosphonium hydroxide, condensation products of melamine and formaldehyde, unsaturated derivatives of phosphoric acids, and other compounds [26, 133]. Fabrics designed for use as decorative and upholstery materials and also carpet fabrics made of polyester fibres are treated with latex of chlorine-containing polymers (polychloroprene, polyvinylidenechloride, vinyl chloride copolymers).

Materials Based on Polyamides

The assortment of polyamide materials is very diverse. In addition to fibres and fabrics, polyamides are used in the manufacture of films, coatings, foams and reinforced plastics. These can replace nonferrous metals and also reactive plastics. The reinforcement of polyamides with fibrous inorganic fillers permits creation of construction materials with high mechanical properties, durability and stability of form at high temperatures. They are relatively nonhydroscopic and sufficiently resistant to oils and lubricants.

The most widely used materials in technology are aliphatic polyamides which have a higher flammability than aromatic polyamides. Different methods are used to reduce the flammability of polyamide materials, depending on their type. Flame retardant fillers can be used, in particular, for polyamide plastics; for fibres and films, additive-type flame retardants can be used with special surface treatment.

It is much more difficult to reduce the flammability of polyamide materials by comparison with other polymers. The effectiveness of flame retardants in polyamides is low; a high concentration is required. The most radical method of flammability reduction in polyamides is through chemical modification and elimination of the aliphatic hydrocarbon fragments from the polymer's structure. This method is carried out as a synthesis of more thermostable polyamides. Monomers with heteroatoms, and with aromatic and heterocycles, are used in the copolymerisation reactions. As a result, polyamidoimides, arylaliphatic polyamides and others are obtained.

The use of reactive flame retardants (diacids, diamines, lactams containing phosphorus or halogen) for the modification of polyamides has not yet been well developed. There are recommendations for the use of phosphorus derivatives of aromatic dicarbonic acids [142, 143], aromatic diamines with chlorine atoms on the ring or with hydrogen atoms completely substituted by chlorine and fluorine atoms in the alkyl groups [144], phosphorus-containing lactams [145], bromoaromatic dicarbonic acids [146] and other reactive flame retardants. However, the necessity of introducing large amounts of such compounds often makes it difficult to obtain high molecular weight products with the required physical and mechanical properties. As a rule, when the percentage of flame retardant increases, the degree of polyamide crystallisability falls and the film and fibre-forming properties of the polyamides are reduced. Such modified polyamides can, however, be used for composite materials.

Direct halogenation of aliphatic polyamides is not practised, since direct halogenation reinforces chemical decomposition of the polymers and increases their solubility in organic solvents. The effectiveness of a halogen in reducing aliphatic polyamide combustibility is low.

At the same time, the surface treatment of articles made from aromatic polyamides (articles that have large surface-to-volume ratios: fabrics, foams) with chlorine, bromine or sulfuryl chloride reduces their flammability noticeably. Fabrics modified in such a way are characterised by only slight shrinkage under thermal influence and by a shorter smouldering period after removal of the flame. In the case of 9-10% chlorine, the oxygen index increases from 28-30% to 35-37% [13]. Inside the flame (1100°-1200°C) the fabric becomes black without being ignited and does not shrink. Cellular materials made from such modified polyamides are well suited to the manufacture of completely carbonised and graphitised articles [12].

Phosphorylation of polycaproamide fibre by dichloroanhydride methylphosphonic acid has been studied in detail [147]. Active hydrogen atoms near the nitrogens of the amide groups are substituted by phosphorus-containing groups and hydrogen chloride is liberated. In order to reduce chemodestruction, the reaction is carried out in an anhydrous organic solvent medium. The retention of a fibrous structure has been observed at a relatively low degree of substitution (9.6% of theoretical) and for a small percentage of phosphorus (2.3%). This amount is insufficient and the fibre remains flammable.

There is interest in studies dealing with the grafting of polyvinyl fluoride [148] or polyvinylidenechloride to polyamide fibres and fabrics by irradiation [149]. These studies have not, however, had any practical application. In the case of composite materials, use is occasionally made of halogen-containing polymer compounds, for instance, polyvinylidenefluoride [150].

Flame retardants of a nonhalogen type, or halogenated ones that do not eliminate hydrogen halide during reprocessing and formation of polymers, are the preferred flame retardants for reducing polyamide material flammability. This is connected with the sensitivity of polyamides to the heterolytic reaction of chain scission. This is why use is made most frequently of aromatic hydrocarbons, adducts of hexachlorocyclopentadiene, combined with oxidisers and metal borates [26].

Flame retardants introduced into the melt of polyamides before spinning and moulding are listed in reference [24]. Phosphorus flame retardants are several times more effective than halogen-containing ones. The present controversial data makes it difficult to judge how the degree of oxidation of phosphorus affects the efficiency of flame retardants. Red phosphorus organophosphates, phosphonates, phosphites and their halogen-containing derivatives have been proposed for polyamide materials.

Phosphorylated phenol-formaldehyde novolacs [151] and linear and cyclic phosphazine compounds have been patented [152]. It was discovered that cyclic phosphazines with alkylated amino groups near the phosphorus atoms affect the oxygen index of polycaproamide fibre. When bromine is present in the organic substituents, the general effect does not increase substantially. For example, if there is an admixture of phosphorus- and bromine-containing flame retardant in the fibre (converted to elements P = 0.82% and Br = 4.5%), the OI changes from 19.3% to 20.3%; in the case of introduction of an analogue that does not contain bromine (P = 0.70%), the OI goes up to 20.0% [153].

Among the nitrogen-containing flame retardants mention should be made of compounds with triazine rings, melamine and its derivatives and products of their condensation.

For the surface treatment of films, fibres or foams made of polyamides, the same is true as for polyester materials: flame retardants that contain reactive groups (methylol, vinyl and others) are used in the main. In addition, inorganic salts are introduced.

Materials Based on Carbamide Resins (Urea-formaldehyde, Melamine-formaldehyde) and Phenol-formaldehyde Resins

Carbamide and phenol-formaldehyde resins are a mixture of oligomer and low molecular weight products of the condensation of urea, melamine or phenol with formaldehyde. The mixing is performed in the presence of basic and acidic catalysts. When hardening occurs under certain conditions, polymers with a three-dimensional network structure are formed. According to GOST 17088-71, the polymer products belong to the group of materials that are ignitable with difficulty or combustible with difficulty. The temperatures of self-ignition are close to 510°, 540° and 600°C for phenol-, urea-, and melamine-formaldehyde polymers respectively.

Carbamide polymers are less combustible than phenol-formaldehyde on account of the larger percentage of nitrogen in their structure. During decomposition, they liberate incombustible gases: ammonia, water and carbon dioxide. However, highly toxic products (hydrogen cyanide, dicyanogen, carbon monoxide) are also formed. Inside a high-temperature flame (1000-1500°C) the carbamide (especially urea-formaldehyde) polymer materials are destroyed; they swell up and form a char layer, which little by little burns out completely.

Phenol-formaldehyde polymer materials are destroyed to a lesser degree by a flame than carbamide polymers, even though the former are more combustible. They usually carbonise first while retaining their initial form. The carbonised product burns out completely. Heterogeneous combustion (smouldering) is observed and continues for some time after the flame has been removed.

During combustion the polymers discussed above liberate relatively small amounts of smoke. Carbamide and phenol-formaldehyde resins are often used in order to reduce the flammability of other more combustible materials.

One of the ways of increasing the flame-resistance of phenol-formaldehyde materials is to modify the structure through preliminary heat treatment. This method is used for the production of different carbon materials (foams, plastics, fabrics, fibres). As the temperature of the preliminary heat treatment of phenol-formaldehyde fibres increases there is a significant rise in their resistance to fire. The combustion time inside a flame at a temperature of 950°C is 0.8 min for fibres heated to 800°C, and 7.35 min for fibres heated to 2600°C [154]. Alkylation of phenolic hydroxyl groups and a reduction in the electron density on the aromatic ring increases the resistance of phenol-formaldehyde materials to oxidation [155]. At the same time, their brittleness falls and decomposition decreases [156].

The chemical modification of carbamide and phenol-formaldehyde resins in order to increase their thermal stability leads to an increase in their resistance to flame. At present a number of studies have been conducted on the synthesis of oligomers containing condensed aromatic and heterocyclic fragments in their structure [23]. It was discovered that the use of reactive flame retardants containing phosphorus, nitrogen, halogen, boron, silicon and other elements makes it possible to reduce the flammability of materials prepared from carbamide and phenol-formaldehyde resins. Boron and phosphorus-containing compounds suppress incomplete combustion of the material after removal of the flame.

In the case of synthesis of phenolic resins, application of the following compounds during copolycondensation has been recommended: trimethylphosphineoxide products of condensation of phenol with vinylphosphate [158], methyl-1,1,2,2-tetrafluoroethylmethylphosphine oxide [159], 4,4'-dioxydiphenylsulfone, and 4,4'-diphenylphenylphosphineoxide [160], halogen-substituted aromatic polyhydroxyls or amino derivatives [161]. The addition of polyhydroxyl derivatives of chlorinated biphenyl during the synthesis of phenol-formaldehyde resins has been recommended. The flame of an oxy-acetylene torch burns through a panel of foam plastics made from this product (thickness, 7 cm) in 2 min, 45 sec. In that time there is no ignition of the foam plastic and only charring occurs. Non-modified foam phenoplastic is destroyed faster in the flame, and the combustible pyrolysis products ignite [162]. In order to reduce the flammability of phenolic polymers, use is also made of nitrogen compounds that react with formaldehyde: urea, dicyandiamide, melamine [163] and other triazine compounds [23].

When polymer products with increased resistance to flame are produced, chloroendic anhydride [164], chloride and hydroxide of tetra(oxymethyl)-phosphonium [165] are used as hardeners.

Also, the methods of direct chlorination or bromination of phenol-formaldehyde resins in an alcohol-water medium [166], and phosphorylation and sulfonation are also practised. Phenol-formaldehyde resins can be treated by chlorophosphocyclopentene [168]. This treatment is applied to the surface of the finished articles, such as to foam phenoplastics.

One of the most practicable ways of reducing material flammability is to use inorganic fillers. These are used in the production of both plastics and foam plastics. Glass fibre, asbestos, porous filler (perlite) [169] and ash (by-products of heat and electric power plants) [170] are introduced into the foaming compound.

When foam phenoplastic is reinforced with cut glass fibre (6-8 weight %) its volumetric shrinkage in an atmosphere of air decreases from four to five times, its bending strength becomes two to three times greater and its adhesion to metal improves [171]. There are cases when flame retardant fillers are used to produce thermoinsulation materials that are combustible with difficulty from phenol-formaldehyde resins. By using aluminium hydroxide, zinc oxide and orthophosphoric acid, it has proved possible to prepare foam phenoplastics that are combustible with difficulty with a low water absorption (3-4%) and improved water resistance [172].

Flame retardants such as ammonium phosphate and polyphosphate, metaborates and borates of metals, organophosphates and phosphonates and their halogen derivatives [160, 161] have been recommended for reducing the flammability of materials created from carbamide and phenol-formaldehyde resins.

Materials Based on Polyurethanes

Polyurethanes are used for the production of foam plastics, rubber technical components, fibres, coatings, construction materials, adhesives and fillers. Rigid and elastic foam plastics occupy a leading position among them. Polyurethane materials burn with pronounced liberation of smoke and toxic products (HCN, CO, isocyanates).

A large burning volume and the increased flammability of polyurethane foam reinforces the necessity of studies aimed at reducing the flammability of these materials [27].

The modification of polyurethanes is viewed as the most promising way of solving this problem. We have already mentioned the chemical change in structure of products obtained from the same initial compounds but with application of different catalysts during synthesis. This method is successfully employed for the creation of rigid incombustible foam plastics made from polymers with carbodiimide and isocyanurate fragments in their structure [26].

The introduction of aromatic and heterocyclic fragments into the macromolecular chain of polyurethanes increases their structural rigidity as a result of an increase in the yield of carbonised residue during combustion of these modified polyurethanes. At the same time, the ignitability and combustibility of these materials decreases. Methods have been developed, for instance, of deriving polyurethanes with benzimide and benzimidazole rings [15, 173].

Modified polyurethanes with decreased flammability were obtained when phenolaldehydes, urea- and melamine-formaldehyde oligomers [174, 175], furyl alcohol [176], aromatic acids and anhydrides were used as compounds with active hydrogen atoms. Polyurethanes with an increased number of crosslinks have been prepared by using multiatomic alcohols, sugars and their oxyalkylated derivatives [177].

The most widespread method of reducing the flammability of polyurethane is the use of flame retardants. Reactive flame retardants are preferable for the preparation of polyurethanes with reduced flammability, since they change the physical and mechanical properties of the materials to a lesser degree. Halogen-, phosphorus- and nitrogen-containing derivatives of isocyanates and of compounds with active hydrogen atoms belong to these flame retardants. Hydrogen atoms inside the aliphatic or cycloaliphatic radicals or in the aromatic nuclei of diisocyanates can be substituted by chlorine, bromine and fluorine [178]. However, it is halogenated diols, polyols or polyetherpolyols that are used most frequently, for example: tetrabromohexandiol, polyetherpolyols based on chlorendic anhydride, tetrabromophthalic acid, dibromo- and tribromopentylglycol and others [26]. Phosphoric compounds are more effective than halogen-containing flame retardants. The range of phosphorus-containing isocyanates is limited. Usually they are obtained in the form of prepolymers when an excess of isocyanate reacts with phosphorus polyols or with diols. The phosphorus polyols and polyetherpolyols used for purposes of reducing the flammability of polyurethanes can be subdivided into phosphate, phosphonate and phosphite types of compounds.

Phosphates are obtained by the direct reaction of phosphoric acid and alcohols, by alkoxylation of esters of pyrophosphoric acid, by oxidation of phosphites, by phosphorylation of polyols, by phosphorus oxychloride and other methods. Phosphonate-type polyols are prepared according to the Michaelis-Arbusov reaction or the Mannich condensation, or by addition of dialkylphosphites to unsaturated compounds. Phosphite polyols are prepared by a reaction of cross-esterification of different polyols by means of triaryl or trialkylphosphates [179]. In spite of numerous studies on the use of phosphorus polyols and polyether polyols it is still not clear what type of compounds are the most effective for reducing polyurethane flammability [180, 181].

A vast number of polyetherpolyols have been proposed for reduced-flammability polyurethanes. These polyetherpolyols contain both halogen and phosphorus. Self-quenching rigid foam polyurethanes that have good dimensional stability both in dry and wet atmospheres have been prepared with up to 1.5% phosphorus and 10-15% halogen in the foamed composition [179]. Halogenated phosphonates containing amide groups are used in order to decrease the flammability of coatings, plastics and foam plastics made from polyurethanes [182].

The additive-type flame retardants, used to reduce the flammability of polyurethane materials constitute a large group. Among the inorganic compounds, those used are ammonium polyphosphates and calcium polyphosphates, red phosphorus, borates and fluoroborates of metals, hexafluorotitanates or ammonium silicates and aluminates [183]. Generally recommended are mixtures of organic, low or high molecular weight halogen-containing compounds, together with oxides and hydroxides of antimony, zinc, titanium, aluminium and other metals. Elastic foam polyurethanes with reduced flammability have been prepared by the introduction into a foam of 2-100 parts by weight of halogen-containing powdered polymer, 30-80 parts by weight aluminium hydroxide and 1-20 parts by weight antimony trioxide based on 100 parts by weight polyetherpolyol.

The use of polyvinyl chloride and chlorinated polyethylene as a halogen-containing polymer [184] has been recommended. Organo- and halogen-organophosphate plasticisers are introduced into elastic foam polyurethanes but have a relatively low effectiveness.

Light mineral fillers in combination with flame retardants and substances that decompose forming noncombustible gases when the material is exposed to a flame, could be used for foam plastics and fire-protecting polyurethane coatings. As a result, swelling and the formation of a charred protective layer occurs. Perlite, ceramic gravel, foam glass granules and foamed alumina are used as fillers. Also, some of the substances that swell the surface layer material under the action of a flame when used as fillers - azodicarbamide, 4,4 bis(benzenesulfonyl)carbazid and sodium bicarbonate - are employed. In general, decomposition activators are also introduced [185]. It is obvious that reducing the flammability of elastic polyurethanes is more difficult than reducing the flammability of rigid ones.

References

1. Andrianov K. A., Polymery s neorganicheskimi glavnimi tsepiami molecul. M.: Nauka, 1962.
2. Neorganicheskiye polymeri, Stone F., Graham V., eds. M.: Mir., 1965.
3. Fabris H. J., Sommer J. G., Rubber Chem. and Technol., 1977, vol. 50, no. 3, pp. 522-564.
4. Korshak V. V., Termostoykie polimery. M.: Nauka, 1969.
5. Li G., Stoffy D., Nevill K., Novie lineiniye polymery. M.: Izd-vo Inostr. lit., 1972.
6. Termo-, zharostoykiye i negoryuchie volonka. Konkyn A. A., ed., M.: Khimiya, 1978.
7. VanKrevelen D. W., Polymer, 1975, vol. 16, no. 8, pp. 615-620.
8. Packham D. J., Packley F. A., Polymer, 1969, vol. 10, no. 8, pp. 559-561.
9. Korshak V. V., Vinogradova S. V., Siling S. A., et al., Dokl. AN USSR, 1970, vol. 195, no. 5, pp. 1113-1116; Visokomolec. soyed., 1979, vol. 21A, no. 2, pp. 288-293.
10. Smirnov R. P., Vorob'yov Y. G., Berlin A. A., Aseeva R. M., Izv. Vizov. Khim. i. khim. tekhnol., 1975, vol. 18, no. 9, pp. 1451-1455.

11. Vanberkel F., Gratjahm H., J. Appl. Polymer Sci., Appl. Polymer Symp., 1973, vol. 21, pp. 67-80.
12. Pat. 3607798 (USA)
13. Hathaway C. E., Early C. L., J. Appl. Polymer Sci., Appl. Polymer Symp., 1973, vol. 21, pp. 101-108.
14. Stannet V., Walsh W. K., Bittencourt E. et al., J. Appl. Polymer Sci., Appl. Polymer Symp., 1977, vol. 31, pp. 201-207.
15. Pat. 3823158 (USA).
16. Pat. 2082651 (France).
17. Pat. 2290459 (France).
18. Pat. 49-43398 (Japan).
19. Pat. 1351954 (Gr. Brit.).
20. Pat. 2140226 (France).
21. Pat. 3644232, 3657161 (USA).
22. Pat. 3673128 (USA).
23. Alaminov K.A., Olygomeri i polymeri ot izacianourovoy kisloty. Autoref., Ph.D. Dis. Sofia, Chem.-technol. Inst., 1978.
24. Pearce E. M., Shalaby S. W., Barker R. H., in Flame retardancy of polymeric materials. Lewin, M. et al., eds. New York, London: Plenum Press, 1975, pp. 239-290.
25. Hirsch S. S., Chem. Eng. News, 1970, vol. 48, pp. 34-36.
26. Ognezashchitniye polymerniye materialy: index for domestic and foreign patents. M: OMVDT BNIPI Teploproekt, 1978.
27. Pushkaryova I. N., Ushkov V. A., Aseeva R. M. et al., Ognezashchishchyonniye polymerniye materialy: Express information. Ser. special construction works. M.: TsBNTI Minmontazhspetsstroy, 1979.
28. Osipova L. B., Barabanova A. V., Khim. prom-st za rubezhom, 1976, no. 6, pp. 3-38.
29. Pat. 3718615, 3816301 (USA).
30. Pat. 3725319 (USA); Pat. 2088266 (France).
31. Pat. 2119407 (France).
32. Pat. 2101453 (France).
33. Pat. 1421334 (Gr. Brit.).
34. Weil E. D., in Flame retardancy of polymeric materials. Kuryla W. C., Papa A., eds. New York: Marcel Dekker, 1975, vol. 3, ch. 3, pp. 185-244.

35. Agra B. V., Vasilyenko E. A., et al., Plastmassy, 1977, no. 2, p. 59.
36. Sanders H. J., Chem. and Eng., 1978, Apr. 24, p. 2237.
37. Pat. 3823097 (USA).
38. Pat. 3829532 (USA).
39. Pat. 3779953 (USA).
40. Pat. 3839266, 3840492 (USA).
41. Pat. 3705128 (USA).
42. Pat. 3326832 (USA).
43. Martin F. J., Price K. R., J. Appl. Polymer Sci., 1968, vol. 12, no. 1, pp. 143-151.
44. Sobolev Y., Woycheshin E. A., in Flammability of Solid Plastics. Hilado, C. J., ed. New York: Technomic Publ. Co., 1974, vol. 7, pp. 263-275.
45. Benbow A. W., Chalabi R., in Report on 5th Conference on Nehorlavost Polymernych Materialov. Bratislava, 1978, 7-9, Nov. 18.
46. Kisselev A. V., Lyghin V. I., Kolloid. zhurn. 1959, vol. 21, no. 2, pp. 581-586.
47. Little L., Infrakrasniye spectry adsorbirovannih molecule. Lyghin V.I, tr. from Eng., ed. M.: Mir, 1969.
48. Aseeva R. M., Berlin A. A., Meshikovsky S. M., Chem. zv., 1972, vol. 26, no. 2, pp. 397-403.
49. Plast Technol, 1976, vol. 22, no. 8, pp. 71-74.
50. Plast World, 1976, vol. 34, no. 7, pp. 46-60.
51. Monte S. J., Poliplasti, 1977, vol. 25, no. 236/237, pp. 59-61.
52. A. s. 373255 (USSR); B. I., 1975, no. 6.
53. Pat. 2635874, 3652488 (USA).
54. Konkin A. A., in Termozharostoykiye i negoryuchie kolokna. Konkin A. A., ed. M.: Khimiya, 1978, pp. 217-340.
55. Artemenko S. E., Vilkova S. P., Tiuganova M. A., Plastmassy, 1976, no. 5, pp. 75-77; 1978, no. 5, pp. 23-25.
56. Khalturinsky N. A., Artemenko S. E., Vilkova S. P., Tiuganova M. A., Plastmassy, 1978, no. 11, pp. 33-35.
57. Zhevlakov A. F. et al., Khim. volokna, 1976, no. 5, pp. 28-29.
58. Eagles T. B. et al., J. Appl. Polymer Sci., 1976, vol. 20, no. 2, pp. 435-448.
59. New Ind. and Appl. Mater., 1974, no. 2, pp. 609-627.

60. Howarth J. T., Sheth S. G., Sidman K. R., Dawn F. S., Flame resistant elastomeric polymer development: Final Report for 1 Apr. 1974 to 1 June 1975. N.Y.: NASA, Little A.D., 1975.

61. Kawaller S. I., Oil Gas J., 1973, vol. 71, no. 4, pp. 78-80.

62. Vandersall H. L., in Fire prevention and suppression. Hilado C. J., ed. N.Y.: Technomic Publ., 1974, vol. 10, pp. 141-184.

63. Lyons J. W., The chemistry and uses of fire retardants. N.Y.: Wiley Intersci., 1970.

64. VanKrevelen D. W., J. Appl. Polymer Sci., Appl. Polymer Symposia, 1977, vol. 31, pp. 269-292.

65. Spilda I., Paulik J., in Nehorlavost polymerniych materials. Bratislava: Dom Techniky CSVTS, 1976, pp. 87-89.

66. Gouinlork E. V., Porter J. F., Hindersinn R. R., in Surface flame spread. Hilado C. J., ed. N.Y.: Technomic Publ., 1973, vol. 5, pp. 86-98.

67. Ravey M., Fishler T., Fire and Materials, 1978, vol. 2. no. 2, pp. 80-83.

68. Batorewicz W., Highes K. A., Fire retardants: Proc. 1974 Intern. Symp. on Flammability and Fire Retardants. Bhatnagar V. M., ed., Westport: Technomic, 1975.

69. Pappova M., Spilda I., Volf J., Riska M., in Nehorlavost polymernych Materialov. Bratislava: Dom Techniky CSVTS, 1978, pp. 91-99.

70. Holve D. J., Sawyer R. W., Polymer Flame Retardant Mechanisms, Report ME-75-2, NBS N USDC-5-9003, Fed. 1975.

71. Hastie J. W., R. Res. NBS, Phys. and Chem., 1973, vol. 77A, no. 6, pp. 733-754.

72. Bostic Y. E. et al., J. Appl. Polymer Sci., 1973, vol. 17, no. 2, pp. 471-482.

73. Robertson A. F., Fire and Materials, 1976, vol. 1, no. 1, pp. 9-13.

74. Johnson P. R., J. Appl. Polymer Sci., 1974, vol. 18, no. 2, pp. 491-504.

75. Shnurova T. E. et al., Khim. prom-st za rubezhom, 1976, no. 1., pp. 18-38.

76. Martirosova A. I. et al, Khom porm-st za ruhezhom, 1976, no. 2, pp. 42-55.

77. Kostandov L. A., ZhVKhO, D. I., Mendeleev, 1976, no. 1, pp. 2-4.

78. Tverskaya L. S., Khim. prom-st za rubezhom, 1976, no. 3, pp. 52-63.
79. Winter H., Chem. Age, 1976, vol. 112, no. 2964, pp. 3-4.
80. Pokrovsky L. I. et al., Plastmassy, 1975, no. 10, pp. 4-8.
81. Kruglov B. I., Plastmassy, 1978, no. 5, pp. 53-55.
82. Chem. Age, 1975, vol. 110, no. 2904, p. 12.
83. Sihrota A. G., Modifikatsiya structury i svoystv polyolephynov. L.: Khimaya, 1969.
84. Schwarz R. J., in Flame retardancy of polymeric materials. Kuryla W.C., Papa A., eds. N.Y.: Marcel Dekker, 1973, vol. 2, ch. 2, pp. 84-133.
85. Pat. 3853811 (USA).
86. Samoylova S. M., Rutner B. R., Yankova S. D. et al., Plastmassy 1976, n. 5, pp. 8-11.
87. Pat. 51-19856 (Japan).
88. Vlasov A. V., Komarov, L. I., Korshak V. V. et al., Dokl. AN USSR, 1970, vol. 193, no. 3, pp. 615-617.
89. Kawase K., Hayakawa S. J., Polymer Sc., 1969, vol. 7A, pp. 3363-3374.
90. Pat. 1495694 (Fed. Rep. Germany).
91. Pat. 50-02785 (Japan).
92. Pat. 47-35553 (Japan).
93. Pat. 45-18469 (Japan).
94. Pat. 48-01089 (Japan).
95. Robertson A. B. et al., in Advance fire retardants. Westport (Conn.), 1972, pt. 1, pp. 64-73.
96. Pat. 48-18099 (Japan).
97. Pat. 3827997 (USA).
98. Pat. 2522653 (Fed. Rep. Germany).
99. Pat. 2236435 (Fed. Rep. Germany).
100. Pat. 2303044 (France).
101. Pat. 3926876 (USA).
102. Pat. 3930104 (USA); Pat. 2300109 (France).
103. Wiesner E., in Nehorlavost polymernych materialov. Bratislava: Dom Techn. CSVTS, 1967, pp. 100-107.
104. Pat. 3931081, 3931101, 3943194 (USA); Pat. 2185641 (France).
105. Pat. 47-29774 (Japan).

106. Pat. 3915910 (USA).
107. Pal G., in Nehorlavost polymernych materialov. Bratislava: Dom Tech. CSVTS, 1978, pp. 69-80.
108. Nagy P., Szöllösi I., Pal G., Fire and Materials, 1978, vol. 2, no. 2, pp. 63-67.
109. Delfosse L., J. Macromol. Sci. Chem., 1977, vol. 11A, no. 8, pp. 1491-1501.
110. O'Mara M. M. et al., in Flame Retardancy of Polymeric Materials. Kuryla W. C., Papa A., eds. N.Y.: Marcel Dekker Inc., 1973, vol. 1, pp. 125-275.
111. Pat. 3792113 (USA).
112. Pat. 3729436 (USA).
113. Pat. 3691127 (USA).
114. Ueno T., Kobunsi Kako, 1971, vol. 20, pp. 526-530.
115. Ignatova G. F., Trizno V. L., Nikolaev A. F. et al., Plastmassy, 1974, no. 7, pp. 39-41.
116. Ilyin S. N. et al., Plastmassy, 1973, no. 10, pp. 76-78.
117. Gorshkov V. S., Berlin A. A., Kotova A. V. et al., Vysokomolec. soyed., 1978, vol. 20A, no. 6, pp. 1368-1372.
118. Pat. 2310252 (Fed. Rep. Germany).
119. Polystyrolniye plastiky: Catalogue/Handbook. Tcherkassy, NIITEKhIM, 1973.
120. Fyjii H., Japan Plastic Age, 1974, vol. 12, no. 2, pp. 15-26.
121. Lindemann R. F., in Flame retardancy of polymeric materials. Kuryla W. C., Papa A. S., eds. N.Y.: Marcel Dekker, 1973, vol. 2., ch. 1, pp. 2-81.
122. Pat. 2127080 (Fed. Rep. Germany).
123. Zhubanov B. A., Nazarova S. A., Karzhaubova R. G., Ghibov K. M., Vysokomolec. soyed., 1976, vol. 18B, no. 3, pp. 150-152.
124. Plast. Ind. New Jap., 1969, vol. 15, no. 8, pp. 93-94.
125. Beylina V. I., Kolosova T. O., Gefter E. L. et al. Plastmassy, 1977, no. 11, pp. 32-34.
126. Pat. 2213293 (France).
127. Pat. 2169575 (France).
128. Pat. 2209780 (France).
129. Eng Pi Chang, Kirsten R., Slagowskii E. L., J. Appl. Polymer Sci., 1977, vol. 21, no. 8, pp. 2167-2180.

130. Benbow A. W., Cullis C. F., in Combustion Institute, European Symposium. London; New York: Acad. Press, 1973, pp. 183-188.
131. Deshpande A. B., Pearce E. M., Yoon H. S. et al., J. Appl. Polymer Sci., Appl. Polymer Symp., 1977, vol. 31, pp. 257-268.
132. Pat. 1437363 (Gr. Brit.).
133. Lawton E. L., Setzer C. J., in Flame retardant polymeric materials. Lewin M. et al., eds. N.Y., L.: Plenum Press, 1975, ch. 4, pp. 193-221.
134. Pat. 47-32430 (Japan).
135. Pat. 1435162 (Gr. Brit.).
136. Pat. 1435824 (Gr. Brit.).
137. Laskevich B., Levandovsky Z., in Nehorlavost polymernych materialov. Bratislava: Dom Techniky CSVTS, 1978, pp. 108-118.
138. Pat. 1438153 (Gr. Brit.).
139. Pat. 1433210 (Gr. Brit.).
140. Pat. 1429038 (Gr. Brit.).
141. Pat. 1447785 (Gr. Brit.).
142. Frunze T. M., Korshak V. V., Kurashov V. V. et al., Izv. AN USSR, OKhN, 1958, no. 6, pp. 783-785.
143. Pat. 48-20040 (Japan).
144. Pat. 3349062 (USA).
145. Pat. 3448086 (USA).
146. Pat. 3929734 (USA).
147. Mukhin B. A., Rogovin Z. A. et al., Khim. volokna, 1973, no. 1, pp. 17-19.
148. Usmanov K. U., Ulchibayev A. A., Syrlibaev T., Izv. Vusov. Khim. i Khim. teckhnologiya, 1969, vol. 12, no. 3.
149. Pat. 2015100 (France).
150. Pat. 49-35332 (Japan).
151. Pat. 3808289 (USA).
152. Pat. 2255337 (France).
153. Struszczyk H., Laszkiewicz B., in Nehorlavost polymernych materialov. Bratislava: Dom Techn. CSVTS, 1978, pp. 25-35.
154. Lin R. Y., Economy J., J. Appl. Polymer Sci. Appl. Polymer Symp., 1973, vol. 21, pp. 143-152.
155. Economy J., Worker L. C. et al., J. Appl. Polymer Sci., Appl. Polymer Symp., 1973, vol. 21, pp. 81-88.

156. A. s. 337993 (USSR) B. I., 1972, no. 15.
157. A. s. 272548 (USSR) B. I., 1970, no. 19; A. s. 260175 (USSR), B. I., 1970, no. 3.
158. A. s. 322347 (USSR), B. I., 1971, no. 36.
159. A. s. 179920 (USSR), B. I., 1966, no. 6.
160. Sunshine N. B., in Flame retardancy of polymeric materials. Kuryla W. C., Papa A. J., eds. N.Y. Marcel Dekker: 1973, vol. 2, ch. 2, p. 201-228.
161. Conley R. T., Quinn D. F., in Flame retardant polymeric materials. Lewin M. et al., eds., N.Y.; L.: Plenum Press, 1975, ch. 8, pp. 337-369.
162. A. s. 295261 (USSR), B. I., 1971, no. 7.
163. Pat. 3907723, 3915905 (USA).
164. A. s. 215485 (USSR), B. I., 1968, no. 2.
165. A. s. 281815 (USSR), B. I., 1970, no. 29.
166. A. s. 247501 (USSR), B. I., 1969, no. 22.
167. A. s. 519442 (USSR), B. I., 1976, no. 24; Pat. 1359683 (Gr. Brit.)
168. A. s. 249640 (USSR), B. I., 1969, no. 22.
169. A. s. 321531 (USSR), B. I., 1974, no. 35.
170. A. s. 462846 (USSR), B. I., 1975, no. 9.
171. Mayzel I. L., Kamenetsky S. P., Kalyinin V. I., in Ulucksheniye svoystv phenolnogo penoplasta/Sb. Trudov VNIPI Teploprogekt, 1976, ad. 43, pp. 69-77.
172. A. s. 389120 (USSR), B. I., 1973, no. 29.
173. Pat. 3719639 (USA).
174. Pat. 2102029 (France); Pat. 1359014 (Gr. Brit.).
175. Pat. 303395 (Australia).
176. Pat. 3865757 (USA).
177. Pat. 1398529, 1330254 (Gr. Brit.).
178. Pat. 3704256 (USA).
179. Papa A. J., in Flame retardancy of polymeric materials. Kuryla W. C., Papa A. J., eds. N.Y.: Marcel Dekker, 1973, vol. 3, ch. 1, pp. 2-133.
180. Papa A. J., Proops W. R., J. Appl. Polymer Sci., 1972, vol. 16, no. 9, pp. 2361-2373.
181. Kresta J. E., Frigew K. C., J. Cell. Plast., 1975, vol. 11, no. 2, pp. 68-73.

182. Pat. 1279205 (Gr. Brit.); Pat. 2149075 (France).
183. Pat. 1273072, 1273071, 1316198 (Gr. Brit.).
184. Pat. 1368931 (Gr. Brit.); Pat. 2182199 (France).
185. Pat. 3455850, 3826740 (USA).

Subject Matter Index

A

Ablation,
- carbonising polymer 169
- non-stationary and stationary 169

Absorptive power,
- polymer materials 74

Activation energies,
- thermal degradation 124

Activation energy,
- elementary reactions 136
- function of pyrolysis temperature 153
- reaction between carbon and steam 177

Addition polymers,
- pyrolysis of 130

Adiabatic self-ignition 19

Alkali metal salts and salts of alkali earth metals,
- inhibitor action of 222

Amidoimides 314

Amorphous carbon 144, 170

Antimony oxides,
- used as synergists of halogen-containing flame retardants 225

Antipyrene,
- flame retardant 329

Antipyrines, 212

Aromatic complex polyethers (polycarbonates, polyacrylates),
- pyrolysis of 158

Aromatic polyamides,
- calculations of bond stability 160
- combustibility 159
- effective activation energy of thermal destruction of 161
- oxygen index 159
- pyrolysis of 159
- resistance to flame ignition of 160
- thermal stability 159, 160

Aromatic polyheteroarylenes,
- flame resistance 163
- thermal stability of 163

Aromaticity index 163

Aromatisation reactions,
- pyrolysis 142

Arrhenius law 14

B

B number 72

Biot number 15

Blue flame zone 76

Breakdown temperature 39

Burning polymer,
- flame of 70

Burning surface,
- shape and size of 279
- shape and structure of 280

Burning wires,
- method of 278

C

Candle flame 63

Candle-type polymer diffusion combustion 109

Carbamide resins 359
- chemical modification of 360

Carbon,
- oxidation and gasification of 170
- oxidised by molecular oxygen 173
- vaporisation of 170

Carbon black,
- formation of 63

Carbon fibres,
- ignition of 27

Carbon monoxide,
- combustion of 208
- ignition of 208

Carbon monoxide oxidation,
- chain process of 208
- mechanism of 208

Carbon oxidation,
- formation of products from 174
- general reactions of 174
- high temperatures 175

Carbon-oxygen complexes 171

Carbon particles,
- ignition of 27

Carbon polyacrylonitrile fibres 146

Carbon skeleton structure,
- formation of 144

Carbon structure,
- model of 172

Carbon vaporisation,
- superficial complexes formed during reactions of carbon with gases 171

Carbonisation,
- inclination towards 165
- polymers 105
- pyrolysis 141

Carbonisation products,
- formation of nonvolatile 127

Carbonised cellulose products,
- X-ray diffraction analysis of 148
Carbonised layer 4
- gasification of 168
- morphological structure of 169
- porosity of 169
Carbonised products,
- polyvinyl chloride and polyvinylidenechloride 144
- transformation of polymers into 150
Carbonised residue,
- ability to form a 166
- yield of, during pyrolysis 149
Cellulose,
- char residue 231
- combustibility of 147
- effect of phosphorus flame retardants on 231
- reducing the flammability of 147
- reduction in the combustibility of 231
Cellulose acetate,
- self-ignition performance temperature of 41
Cellulose-based char materials 148
Cellulose dehydration,
- presence of phosphorus compounds 231
Cellulose materials,
- incomplete combustion of 48
Cellulose thermal degradation,
- decomposition products of 147
Chain-reaction self-ignition 13
Char,
- cellulose 173
- chemisorption of oxygen 173
- formation of soot deposited on 173
- paramagnetic centres 173
- physical adsorption of oxygen 173
- polyvinyl benzene 173
- polyvinylchloride 173
- polyvinylidene chloride 173
- thorough combustion of 173
Char formation,
- contribution of substituent groups to 149
Char layer,
- characteristics of 168
- efficiency as a heat barrier 263
- oxidation of 48
- participation in combustion reactions 168
Char number,
- bakelite "A"-based polymer 154
Char numbers 165
- calculated 166
- experimental and calculated values 168
Char residue,
- correlation between the yield during pyrolysis and the oxygen index of the polymers 165
- determination of the thermochemical yield of 155
- thermochemical yield of, during pyrolysis 154
Char yield 148
- novolac oligomer 155
- pyrolysis 155
Chars,
- brown coal 177
- phenol-formaldehyde polymer 154
Chemical energy 72
Chemical modification,
- in bulk 315
- in bulk with a view to reducing flammability 316
- polymers to reduce flammability 315
- superficial 315
Chemical processes,
- combustion of polymeric materials 119
- condensed phase 120
Chemical reaction,
- initiation time 36
Chemical transformation reactions,
- polymer materials at high temperatures 126
Chemoinactivation 241
Coatings,
- based on magnesium oxychloride 327
- based on mineral binding materials 327
- heat insulating 327
- nonswelling 327
- swelling 327
- to reduce the flammability of fibrous, film and cellular polymer materials 327
Combustibility 90
- chemical aspects of the reduction in 267
- how structural factors affect the 264
- methods of testing 295
- polymer materials 107, 110
- polymers 164
- predicted on the basis of chemical structure 166
- relative determination of 66
Combustibility of polymer materials,
- reduction in 212
Combustion,
- and motion 72
- chemistry of 119
- conversion reactions in the condensed phase 62
- counterflow of gaseous oxidiser 106
- critical conditions of 8
- definition of 1
- determination of the capacity of materials to propagate 294

- heterogeneous 3
- homogeneous 3
- mechanism and kinetic principles 278
- methods for the study of 277
- methods for the study of the processes that occur within the surface layers during 288
- nonvolatile condensed substances 4
- PMMA spheres 85
- spherical polymer particles 103
- theoretical studies of 63
Combustion characteristics,
- determination of 277
Combustion intensity 66
Combustion models,
- quasi-unidimensional 8
- three-dimensional 8
- two-dimensional 8
- unidimensional 8
Combustion of polymer materials,
- disappearance of the condensed phase during 124
- in an oxygen atmosphere 71
- modelling 304
Combustion of polymers,
- formation of condensed products during 194
- formation of soot 194
Combustion process,
- influence of gravitational forces 111
- mathematical analysis of 103
- mathematical description of 8
- nature of the metals in the recombination of active centres in the 224
Combustion products zone 60
Combustion rate 72
- large-sized polymer samples 83
- polymer materials 61
Combustion rate constants,
- spherical particles of polymer materials 105
Combustion regime 28
Combustion retardant agents,
- use of inorganic compounds 212
Combustion source diameter 6
Combustion tests,
- full-scale 304
- small-scale 304
Combustion wave 3
- leading zones 9
- methods used for the determination of temperature within 282
- structure of 3, 60
- when carbonised polymers burn 169
Complete combustion,
- gaseous fuel 35
Composite materials 324
- modified polyamides used for 357
Composites 323
Condensed substance combustion,
- nongaseous during combustion 4
- vaporised during combustion 4
Convectional heat exchange 73
- in a gaseous phase at the edge of the flame front 94
Convectional thermal flow 36
Cool-flame,
- secondary 50
Cool flame ignition 42
Cool-flame stage 42, 61
Cool-flame zone 42, 61
Cotton,
- self-ignition performance temperature of 41
Critical combustion conditions 8
Critical overload value 112
Critical pressures 110
Critical self-ignition 23
Critical temperature 23
Crosslinking reactions,
- pyrolysis 141

D
Damkeler number 87, 94, 334
Damkeler's number 110
Degradation,
- chain transfer 131
- initiation 131
- kinetic models 132
- linear polymethylmethacrylate and network polymer 122
- reverse polymerisation 131
- role of oxygen 120
- termination 131
- thermal-oxidative 120
Degradation of polymer materials,
- during diffusion combustion 62
Degradation of polymers,
- high-temperature, during combustion 4
Degradation temperature 39
Degree of flammability,
- evaluation of 290
Dehydration reactions 153
Dehydrochlorination,
- effective activation energy of 143
- polyvinyl chloride 143
Depolymerisation,
- mechanism of 131
Destructive degradation 39
Diffusion combustion,
- fuel drops 104
- mass and heat transport 66
- polymer materials 7, 61

Diffusion combustion,
- polymer spheres 104
- reactions of condensed phases during 63
- theory of 7
Diffusion combustion model 37
Diffusion combustion stability,
- effect of the size of polymer samples 112
Diffusion flame,
- analysis of 106
- composition of gaseous products of polymethylmethacrylate 179
- stability of 108
Diffusion flame spreading,
- model of 99
Diffusion flames,
- calculated adiabatic temperatures of 76
- composition of the gas within 178
- geometric configurations of 278
- maximum temperature of 77
- measured maximum temperatures of 76
- optical properties 74
- polymers 178
- polyolefins in a counterflow of oxidising gas 185
- sampling and analysis of gases from 191
- stability limitations on 108
- stability limits on 110
Diffusion flames of polymer materials,
- optical properties along the height of 83
Diffusion regime,
- combustion 5
Dispersion,
- particles 5
- polymer particles in the gaseous phase 122
Dispersion processes,
- during the combustion of the decomposing condensed substances 4
Disruption of combustion 112
Downwards combustion,
- vertical polymer samples 106

E

Effective activation energy,
- phenol-formaldehyde polymer pyrolysis 151
- self-ignition 41
Effective heat of vaporisation 67
Efficiency coefficients,
- bromine and chlorine atoms in the reduction of polyolefin flammability 343
Elementary combustion model 65
Elimination rates,
- hydrogen halides 142
Energy of formation,
- carbon compounds 192
Environmental flow,
- hydrodynamic picture of 72
Epoxy plastics,
- flammability of 325
Epoxy polymer,
- analysis of the products along the surface of 191
Epoxy polymers,
- ignition index for 229
- metal-containing compounds 228
- to reduce the combustibility of 228
Epoxy resin,
- composition of gases during candlelike combustion of 190
- concentration profile of products along the axis of the diffusion flame 190
- diffusion flames of 190
Evaporation enthalpy 67
Exothermal reaction,
- kinetic parameters of 17
Exothermal reactions,
- self-acceleration of 13
Experimental methods,
- study of the combustion of polymer materials 277
External heating 28
Extinction criteria 110
Extinction rates,
- burning polymer spheres 122

F

Fillers,
- carbon fibre 325
- common organic fibrous 325
- inorganic-type 322
- modification of 324
- modified by organotitanates 324
- modified organic 322
- multifunctional 322
- noncombustible thermostable organic 322
- thermostable fibrous 326
- used in order to reduce the flammability of polymer materials 322
Fire hazard,
- combustible and polymer materials 303
Fire index,
- methods for calculating criteria of flammability 302
Fire modelling 277
Fire prevention,
- international organisations 289
- norms and regulations 291

Fire simulation 277
Fire tests,
- modelling full scale 9
- modelling large scale 9
Fireproof coatings 312, 326
- swelling type, form carbonised foam layers 327
Fires,
- optical properties of 83
Flame,
- appearance of 12
- chemoionisation of hydrocarbon fragments inside a 202
- concentration profiles of products inside a 284
- definition of 2
- effect of flame retardant extinction 334
- formation of charged particles in a 202
- front of 6
- fuel gas composition inside the 191
- heat dissipation by 68
- heterogeneities inside a 281
- hydrogen 206
- inhibition of a 215
- maximum temperature in the 181
- measuring the absorption and radiation spectra of 285
- model of 94
- of burning polymethacrylate 181
- shape and structure of 280
- spherical 103
- unidimensional 103
Flame analysis,
- methods of 285
Flame characteristics,
- temperature profile 281
Flame diffusion,
- de Ris' theory of 86
- model for 85
- over the surface of a polymer material, theoretical models of 91
- over the surface of thermally thick samples 97
- rate of 299
- two-dimensional 85
Flame diffusion index 299
Flame diffusion over the surface of a polymer material,
- change in character of the gaseous flow for a change in the orientation of 99
Flame diffusion rate,
- over the surface of polymethylmethacrylate sheets 100
Flame extinction 106
- limiting conditions of 107
Flame front 2
Flame inhibition,
- by metal-containing compounds 223
- by sodium salts 222
- phosphorus compounds 236
Flame luminosity 195
Flame properties,
- experimental methods applied in determining 6
Flame quenching 87
- analysis of 88
Flame reaction zone 60, 63
Flame reactions,
- formation of carbon particle nuclei 201
- in the zone of 201
- participation of phosphorus compounds 237
Flame retardant,
- affects the stability of a flame 335
- effect on the process in the C phase during the combustion of polymers 336
- effectiveness determined from a change in mass-transfer number B 334
- mixtures 336
- nonreactive type 214
- reactive type 214
Flame retardant action,
- halogen-containing organic compounds 214
- mechanism of 214, 317
Flame retardant effectiveness,
- activation energy of the gas-phase combustion reaction as a quantitative measure of 335
- criteria for evaluating 329
- measure of 329
Flame-retardant fillers,
- aluminium hydroxide, ammonium pyrophosphate and hydrated metal carbonates 323
- aluminium hydroxide and calcium carbonate 344
- effectiveness of 323
Flame retardants 212, 317
- action mechanism of 213
- additive-type 318, 340
- additive-type, used to reduce the flammability of polyurethane materials 363
- blending with nonpolar polymer substrates 342
- capable of polymerisation 348
- classification of 214
- effect of different metal oxides as synergists 227
- effect on self-ignition temperature 45

Flame retardants,
- effect on the rate of stationary combustion, during convective heat transfer and under conditions of flame quenching 334
- effectiveness in polyamides 356
- efficiency coefficient a of 331
- fluoroelastomer coatings 327
- for reducing the flammability of carbamide and phenol-formaldehyde resins 361
- halogen-containing 268
- inert 318
- inert or additive-type 312
- inorganic derivatives of phosphoric acids 233
- metal-containing compounds 218
- metallic compounds 271
- neoprene-based coatings 327
- nonlinear nature of the relationship between the oxygen index and the concentration of 333
- organic derivatives of phosphorus-containing acids 234
- phosphorus and its compounds as 230
- phosphorus-containing 270
- plasticising effect of 320
- polymeric halogen-containing 321
- reactive 312, 318, 321, 340
- reactive phosphorus 349
- stable 342
- use of 312
- use of metallic complexes with organic ligands, containing phosphorus, nitrogen, oxygen and other atoms with unshared electrons 322
- used as fillers 320
- volatile metal halides 273
- with multifunctional effects 319
- with polyolefins 341

Flame retarder,
- effectiveness of 107

Flame spread,
- directions of 91
- orientational effects 111
- over a vertical surface 99

Flame spread upwards,
- theoretical analysis of 102

Flame spreading 75
- on the surface of polymer materials 90

Flame temperature 76

Flameless combustion 47

Flames,
- aura 64
- dark zone 64
- ionic process inside 203
- laminar 5
- luminous reacting layer 64
- luminous zone 64
- mixed or premixed 5
- monomer-polymer degradation products 268
- turbulent 5

Flaming reaction,
- gaseous phase, kinetic parameters of 110
- temperature coefficient of 107

Flammability,
- aliphatic polyamides 356
- aromatic polyamides 356
- characteristics of copolymers of vinyl chloride 345
- characteristics of nonplasticised PVC 345
- classification of materials 295
- dust particles from a polymer material 104
- effect of flame retardants on 329
- estimation methods for 288
- estimation of 277
- materials 53
- methods for the evaluation of 289
- plasticised PVC 345
- polymer materials used in electronics and instrumental fabrication 297
- polymeric materials 119
- prediction of 304
- reduction of 309
- reduction of, by the use of mixtures of various flame retardants and their synergists 319
- standard methods of testing 298
- testing of 289

Flammability properties of polymer materials,
- analysis of 82

Flammability reduction,
- chemical methods 309
- physical methods 309

Flammability testing 292
- cellular plastic materials 291
- fabrics 296
- floor coverings 290
- plastic materials and foam plastics 296

Flashing 12

Flat flame,
- device for the generation of 279

Fluctuating flame 50

Foamed polyurethane,
- incomplete combustion of 48

Focus ignition 11

Fog,
- observed in an acetylene flame 201

Forced ignition 11

Formaldehyde,
- oxidative degradation of 211

Formation energy,
- carbon compounds 192
Fourier numbers 306
Frank-Kamenetskii parameter 15, 18
Froude number 304
Fuel 5
- oxidation reaction of 64
Fuel content,
- combustible and polymer materials 303
Fuel degradation products,
- flow of 64
Fuel diffusion rate,
- over the horizontal surface of a thermally thin material 96
Fuel gases,
- generation of 62
Fuel to oxidiser ratio 76
Fusion enthalpy 67

G
Gas-dynamic characteristics,
- in the case of diffusion combustion 287
Gas phase,
- chemical process in the 177
Gas phase combustion,
- inhibition of 217
Gas phase combustion reaction,
- Arrhenius equation for 88
Gas phase flame reactions,
- low-molecular components 119
Gas phase ignition,
Gas phase ignition 48
- theoretical analysis of 53
Gas phase reactions,
- kinetics and mechanism of degradation of different carbon compounds 193
Gaseous flow rate,
- near the edge of a polymer flame 287
Gaseous phase,
- chemical processes in the 265
Gasification,
- polymer material 34
Gasification of carbon,
- carbon dioxide 176
- presence of steam 176
Gasification temperature 39
Glass carbon 170
Graphite 144, 170
Graphitisation 144
Grashof number 23, 305

H
HBr,
- inhibitor action of 217
HCl,
- inhibitor action of 217
Heat,
- liberated during combustion 299
Heat absorption,
- rate of 16
Heat balance 108
Heat emission 73
Heat energy 72
Heat exchange 20
- between a solid substance and a gas 73
Heat flow 18
Heat losses,
- through the condensed phase 88
- through the gaseous phase 88
Heat of combustion 67, 263
- polymer materials 69
Heat of gasification,
- polymer materials 69
Heat of polymerisation,
- criterion of polymer behaviour during heating 127
Heat of reaction 12
Heat transfer 12, 20
- acceleration parameter 103
- by heat conductivity over the surface of cellulose samples 101
- from the flame 96
- gaseous and condensed phases 100
- initiated by a purely mechanical motion process in the hot polymer melt 78
- laboratory tests 80
- through heat conductivity 73
- through heat conductivity, convection, radiation 73
- through heat conductivity in the condensed phase 94
- use of external radiation 82
Heat transfer coefficient 67, 73
Heat transfer mechanism,
- by means of radiation 102
- determination of 75
Heat transfer parameters,
- polymer combustion 79
Heat wave 17
Heating,
- Initiation by radiant flow 34
Heterochain polymers,
- thermal degradation of 139
Heteroelement polymers,
- synthesis of 313
Heterogeneous ignition 32
- carbon 33
- induction period of 31
- solid combustion materials 28

Heterogeneous, ignition
- thermal model of 28
Heterogeneous ignition model 28
Heterogeneous oxidation,
- carbon 170
- polymer materials 50
Heterogeneous reaction,
- carbon and gases 170
- on the surface 30
- transition from a kinetic to a diffusion regime 30
High-temperature degradation,
- condensed substances 120
Hissing surface zone 60
Homogeneous combustion,
- vaporisation products in the gaseous phase 4
Hydrocarbon flames 202
Hydrocarbon fragments 63
Hydrocarbons,
- active centres during the oxidation of 209
Hydrodynamics,
- polymer combustion 79
Hydrogen,
- combustion of 205
Hydrogen combustion,
- elementary reaction of 207
- kinetic characteristics 207
- mathematical modelling of the chemical process of 206
Hydrolytic stability,
- heterocycle polymers 150
Hypergolic heterogeneous ignition 27
Hypergolic ignition 28

I

Ignitability,
- comparative analysis of 165
- determination of 292
- materials 53
- measure of 110
- oxygen index (OI) method to test 293
- polymer materials 107, 110
Ignition,
- asymptotic method to analyse 54
- condensed phase models 22
- criteria for 13
- critical conditions of 8
- destabilisation of 53
- flow of hot gas 20
- gaseous phase 33
- heterogeneous regime of 28
- incandescent material 21
- induction period 39
- induction period of 35
- limitations on 39
- models in the gaseous phase 37
- models in the gaseous state 34
- PMMA 52
- polymers 44
- radiant energy flow 22
- solid fuel 37
- temperature dependence 39
- thermal theory of 17
- through radiant heat flow 51
- time delay 39
- under the influence of a radiant thermal flow 43
Ignition characteristics,
- experimental studies 38
Ignition criterion 30, 37
Ignition induction period,
- polymers 43
Ignition limits,
- calculated 55
- study of 51
Ignition mechanism,
- effect of different admixtures on 51
- solid phase 21
Ignition model,
- unidimensional, of a semi-infinite combustion plate 33
Ignition number 30
Ignition parameters 35
Ignition process,
- gaseous phase 39
Ignition source,
- energy of 53
Ignition temperature 20
- determination of 292
- effect of heating rate 261
- polymers 46
Ignition time delay 21
- calculations of 55
Incombustible polymers,
- synthesis of 313
Incomplete combustion,
- causes of 68
- initiation of 47
- initiation period for 50
- mechanism of 48
- model of 49
Induction period 32
- ignition 13
- self- ignition 13
Inert flame retardants,
- halogen-containing compounds 320
- inorganic compounds: oxides, hydroxides and metal salts, derivatives of phosphoric acids and halogen-containing compounds 318

Inert flame retardants,
- organic: borates, fluoroborates of ammonium, alkali and alkali earth metals 318
- organic derivatives of phosphoric acids 320
Inhibitor action mechanism,
- metal-containing compounds 220
Initiation of combustion,
- mathematical description of 27
Inorganic polymers,
- synthesis of 313
Interface,
- solid fuel and gaseous oxidising agent 27
Intermolecular crosslinking actions,
- pyrolysis 149
Intermolecular cyclisation,
- statistical theory 146
Intramolecular cyclisation,
- pyrolysis 142, 149
Isomerism,
- heterocycle polymers 150

K

Kinetic chain length 133
Kinetic regime,
- combustion 5
Kinetics and mechanism,
- phenol-formaldehyde polymer pyrolysis 151
Kirchhoff's law 74

L

Ladder and three-dimensional network structures 150
Ladder polymers 145
Laminar diffusion combustion,
- chemical transformation during 123
- polymer materials 64
Laminar diffusion flame,
- maximum temperature of 86
Laminar diffusion flame spread,
- model of 109
Laminar diffusion flames,
- condensed substances 5
Laminar flame structure 6
Leading zone 61
Lewis number 67, 107
Lewis parameter 29
Limiting oxygen concentration 109
Limiting oxygen index 109
Linear pyrolysis (LP),
- kinetic parameters of 123
Linear pyrolysis method 120
Linear pyrolysis rate (LP rate) 121
Linear rate of combustion,
- determination of 278
Luminous flames,
- soot formation 69

M

Macroradical 130
Macroradical degradation,
- activation energy of 138
Macroscopic combustion characteristics 7
Macroturbulence 6
Mass flow of fuel,
- inside the flame zone 100
Mass rate of polymer combustion 84
Mass transfer coefficients 67
Mass transfer parameter, '
- for combustion in air 71
Mass transfer parameters,
- in the combustion of polymer materials 72
Maximum adiabatic temperature of combustion 88
Maximum flame temperature,
- polymer materials 76
Melamine-formaldehyde polymers,
- char layer 359
- temperature of self-ignition 359
- toxic products 359
Mesophase formation,
- carbonisation 144
Metal-containing compounds,
- effectiveness of combustion retardation 219
- mechanism of flame process inhibition 219
Methane,
- combustion of 209
- high temperature combustion of 209
Method of successive approximations 26
Michelson's equation 284
Microturbulence 6
Modelling,
- atmospheric pressure 304
- combustion process 9
- elevated pressure 305
- geometric similarity 305
- hydrodynamic similarity 305
Modelling methods,
- based on similarity theory 304
- full-scale combustion of polymer materials 304
modelling of fires 82
Molecular weight distribution,
- during degradation 131
Multi-phase combustion 60

N

Neutron activation analysis,
- detection of oxygen in burning polypropylene 125
Newton formula 73
Newton's law 20
Nitrocellulose,
- forced ignition of 25
- ignition characteristics of 26
Nitrocellulose degradation 23
- kinetic parameters of 26
Nitrosofluorocarbon elastomers 313
Nomex,
- treated with a mixture of phosphorus oxychloride and bromine oxychloride followed by gaseous bromine 315
- treated with chlorine 315
- treated with sulfuryl chloride 315
Nonstationary temperature distribution 17
Nylon 6,6,
- self-ignition performance temperature of 41

O

Oligoarylenes,
- pyrolysis of 164
Oligoether-based polymer,
- self-ignition of 41
Optical methods,
- to study shape and size of the flame and the burning surface 280
Organic polymers that liberate primarily incombustible gases during decomposition,
- synthesis of 313
Oscillating cool-flame 51
Oxidant,
- flow of 64
Oxidation,
- carbon 171
Oxidation of graphite,
- oxidation rate constants 172
Oxidation reactions,
- pyrolysis products 180
Oxidative transformations,
- polymer under combustion conditions 125
Oxidiser 5
Oxidiser flow 106
- bidirectional 107
Oxygen,
- effect on characteristics of polymer combustion 82
- effect on polymer decomposition 259
Oxygen concentration 109
Oxygen index 8, 88, 106, 110
- homopolymers and copolymers of styrene 348
- polyethyleneterephthalate 352
- polyvinyl chloride 347
- thermoplastic polymers 332
Oxygen index (OI),
- dependence on the concentration of flame retardant 330
Oxygen indexes,
- experimental and calculated values 168

P

Performance time,
- chemical reaction inside the flame 99
PET destruction,
- radical process of 157
Phase contact area 3
Phenol-formaldehyde,
- carbonised product 359
- heterogeneous combustion (smouldering) 359
Phenol-formaldehyde polymer,
- pyrolysis products 152
Phenol-formaldehyde polymers,
- pyrolysis reactions 151
Phenol-formaldehyde resins 359
- chemical modification of 360
- reducing material flammability with inorganic fillers 361
Phenolic polymer,
- combustion of 104
Phosphorus and its compounds,
- for the production of low-flammability polymeric materials 229
Phosphorus compounds,
- inactivation of carbon oxidation active centres by 240
Phosphorus-containing polymer systems,
- nonvolatile residues after pyrolysis and combustion of 233
Phosphorus flame retardant action,
- mechanism of 270
Phosphorus flame-retardants,
- action mechanism of 229
- unique ability to extinguish the incomplete combustion reactions of the carbonised residues of polymeric materials 240
Phosphorus oxidation,
- mechanism of 235
Photochemical degradation reaction,
- heat of 35
Pilot ignition 51
PMMA,
- heat transfer from the flame to the surface of 82
- ignition time delay for 52

- kinetic parameters of the heterogeneous reaction of 33
- mechanism of reactive ignition of 51
- pilot ignition of 51
- quenching of flames 89
- rate of flame diffusion over the surface of 81
- rate of thorough combustion 85
- relationship between ignition time delay and energy of the radiant flow 53
- vaporisation 55

Polmer ignition temperature 44

Poly-ε-caproamide,
- thermal depolimerisation of 140

Poly-tert-butylmethacrylate,
- thermal degradation of 139

Polyacrylonitrile,
- cyclisation of 145
- self-ignition performance temperature of 41

Polyacrylonitrile cyclisation,
- effective activation energy of 145

Polyacrylonitrile degradation,
- volatile products of 145

Polyamide,
- flame retardants of a nonhalogen type 358
- induced self-ignition time 40

Polyamide (nomex),
- self-ignition performance temperature of 41

Polyamide (nylon 6,6),
- activation energy 124
- self-ignition temperature of 41

Polyamides 356
- flame retardants for the surface treatment of films, fibres or foams 359
- flame retardants introduced into the melt of 358
- flammability reduction through chemical modification and elimination of the aliphatic hydrocarbon fragments 357
- nitrogen-containing flame retardants 359
- oxygen index of 358
- phosphorus flame retardants 358
- phosphorylation of 357
- rate constant for the degradation of 140
- surface treatment with chlorine, bromine or sulfuryl chloride 357
- use of reactive flame retardants for the modification of 357

Polybenzimidazoles 314

Polybenzimidazopyrrolones 314

Polybenzoxazoles 314

Polycarbonate,
- combustion of 104

Polycondensation polymers,
- thermal degradation of 139

Polyene fragments,
- intramolecular and intermolecular reactions of 144

Polyester,
- fibres, fabrics and films surface-treated with reactive flame retardants 356
- reduction in combustibility of 353

Polyesters 352
- additive-type flame retardants 354
- direct bromination or chlorination of 353
- flame retardants 355
- phosphorus and its compounds used in reducing the combustibility of 354
- reactive flame retardants for 353
- reduced flammability through the use of halogen-containing or phosphorus-containing diols or by using dibasic acids and their esters 353
- self-extinguishing 354

Polyethers,
- pyrolysis of 156

Polyethylene,
- activation energy 124
- candle-type combustion of 78
- changes in enthalpy, entropy and free energy of pyrolysis 128
- chlorinated or brominated 340
- chlorosulfonation of 340
- composition of gases in the candlelike flame of 181
- concentration profiles of gaseous products in a diffusion flame of 182
- diffusion flame of 65
- diffusion flames 181
- flames of 63
- fluorinated 340
- formation of soot inside the flame 182
- stationary combustion of 79

Polyethylene combustion,
- composition of gases 183

Polyethylene diffusion flame,
- concentration profiles of gases 183

Polyethyleneterephthalate,
- combustion 352
- concentration profiles of gases along the axis of a plane (horizontal) diffusion flame of 188
- degradation products 156
- effective activation energy of PET pyrolysis 156
- laminar diffusion combustion of 187
- oxygen index 353
- pyrolysis of 156
- pyrolysis products of 187
- self-ignition performance temperature of 41

Polyformaldehyde,
- composition of gases inside the diffusion flame of 186
Polyheteroarylenes,
- thermal stabilities of 161
Polyhexamethyleneadipylamide,
- pyrolysis of 140
Polyhexazocyclanes 314
Polyimides 314
- analysis of the pyrolysis products of 162
- effective activation energy 162
Polymer combustion,
- chemistry of 7
- fire development 85
- initiation of 2
- large-scale tests 85
- mechanism of 3
- small-scale tests 85
Polymer combustion products 63
Polymer diffusion flames,
- optical characteristics of 83
Polymer flammability,
- role that the halogens play in the reduction of 269
Polymer foam materials,
- ignitability, combustibility and the ability to form smoke 165
- relative toxicity of the gaseous products 165
- thermochemical yield of the char residue 165
Polymer gasification,
- temperature of 108
Polymer ignition,
- gaseous phase 108
Polymer ignition by radiant flow,
- gas phase model of 53
Polymer material,
- ignition of 37
Polymer pyrolysis,
- effective kinetic parameters of 85
Polymer vaporisation,
- during the action of a stationary radiant flow 55
Polymers with a minimum organic fuel component 313
Polymers with aromatic and heterocyclic structures,
- synthesis of 313
Polymethacrylate,
- high-temperature pyrolysis of 181
Polymethylmethacrylate,
- activation energy 124
- activation energy of pyrolysis 138
- active pyrolysis zone for 120
- candle-type combustion of 78
- combustion of 104
- depolymerisation mechanism of 133
- diffusion combustion of 110
- diffusion flame 178
- flames of 63
- gas phase temperature profiles during the spreading of a laminar flame over the surface of 283
- heterogeneous ignition of 31
- ignition of 27
- temperature profiles of 75
Polyolefin,
- chemical modification 340
- methods used most frequently to reduce flammability in 340
- oxygen index of 228
Polyolefin-based materials 339
Polyolefin diffusion flames,
- composition of gaseous in 184
Polyolefin flame,
- maximum temperature inside the 185
Polyolefines,
- during the oxidation of 51
- self-ignition 42
Polyolefins,
- combustion of 183
- graft copolymerisation with reactive flame retardant monomers 341
- high molecular weight halogenated flame retardants for 343
Polyoxymethylene,
- activation energy 124
- candle-type combustion of 77
- concentration profile of gases inside the diffusion flame of 186
- diffusion flame near the melt surface 186
- effective activation energy of polymer thermal degradation 140
Polyperfluoroalkylenetriazines 313
Polyphenylenes 314
Polyphosphazines 313
Polypropylene,
- activation energy 124
- analysis of the degradation of volatile products 134
- combustion profiles 184
- composition of gases in the candlelike flame of 181
- containing chlorine 340
- effect of flame retardants 45
- products of decomposition 136
- qualitative composition of the products near the surface of burning 185
- reduction in flammability by using bromoaromatic compounds and by using phosphoric acids 343

- self-ignition of 42
- stationary combustion of 79
- substances containing halogen and phosphorus simultaneously in 343
Polypropylene diffusion flame,
- concentration profiles of gases 184
Polypropylene flame,
- relationship between concentration of oxygen in surrounding atmosphere and maximum temperature 78
Polypyromellitimide,
- thermal oxidative degradation of 163
Polyquinoxalines 314
Polysiloxane 313
Polystyrene,
- active pyrolysis zone for 120
- chemical modification of, by direct chlorination or bromination 350
- ignition delay for 55
- relationship between mass rate of thorough combustion and concentration of oxygen 84
- relationship between mass rate of thorough combustion and the intensity of the external heat flow 84
- relationship between the change in oxygen index and the concentration of flame retardant 352
- self-extinguishing 351
- self-ignition 42
- surface pyrolysis reaction of 56
Polytetrafluoroethylene 313
- changes in enthalpy, entropy and free energy of pyrolysis 128
- concentration profiles of gases along the axis of a diffusion flame of 189
Polytetrafluoroethylene(PTFE),
- composition of gases in flaming combustion 189
Polyurethane,
- activation energy 124
- modification with a view to reducing flammability 316
- use of flame retardants 362
Polyurethanes,
- chemical modification of 311
- elastic foam with reduced flammability 363
- flammability of 361
- liberation of smoke and toxic products 361
- modification of, aimed at reducing the flammability 362
- noncombustible rigid foam 316
- phosphorus polyols and polyetherpolyols used for purposes of reducing the flammability of 363
- self-extinguishing rigid foam 321
- self-quenching rigid foam 363
- use of light mineral fillers in combination with flame retardants and substances that decompose forming noncombustible gases 364
Polyvinyl chloride,
- flaming combustion of 195
- kinetics and mechanism of pyrolysis 143
- self-ignition 42
- smouldering and flaming combustion of 195
Polyvinyl chloride-based materials 344
Polyvinyl nitrate,
- combustion rate 65
Pre-flame zone 61
Pre-ignition time 40
Precombustion process 41
Preflame zone 63
- pyrolytic reactions of organic compounds inside the 191
Preflaming zone,
- carbon monoxide and carbon dioxide in the 180
- decomposition of gaseous products 179
Profilometer 85
PVC,
- chemical modification of 345
- chlorinated 345
- use of flame retardant fillers 346
Pyrolysis,
- carbochain polymers 131
- carbonisation of the polymer during 48
- cellulose 146
- characterised by the presence of aromatic carbocyclic and/or heterocyclic segments in the principal macromolecular chain 149
- chemical reactions taking place during 126
- deuterated and regular polyvinyl chloride 143
- formation of carbon during 128
- heterochain polymers 131
- kinetic parameters of 137
- nature of active centres in 129
- of a polymer 35
- phenolphthalein-based polyarylates 175
- polymers 149
- polymers that react through intermolecular processes involving the elimination of atoms or functional groups 141
- polymers which degrade by depolymerisation 131
- polyvinyl methyl ketone 146
- polyvinylidene chloride 144
- result of the fracture of principal chain bonds 130

Pyrolysis,
- subdivision of polymers into specified groups 127
- surface initiation temperature 36
- thermodynamic parameters of, for polyethylene and polytetrafluoroethylene 128
- various concentrations of oxygen 259
Pyrolysis enthalpy 67
Pyrolysis process,
- inside flames 200
Pyrolysis reaction,
- heat of 35
Pyrolysis resistance,
- polymer materials 49
Pyrolysis zone 120
Pyromellitic acid dianhydride 163

Q
Quenching,
- diffusion flames of fuel liquids 89
- theory of 89
Quenching of flaming combustion,
- effect of radiation on 88
- mechanisms for 268
Quenching of polymer materials,
- limits of 89
Quinoxalineimides 314

R
Radiant energy 72
- flow of 74
Radiant heat transfer 73
Radiant power values,
- hydrocarbon flames that do not form soot 79
Radiant thermal flow 36
Rate,
- combustion 278
Rate constants,
- bond breakage 133
Rate of carbon oxidation,
- atomic oxygen 177
Rate of combustion 61
- determination of 90
- linear or normal 62
Rate of depolymerisation 133
Rate of flame diffusion 94
- dependence on oxygen concentration 96
- dependence on pressure 96
- for indefinitely thick layers 93
- over the horizontal surface of cellulose 97
- over the surface of polymer materials 81
- over the surface of polystyrene and polymethylmethacrylate, effect of the oxygen concentration and pressure 97
Rate of flame radiation,
- large-sized polymer samples 83
Rate of flame spread 86
- for thermal thin material layers 93
Rate of heat transfer 12
Rate of heating 17
Rate of linear pyrolysis,
- impact-resistant polystyrene 124
- unsaturated polyether 124
Rate of mass change,
- stationary combustion process 279
Rate of polymer combustion,
- effect of an external heat source on 83
Rate of polymer material extinction 107
Rate of reaction 12
Rate of vaporisation,
- linear polymethylmethacrylate and network polymer 122
Reactive flame retardants,
- containing phosphorus, nitrogen, halogen, boron, silicon and other elements to reduce the flammability of carbamide and phenol-formaldehyde resins 360
- halogen-containing 349
- halogenated ethers of phosphoric acids 321
- preparation of polyurethanes with reduced flammability 362
- used for the chemical modification of styrene polymers and copolymers 348
- with phosphate, phosphonate and phosphite groups 321
Redox process,
- inside the flame front 6
Reduced flammability,
- copolymers derived from ethylene and esters of phosphoric acid 340
- design of polymer materials with 310
- impact-proof styrene plastics with 350
- polymer materials with 338
Reflective power 74
Retardation of combustion,
- hydrogen or carbonic flames 219
Reynold's number 6, 306

S
Scale temperature 19, 30
Self-accelerating chemical reaction 11
Self-accelerating exothermal oxidation,
- in the gaseous phase 91
Self-accelerating exothermal reaction 27
Self- acceleration 11

Self-extinction time,
- polyvinyl chloride 347
Self-ignition 28, 39
- activation energy of 40
- concentration dependence 42
- determination of 23
- initial conditions of 16
Self-ignition heating 17
Self-ignition initiation time,
- function of temperature 40
Self-ignition temperature,
- determination of 292
- polymers 44, 46
- unsaturated oligoether-based polymer materials 47
Self-ignition time delay 24
Semenov number 16
Shift in the combustion front,
- techniques applied to monitor 278
- visual observation 278
Siloxane polymers with carborane groups 313
Single-stage combustion regimes 65
Smoke,
- filler reduces formation of 323
- formation of 267
- specific optical density of 301
- susceptibility of a polymer material towards formation or 71
- toxicity of 267
Smoke density,
- critical value of 301
Smoke formation,
- during the combustion of polymers 194
- chemical aspects of the reduction in 267
- during flaming combustion 195
- quenching of 241
- retardation of 241
Smoke-formation capacity,
- determination of 300
- determined with the aid of optical methods 300
Smoke liberation,
- during combustion 212
Smoke opacity,
- index of 301
Smouldering 47
- effect of additives 50
- incomplete combustion 177
- unstable 50
Solid fuel combustion,
- model for 61
Solid fuel combustion wave,
- structure of 61
Soot,
- particle size distribution 201
- quenching the formation of 244
- reactions leading to formation of 199
- susceptibility of a polymer material towards formation of 71
Soot formation,
- acceleration of 200
- chemical process that lead to 204
- during diffusion combustion 196
- influence of different metal-containing systems on 242
- mechanism of 267
- radical-chain nature 197
Soot particles 63
- formation of 64
- nuclei of 201
Spolding number or B number 66
Spread of a laminar diffusion flame,
- model of 92
Spreading of a flame,
- in a gaseous oxidising atmosphere 91
- over the surface of a polymer material, analytical models of 91
Sreznevsky,
- law of 85
Sreznevsky constant 104
Sreznevsky's law 104
Stable combustion process 108
Stable diffusion combustion of polymers 88
Stationary combustion,
- atmosphere of gaseous oxidiser 120
- effect of oxygen 124
Stationary combustion process 12
Stationary diffusion combustion,
- effect of the scale of the combustion source 79
- polymer materials 79
Steady-state combustion 2
Stefan-Boltzmann law 73
Styrene polymers and copolymers 347
Sublimation 60
Surface complexes,
- gas and carbon 172
Surface modification,
- to change the flammability of materials 315
Surface temperature 18, 32, 121
- burning polymer 76
Surface temperature at the ignition point,
- determination of 44
Synergists,
- with polyolefins 341

T

T° flash 12
T° ignition 12
T° self-ignition 12

Temperature characteristics of combustion,
- errors during determination of 282
Temperature distribution,
- gaseous state 29
- reacting system 75
- solid state 29
Temperature field,
- combustion wave zone 75
- inside the front edge of the flame 86
Temperature field measurements,
- during downwards flame diffusion over thermally thick layers of polymethylmethacrylate 100
Temperature gradient,
- near the surface, critical limit value of 108
Temperature gradients 75
- near the surface 120
Temperature index 8
Temperature measurement,
- optical 283
Temperature of gasification 70
Temperature profiles,
- during combustion 282
- gaseous phase during flame spread 76
- preflame zone 75
Testing,
- flammability characteristics of polymer materials 289
Thermal balance,
- stationary combustion of polymer materials 80
Thermal combustion,
- nonstationary theory of 16
Thermal conductivity 29
Thermal degradation 63
- chain nature of 129
- influence of an external heat source 120
- kinetic parameters of 120, 123
- kinetics and mechanism 130
- polymers in a vacuum 123
Thermal energy,
- transfer of 91
Thermal explosion,
- characteristic of 2
- dimensionless parameter of 24
- nitrocellulose 24
Thermal explosions,
- stationary theory of 14
Thermal flow,
- rate of 36
Thermal inductance 11
Thermal oxidising stability,
- relationship between the combustibility of polymers and their 260
Thermal radiation flow,
- polyethylene and polypropylene flow 79
Thermal self-acceleration 17
Thermal stability,
- heterocycle polymers 149
- hydrocarbons at high temperatures 193
- modification of polymers to increase their 311
- polymers 149
- under high heating conditions 261
Thermal treatment,
- cellulose fibres, polyacrylonitrile and other polymers to reduce their flammability 317
Thermo-oxidation degradation 63
Thermo-oxidative decomposition,
- solid phase 14
Thermodynamic stability,
- carbon compounds in a gaseous state 192
Thermodynamics,
- polymer pyrolysis processes 128
Thermokinetic limit 31
Thermokinetic parameter 31
Thermostable polymer materials,
- base of polyheteroarylenes 162
Thermostable polymers,
- combustion of 315
Thorough combustion,
- mass rate of 85
Thorough combustion of polymer materials,
- mechanism of 85
Time of mass transfer,
- inside the flame zone 100
- stationary medium 100
Toxicity,
- combustion products 302
- lethal concentration 302
- specific index of 301
Toxicity of polymer combustion products,
- determination of 301
Truton's law 68
Turbulence 5
Turbulent diffusion flames,
- analysis of 305
Turbulent flames 5

U

Urea-formaldehyde polymers,
- char layer 359
- temperature of self-ignition 359
- toxic products 359

V

Vaporisation enthalpy,
- for degradation products of hydrocarbon polymers 68
- for linear aliphatic hydrocarbons 68
- for unsaturated oligoethers derived from alkylene glycols and hydroxyalkylene glycols 68

Vaporisation temperature 67
Vapour-smoke gaseous zone 61
Volatile product liberation 135
Volatile products,
- formation of 134

Volume ignition 12

Y

Yield of char residue,
- high-temperature treatment 150